The Unfolding Universe

A Stellar Journey

The Unfolding Universe

A Stellar Journey

Lloyd Motz
and
Jefferson Hane Weaver

Plenum Press • New York and London

Library of Congress Cataloging in Publication Data

Motz, Lloyd, 1910–
    The unfolding universe: a stellar journey / Lloyd Motz and Jefferson Hane Weaver.
        p.   cm.
    Includes bibliographical references and index.
    ISBN 0-306-43264-1
    1. Astronomy. I. Weaver, Jefferson Hane. II. Title.
QB43.2.M674   1989   89-8782
520—dc20   CIP

Printed in the United States of America

To Minne and Shelley

Preface

There is hardly an area in our lives that has not been influenced by or does not owe something to astronomy. It has been the friend of the sailor exploring uncharted seas, the inspiration of the poet and artist, the source of mythology and astrology, the great arena in which theology was born and evolved, and the intellectual domain of the philosopher, to name but a few human activities that astronomy has enriched. The greatest chronicler of them all, William Shakespeare, conveyed his belief in the importance of astronomy to romance and music in *The Merchant of Venice*, where he had Lorenzo exhort Jessica to contemplate the heavens while the "sounds of music" crept in their ears and the "soft stillness and night" became "the touches of sweet harmony."

Astronomy is the queen of the sciences because it bore some of our most important scientific principles; so now more than ever the stars, and the universe itself, are the ultimate laboratories where the high-energy particle physicists seek answers to some of the most perplexing problems about the nature of space and time and the origin of the universe. Indeed, astronomy has enabled us to grapple with some of the so-called "ultimate questions" about the nature of existence that scientists used to leave to philosophers and metaphysicists. We now have astronomical evidence that the universe originally began with a gigantic explosion that scattered the primordial ingredients of the cosmos thoughout space.

In this book we have tried to communicate some of our feeling that the study of astronomy is a great adventure that, more than any other intellectual pursuit, reveals the relationship of all things to each other and the present to the past and the future. This sentiment was perhaps most beautifully expressed by the poet William Blake, who saw the "world in a grain of sand, heaven in a wild flower and infinity in an hour." While it is doubtful that Blake knew much about astronomy or had even an inkling that all things might consist of fundamental building blocks that we now call atoms, his poem does suggest his sense that all things are related to each other in some subtle yet pervasive manner. Similarly, Albert Einstein's lifelong search for a unified field theory that would somehow explain all known physical phenomena in the universe was motivated by his belief that the universe is ultimately homogeneous on the subatomic level.

Recent exciting discoveries in astronomy, such as quasars, pulsars (neutron stars), X-ray sources, background cosmic radiation, black holes, interstellar organic molecules, and the origin and evolution of the universe, to name the most spectacular ones, have had an important although not always desirable effect on the exposition of astronomy and the writing of astronomy books. The emphasis both in popular articles and in recent books on astronomy has shifted to such an extent from classical astronomy to the current dramatic discoveries that the general reader may be left with a feeling of discontinuity and a sense that astronomy consists of a series of amazing but unrelated tableaux. It is very difficult, if not impossible, for the person not trained in science to see, without proper guidance, how the various astronomical phenomena now being studied so assiduously are related to each other and to the great body of astronomical knowledge and principles that has been built up over the ages. This book supplies that guidance in a simple, nonmathematical way so that the reader can see just how the modern great developments in astronomy have emerged from classical astronomy and how they are related to each other and to the two great theories of physics—the theory of relativity and the quantum theory—that lead us in our exploration of the universe.

The first eight chapters of this book deal with the major bodies in the solar system and their dynamics. We present the properties of Earth, the moon, the planets and their satellites, the asteroids, the comets, and the sun in a purely descriptive way that any inquisitive reader can understand and absorb as easily as any well-written description of the geographical and topographical features of Earth. Since the dynamics of the solar system are determined essentially by the gravitational force of the sun on the other bodies in the solar system, we introduce Newton's law of gravity and his laws of motion in a simple, direct way with a minimal use of elementary algebra. We discuss the way the orbits of the planets and the motion of the planets (Kepler's laws of planetary motion) are related to Newton's law of gravity. Since a complete understanding of the motions of the planets requires some knowledge of the theory of relativity, the special and the general theories of relativity are described in Chapters 4 and 5.

In Chapters 9 through 13 we examine the general properties of the stars and describe correlations among these properties, such as the Hertzsprung–Russell diagram. Since one must know something about the characteristics of radiation and the structure of the atom to understand how the astronomer obtains his results, we introduce and explain the principles of thermodynamics, the laws of radiation, and atomic theory as they are needed. Chapters 14, 15, and 16 are devoted to the structure and evolution (from birth to death) of stars, and the conditions that determine whether a star will end its life as a white dwarf, a neutron star, or a black hole. The three remaining chapters concern the motions of stars, the structure of our galaxy (the Milky Way), the properties of galaxies in general, and the structure and evolution of the universe (cosmology).

We have tried to write an accessible exposition of the great astronomical and astrophysical principles, presented in their simplest form so as to appeal to the general reader who comes to astronomy because it is the only science that offers an insight into our relationship to the universe. By showing that astronomy has both observational and theoretical features, we demonstrate that this subject presents some of the most spectacular images ever imagined as well as some of the most profound physical principles our intellects have yet uncovered.

Lloyd Motz
Jefferson Hane Weaver

Contents

The
Unfolding Universe

A Stellar Journey

CHAPTER 1

The Earth

Sooner let earth, air, sea, to chaos fall,
Men, monkeys, lap-dogs, parrots, perish all!

—ALEXANDER POPE, *The Rape of the Lock*

Owing to the existence of intelligent life on Earth, our planet is by
far the most interesting of the nine planets in the solar system.
Moving around the sun in an almost circular orbit that lies between
those of Venus and Mars, Earth receives just the right amount of
energy from the sun every second to produce the optimum conditions for life on its surface. In itself, the proper distance of Earth
from the sun is not sufficient to produce intelligent life; its size,
mass, and chemistry, the tilt of its axis of rotation with respect to
the plane in which it moves, the rate at which it spins, the area of
dry land compared to the area covered by oceans, and other minor
characteristics have helped or hindered to a greater or lesser extent
the emergence of life and the evolution of intelligent life on this
planet.

Long before Columbus sailed to the Western continents and
Magellan circumnavigated the globe, people knew that Earth is
round. The ancient Greeks, hundreds of years before the birth of
Christ, observed that the shadow of Earth cast on the moon during
a lunar eclipse had a round rim, and concluded that Earth is a sphere.
They also observed that the direction of the vertical line from Earth
to the sky changes as one moves either due north or due south, and
in 250 B.C. the great Alexandrian librarian Eratosthenes calculated
Earth's circumference by noting that the vertical (a plumb line) at
Alexandria differs in direction by 7 degrees from the vertical at
Syene, which is about 500 miles south of Alexandria. "Eratosthenes
estimated that Alexandria was 5,000 stades due north of Syene, so
that the circumference of the earth was 250,000 stades. Estimates

of the stade vary, but if ten stades be taken to the mile, his polar diameter of the earth comes out only fifty miles short of our modern value.''[1] Eratosthenes could calculate this quantity because he knew that 7 degrees is just a little less than 1/50th of the 360 degrees in a circle. Accordingly, it was a simple operation to multiply the 5000 stadia by 50 to obtain the diameter of 250,000 stadia.

That Eratosthenes devised a method yielding a calculation that would not be approached in accuracy for centuries is not so surprising when one considers that the ancient Greeks were obsessed with what they imagined to be the geometric configuration of the universe. Unlike the idle speculations by Greek philosophers such as Plato about the geometric basis of the universe (Plato considered the true elements of the material world not to be the four elements earth, air, fire, and water but "two sorts of right-angled triangles, one of which is half a square and the other half an equilateral triangle"[2]), Eratosthenes's work showed the advantages of developing more systematic treatments of the relationships between angles and chords, and his work helped give birth to what became known as trigonometry.

Eratosthenes (ca. 276–194 B.C.) was a native of Cyrene who had spent his earlier years in Athens where he displayed a talent for both mathematics and the arts. Although little is known about his life, he appears to have been a good friend of Archimedes and Aristarchus, so it is not surprising that Eratosthenes should have developed a clearheaded approach to mathematics that was not muddled by mysticism or numerology. It is not known what or who convinced Eratosthenes that Earth is spherical (even though very few people at that time believed that it was flat), but it is likely that his conversations with Aristarchus, who propounded the first heliocentric system, influenced him greatly. To suggest that Earth is both spherical and mobile (and thus like the other known planets) was as revolutionary as saying that Earth moves around the sun, because the Greeks believed that Earth and the heavens were fundamentally different in constitution and subject to different physical laws. Earth was considered to be fixed and base, the stars in the heavens mobile and more perfect, moving in perfect circles overhead.

Eratosthenes immersed himself in all areas of knowledge while in Greece and developed an impressive reputation as a teacher and scholar in such diverse areas as poetry, mathematics, astronomy, and history. Although his multitude of academic talents caused him to be in great demand as a tutor, his first love was mathematics and he is chiefly remembered for his work in this subject. He collected and assessed the work of earlier geographers and astronomers, "who had evolved the conception of the earth as a globe, with two poles and an equator, and made a map of the known earth, marked with lines of latitude and longitude, and delineating five zones, two frigid, two temperate, and one torrid."[3] He drew the fundamental line of longitude through Alexandria and Syene and the fundamental line of latitude through the Strait of Gibraltar and the island of Rhodes.[4] He imagined Earth's surface to be mostly ocean, surrounding a single land mass that straddled his "equator" for the modern equivalent of nearly 8000 miles. It is fair to say that Eratosthenes had a much more accurate vision than Christopher Columbus of the vast distance one would have to travel by sailing westward from Europe to Asia, because he despaired that such a trade route would ever be practical. In any event, Eratosthenes's reputation as a scholar became so great that he was hired by Ptolemy III, the ruler of Egypt, to come to Alexandria to tutor his son and become the librarian of the university there. Only after he had arrived in Egypt and noticed that the bottom of a well could only be seen when the sun was directly overhead was he inspired to think of the brilliant method for measuring Earth that has immortalized his name.

Today, very accurate geodetic measurements show that Earth is not a perfect sphere but rather an oblate spheroid that is flatter at the poles than at the equator. At the equator the distance between two points, whose latitudes differ by 1 degree, is 68.7 miles, whereas at the poles the distance between two such points is 69.4 miles. From an analysis of the motion of and the data gathered by artificial satellites, we know that the flattening at the two poles is not quite the same; Earth is slightly flatter at the south pole than at the north pole so that it has a somewhat (but very slight) pearlike appearance.

Because Earth is flatter at the poles than at the equator, the polar diameter is shorter than the equatorial diameter; a line drawn from the North Pole through the center of Earth to the South Pole (the polar diameter) is about 29 miles shorter than the diameter of Earth at the equator. The equatorial diameter is 7929 miles whereas the polar diameter is 7900 miles; the average or mean diameter is 7918 miles. The complete Earth as a sphere was observed and pho-

tographed for the first time by a Soviet satellite and then by the
Apollo astronauts on their journey to the moon. In these photographs features such as Africa, Asia, Arabia, the cloud cover, and
the Antarctic ice and snow caps can be clearly seen. As observed
from the moon, the full disk of Earth is not always visible; only that
part of the sun-lit portion of Earth's disk that reflects the sunlight
to the moon can be seen by an observer on the moon. Thus, to an
observer on the moon, Earth is seen in various phases just the way
the moon is seen in different phases by an observer on Earth.

In a photograph taken by the Apollo 8 mission, Earth is shown
rising above the moon's horizon; slightly more than half of it can
be seen. If Earth were seen in its full phase from the moon, it would
appear about 4 times larger but 75 times brighter than the full moon
does as seen from Earth. The reason for this, even though the surface area of Earth is only 16 times larger than that of the moon, is
that Earth's atmosphere, clouds, and large water areas taken together reflect about 5 times as much sunlight as the moon's surface
does. The distinct bluish color of Earth is caused by our atmosphere,
which reflects the blue rays of sunlight more effectively than it does
the red and yellow rays. This is why the sky looks blue to us on
Earth. A photograph of Earth also shows centers of atmospheric
disturbances, which are the beginnings of tornadoes, hurricanes, or
storms; such disturbances arise from the rotation of Earth, the flow
of winds, and the temperature variations over Earth's surface.

The difference in appearance (aside from size) between Earth
as seen from the moon and the moon as seen from Earth is caused
by our atmosphere and Earth's vast oceans. The atmosphere, consisting of nitrogen (80%), oxygen (about 20%), some carbon dioxide,
and some water vapor, extends out to about 15,000 miles above
Earth's surface. Most of it (about 90%), however, is concentrated
in the first 10 miles. The ozone (the triple oxygen molecule) layer,
which protects us from the ultraviolet solar rays, lies 30 miles above
Earth's surface, and the ionosphere (a thin layer of ionized atoms
which reflect radio waves) lies 70 miles above us. Starting some
5000 miles above us, and extending outward some 10,000 miles beyond that, are the famous Van Allen belts, belts of very rapidly
moving protons (hydrogen nuclei) that are trapped in Earth's magnetic field. The variations of pressure, density, and temperature in
Earth's atmosphere are quite complex.

The present physical and chemical properties of the atmosphere are quite different from what they were when Earth was formed some 5 billion years ago. Indeed, our present atmosphere is a second atmosphere, which was formed as a result of volcanic activity after Earth had lost its first atmosphere. The initial atmosphere consisted of hydrogen, helium, ammonia, methane, water vapor, and carbon dioxide. This atmosphere was driven off by the solar wind and the high temperature of Earth at that time. Following the loss of this atmosphere, Earth continued moving around the sun for millions of years as a hot, rocky structure, devoid of atmosphere. After this sterile period, intense volcanic activity released great quantities of water vapor, carbon dioxide, and nitrogen compounds. Most of the carbon dioxide was ultimately absorbed by the rocks to form carbonates, and most of the water vapor condensed to form our present oceans. There was no oxygen in this second atmosphere when it first formed; the oxygen now present could only have originated from some form of plant or bacterial life that required no oxygen and that synthesized the carbon in carbon dioxide into its own carboniferous structure and released free oxygen.

The most remarkable aspect of our atmosphere is the almost complete absence of hydrogen and helium, which are the most abundant elements in the universe. This situation is due to the low speed of escape from the surface of Earth as compared to the speed of escape from the surfaces of Jupiter and Saturn, which are almost all hydrogen and helium. The speed of escape from the surface of any sphere depends only on the mass of the sphere (how much matter it contains) and its radius. The more massive a sphere, the larger is the speed of escape from its surface, and the larger its radius, the smaller is this speed of escape. We have discussed Earth's diameter, so we now consider its mass, which can be determined by using Newton's law of gravity. Since Earth's gravitational pull on any object, such as an artificial satellite circling Earth, depends on Earth's mass and the distance of the object from the center of Earth, its mass can be calculated if the motion of a body moving at a given distance from the center of Earth can be accurately observed. This has been done for artificial satellites moving in various orbits, and from such observations Earth's mass is known to be 5.98 thousand trillion trillion grams. One gram, the unit of mass, is about $\frac{1}{450}$ of a pound.

From Earth's mass and radius, the speed of escape from Earth is found to be about 7 miles per second or about 25,000 miles per hour. At the average temperature of Earth's atmosphere, hydrogen molecules and helium atoms travel fast enough to escape from Earth whereas the molecules of oxygen, nitrogen, water vapor, and carbon dioxide do not.

It is fortunate for us and life in general that Earth's mass and radius are just about what they are. If the mass were much smaller or the radius much larger, the speed of escape would be smaller and Earth, like the moon, could have retained no atmosphere. On the other hand, if the mass were much larger or the radius much smaller, the hydrogen and helium would not have escaped and there would be no free oxygen, nitrogen, or carbon. Our atmosphere (like Jupiter's) would then have been quite noxious.

When Earth was formed some 5 billion years ago by the gravitational accretion of the debris that was left in the wake of the formation of the sun, a large amount of heat was generated by the vast number of particles that fell in toward each other and coalesced. To this was added the heat of meteorites striking the surface of the forming Earth and the heat from the radioactive decay of uranium and thorium. Some of the heat escaped into space but most of it remained trapped inside, raising Earth's temperature to about 6000°C. At this temperature, Earth's interior became molten, and liquid drops of iron, nickel, and other such elements sank to the center to form the present liquid core of our planet. The lighter elements such as aluminum and silicon floated to the top to form the rocky crust of Earth. During this iron and nickel fallout, the heavy radioactive elements like uranium were squeezed out of the interior and forced into the crust and mantle, where they are now found.

After the crust and mantle formed, the heat in the deep interior could no longer get out. The steady outward flow of heat from the crust that we now observe is generated by the radioactive decay of uranium and thorium. This heat melted the rocky material right under the crust so that the crust, which is lighter than the molten material in the mantle, floats on the mantle.

When the oceans and land masses formed, the continents as we know them today did not exist; instead, all the dry land formed one continuous land mass around the North Pole surrounded by water. The continents were formed later as the result of the motion of the continental plates, which float on the underlying heavier material of the mantle. The continents at present look very much like pieces of a jigsaw puzzle. It is easy to see that the eastern coastlines of North and South America match the western and northern coastlines of Africa, Europe, and Greenland. Geologists have demonstrated that if all of the continents were brought together, they would fit into a single structure quite well. Moreover, it has long been known that the rocks along the eastern coastlines of the American continents match those along the western coastlines of Africa, Europe, and Greenland.

Since the rocks of our continents (e.g., mountain ranges) are about 25% less dense (less heavy volume for volume) than the rocks of the upper mantle, they float about like icebergs in the ocean. The energy required to move them comes from the decay of the radioactive elements in the crust and mantle. For hundreds of millions of years the solid land mass, consisting of all the continents, remained intact, but in time, as great amounts of energy accumulated beneath this land mass, vast forces were brought into play. These forces set the land mass and the entire outer crust in motion. Since the forces acting on the bottom of the land mass were not everywhere the same, different parts of the land mass moved at different speeds, resulting in a shearing action that broke the land mass into continents and caused the latter to drift away from one another.

Today the continents are still moving at different speeds and in different directions so that they press against each other along various lines. This differential motion gives rise to mountain ranges, earthquakes, volcanoes, and ridges in the ocean basins.

We have already noted that atmospheric disturbances are generated by the variations in temperature and pressure and the motions of large air masses. These motions or jet streams are caused by the daily spin or rotation of Earth. As seen from the fixed stars (relative to the fixed stars), Earth rotates once every 23 hours, 56 minutes, and 4 seconds. Astronomers call this length of time a sidereal day, and they use a special clock, called a sidereal clock, to keep track of Earth's rotation and the rising and setting of the stars. Relative to the sun (as seen from the sun), Earth rotates once every 24 hours or once a day. This length of time (exactly 24 hours) is called a solar day. It is the length of our ordinary day as measured by an ordinary or solar clock. A sidereal clock runs about 4 minutes faster every day than an ordinary or solar clock does. This means that, as measured by an ordinary clock, the stars rise about 4 minutes earlier each day. Owing to this shift, the sky changes its appearance from month to month and different constellations are seen in the night sky in the different seasons.

The reason for the difference in the length of the sidereal day (the actual time it takes Earth to rotate once) and the solar day (the actual time between two successive sun risings) is that Earth revolves around the sun in the same direction (eastward, in both cases) as it rotates on its axis. It moves in a large orbit around the sun at a speed of about $18\frac{1}{2}$ miles per second. This motion changes the direction from the sun to Earth (eastwardly) by about 1° per day, and this 1° eastward shift in the sun's direction is why the sun rises about 4 minutes later each day than do the stars, whose directions from Earth hardly change at all. Earth revolves completely around the sun once every 365.2563 (slightly more than $365\frac{1}{4}$) days, and during that time it rotates on its axis $366\frac{1}{4}$ times.

Owing to Earth's rotation, a point on the equator moves at a speed of slightly more than 1000 miles per hour, whereas points on the surface north and south of the equator move more slowly. Near the poles there is hardly any motion at all. Because of this difference in the speeds of rotation, the warm air masses rising from the equator and drifting northward and southward move rapidly to the east relative to the cooler air masses of the northern and southern temperate zones. This warm air collides with the colder air masses of the temperate zones, generating hurricanes and causing other atmospheric disturbances.

Earth's rotation has had another important effect: it has changed Earth's shape from that of a perfect sphere to that of the oblate spheroid we have been discussing. If a perfect sphere of matter is set spinning, it becomes distorted because the spinning motion produces a centrifugal force that is directed perpendicular to (at right angles to) and away from the axis of rotation. This force is largest at the equator and zero at the poles so that a bulge is formed around the equator. The faster the rotation, the greater is this centrifugal force and the more pronounced the distortion. In a massive body like Earth (which is rotating very slowly), the very weak outward centrifugal force and the much stronger inward gravitational force combine to give Earth its present shape. The weight of any object on Earth is the result of the combination of these two forces acting on the object.

That Earth rotates on its own axis can be demonstrated by the counterrotation of a pendulum suspended by a long string. Since

Earth rotates under the pendulum, the plane in which the pendulum is oscillating appears to turn in a westward direction. This phenomenon was first demonstrated by the amateur French scientist Jean Foucault in 1851.

For thousands of years, it was thought that Earth was at the center of the universe and that the moon, the sun, the planets, and the stars revolved around it. The revolutionary idea that Earth is not the centerpiece of the universe is not new, however; it was proposed more than two millennia ago by the Greek philosopher and astronomer Aristarchus of Samos (310–250 B.C.), who argued that the sun is at the center of the solar system and that Earth revolves around it in a perfect circle. The manuscript describing this theory, if in fact it was ever recorded, has not survived, but the writings of others such as Archimedes and Plutarch have attributed this theory to Aristarchus; it apparently gained some notoriety during its author's lifetime as more than one of Aristarchus's contemporaries said that he should have been indicted for impiety. Although Aristarchus was the only one among the early Greek astronomers who was courageous enough to assert that Earth moves, he did impress many of his fellow scientists with his surviving work, *On the Sizes and Distances of the Sun and the Moon*, which contains the first scientific effort to measure the distances of the sun and moon from Earth: "Aristarchus supposed that at half moon, the configuration of the sun, moon, and earth, forms a right-angled triangle, from which the relative distances of the sun and moon can be determined by measuring the angular separation of the sun and moon from the earth. He measured the angle as eighty-seven degrees, from which he calculated that the sun was nineteen times more distant from the earth than the moon, though actually the angle, and thus the ratio of the distances of the sun and the moon from the earth, are larger. As the moon, on the average, just covers the sun during a solar eclipse, Aristarchus concluded that the diameter of the sun is nineteen times larger than that of the moon. From lunar eclipses he estimated that the breadth of the earth's shadow, and thus the approximate diameter of the earth, equals three moon diameters. Thence, he argued, the diameter of the sun must be six to seven times as great as that of the earth."[5] Although we now know that the moon's diameter is 2160 miles, making it a little more than a fourth as wide as Earth, it is striking that with the primitive tools available to him Aristarchus made such an accurate calculation. His underestimation of the distance of the sun, however, led him to underestimate its diameter greatly; we now know to be about 864,000 miles—a little more than 100 times that of Earth. That Aristarchus's method was somewhat inaccurate is less important than the fact that he relied solely on observational evidence and mathematical calculations to derive his results. His approach was purely physical and did not involve nonsensical speculations about mythologies or metaphysical abstractions.

Despite the ease with which Aristarchus's theory explained the apparent movement of the stars in the evening sky, it was neglected and nearly forgotten for almost 19 centuries until it was revived by the Polish astronomer Nicolaus Copernicus. That Earth revolves around the sun can be demonstrated by noting that the stars appear to shift back and forth by minute amounts as Earth moves from one side of the sun to the other. Moreover, Earth's motion causes the path of the light from the stars to appear to be slanted in the direction of Earth's motion just as the paths of falling raindrops appear to be slanted away from the vertical when we are running. Finally, Earth's motion causes the light from stars which we are moving toward to appear somewhat bluer than is normal and the light from stars which we are moving away from to appear somewhat redder than is otherwise the case.

Earth moves around the sun in a large ellipse that brings it to within 90.5 million miles of the sun in January and out to 95.5 million miles in June so that its mean distance from the sun is 93 million miles. As Earth travels around the sun, its axis of rotation (the line passing through the North and South Pole) points very nearly in the same direction at all times. The words "very nearly" here are very important, for the direction of the axis is not fixed but changes very slowly. This slow change is continuous and measurable. In fact, it was first detected by Hipparchus, the most famous of the early Greek astronomers, in 150 B.C. and is now called the precession of the equinoxes. This precession of the axis is caused by the torque exerted by the sun on the equatorial bulge of Earth. The earth is thus like a huge spinning top, which we will discuss shortly.

Hipparchus was the most important astronomer of the ancient world and his work greatly influenced Ptolemy's *Almagest*, which entrenched the geocentric theory of the universe for nearly 1500 years. Except for one relatively unimportant book, all of Hipparchus's works have been lost, and only through the writings of his followers such as Ptolemy can we assess his contributions to astronomy and mathematics. Little is known about his life but it is believed that he was born on the island of Rhodes, where he later built an observatory and did his most important work. Although not a member of the Alexandrian school, he kept in contact with the prominent mathematicians and astronomers there and during his occasional travels to Egypt made some celestial observations there. His contributions to astronomy were extensive but he also developed methods that helped give a more rational basis to astronomy: "He invented or greatly developed a special branch of mathematics [trigonometry], which can be applied to geometrical figures whether on a plane or on a sphere. He made an extensive series of careful observations with all the accuracy that his instruments would permit. He systematically and critically used old observations for comparison with later ones to discover astronomical changes too slow to be detected within a single lifetime. Finally, he systematically employed a particular geometrical scheme (that of eccentrics, and to a less extent that of epicycles) for the representation of the motions of the sun and moon."[6] Hipparchus also discovered through his own observations "the precession of the equinoxes, that is, the slow change in the times of occurrence of the spring and fall equinoxes and the summer and winter solstices."[7]

Although Ptolemy's *Almagest* is often remembered by most historians as an imposing obstacle that delayed the widespread acceptance of the heliocentric theory, it deserves praise as a historical record for it recalls that Hipparchus determined the duration of the year to within 6 minutes of its true value, calculated the inclination of the ecliptic to the equator to within 0.06 of a degree of arc, derived quite accurate figures for the lunar parallax and the solar orbit, determined the perigee and mean motion of the moon and the sun and the extent of the shifting of the plane of the moon's motion, and compiled a list of over a thousand stars (which represented one of the earliest efforts to catalogue the evening sky).[8] "The mathematics that Hipparchus created to survey the earth and heavens has been used since his times to handle numerous practical problems. Surveyors, navigators, and map-makers employ it constantly. Indeed, through the power of Hipparchus' methods and other mathematical methods not detailed here, the Alexandrian Greeks set up map-making as a science. Their maps offered the best knowledge of the earth up to the time of the great explorations in the fifteenth and sixteenth centuries. Very fortunately for later generations the astronomer Ptolemy summarized all of the geographical knowledge amassed by the ancient world in his *Geographia*, a work in eight volumes."[9]

It is unfortunate that Hipparchus did not try to develop Aristarchus's heliocentric theory. His lack of interest in a sun-centered universe was due in part to the theory's lack of agreement with existing Babylonian astronomical records. Hipparchus also realized that to cast Earth adrift in the heavens would imply that it was as common as the other planets—a view that did not enjoy great popularity in the ancient world. Despite this oversight, Hipparchus did more than anyone else until the time of Copernicus to advance astronomy; the great French historian Delambre, who was never reluctant to criticize the work of any scientist, offered a worthy tribute: "When we consider all that Hipparchus invented or perfected, and reflect upon the number of his works and the mass of calculations which they imply, we must regard him as one of the most astonishing men of antiquity, and as the greatest of all in the sciences which are not purely speculative, and which require a combination of geometrical knowledge with a knowledge of phenomena, to be observed only by diligent attention and refined instruments."[10]

Returning now to the spinning Earth, we note that the axis of rotation of Earth is not perpendicular (at right angles) to the plane in which our orbit lies but rather is tilted about $23\frac{1}{2}°$ away from the perpendicular (actually $23°27'8''$) to this plane. This tilt together with Earth's motion produces our change of seasons. On June 21, when the North Pole is tilted toward the sun, summer begins in the north temperate zone. Six months later on December 21, when the North Pole is tilted away from the sun, winter begins in the north temperate zone. On March and September 21, when neither pole is tilted toward the sun, we have spring and fall in the north.

As noted previously, the time it takes us to revolve around the sun once (the revolution of Earth relative to the fixed stars) is 365.2564 days. This interval, which is not the length of the year as measured from the beginning of any season to its next beginning, is called the sidereal year. The year of the seasons (from winter to winter or summer to summer) is called the tropical year. Its length is 365.2422 days (somewhat shorter than $365\frac{1}{4}$ days). The length of the tropical year is about 20 minutes shorter than the length of the sidereal year because of the precession of Earth's axis, as discussed previously. The calendar follows or keeps track of the tropical and not the sidereal year. If the calendar were based on the sidereal year, the seasons associated with the months June and December would be interchanged every 13,000 years.

The earliest practical calendar was introduced by Julius Caesar (the Julian calendar) in 42 B.C. In this calendar the length of the year is assumed to be exactly $365\frac{1}{4}$ days. The $\frac{1}{4}$ day was disregarded for three successive years and was then taken into account by making the fourth year 366 days long (the leap year). This calendar ran slightly ahead of the true year and by the year 1582 it was in error by more than 12 days. In that year, Pope Gregory XIII proposed a new calendar (the Gregorian calendar) in which three leap years are left out of the calendar every four centuries. Only century years that are divisible by 400 are made leap years (366 days) in the Gregorian calendar.

CHAPTER 2

The Moon

To behold the wandering moon,
Riding near her highest noon,
Like one that had been led astray
Through the heav'n's wide pathless way
And oft, as if her head she bowed,
Stooping through a fleecy cloud.

—JOHN MILTON, *Il Penseroso*

The moon, at a mean distance of 384,404 km (240, 252 miles) from
Earth, is the closest and the most thoroughly studied celestial body.
It is the only body outside Earth that we have tread on, and will
probably remain so for many years. As the moon moves in its elliptical orbit around us, its distance from the center of Earth varies
from a minimum of 356,400 km to a maximum of 406,700 km. As
seen from Earth this orbital motion of the moon makes it appear to
move about 13° eastward (opposite to its apparent westward rising
and setting motion, which is due to the rotation of Earth) each day.
Owing to this apparent eastward motion, which, as noted, stems
from the actual revolution of the moon around Earth, the moon
rises about 52 minutes later each day for any observer on Earth.

Since the moon is visible on Earth only because it reflects the
sunlight that strikes it, the appearance of the moon (the phase) at
any particular time depends on the fraction of its surface (illuminated by the sun) that reflects light to Earth. This, in turn, depends
on the angle formed by a line from Earth to the moon and a line
from Earth to the sun. This angle is called the elongation of the
moon and it varies from 0° when the moon and sun are in the same
direction to 180° when the moon and sun are opposite each other
in the sky. When the angle is 0° (the sun and moon, which then rise
and set together, are said to be in conjunction) the moon cannot be

seen from Earth because it reflects no sunlight to Earth. It is then in its new phase (new moon). When the elongation is 90° (the moon is said to be in quadrature), we see the half-moon, and when the elongation is 180° (the moon and sun are in opposition) the moon is full and it rises just as the sun sets.

The moon revolves around Earth once every 27.321 days, but this period, which is called the sidereal period of the moon's motion, is not the calendar month. The calendar month, which astronomers call the synodic month, is the period from new moon to new moon. Its length is 29.5306 days. The reason the synodic month is longer than the sidereal month is that Earth and moon together reolve around the sun while the moon is revolving around Earth. Because of our motion around the sun, the sun, as described in Chapter 1, appears to move 1° eastward each day. This apparent eastward motion of the sun increases the interval of time between two equal phases (new moon to new moon) compared to what it would be if Earth and moon were fixed relative to the sun: as the sun moves to the east, it makes the elongation of the moon at any moment somewhat smaller than it would be if the sun were fixed in the sky.

The plane of the orbit in which the moon revolves around Earth does not coincide with the plane of the ecliptic (the plane in which Earth's orbit around the sun lies) but rather is tilted 5°8′ with respect to the ecliptic. If the plane of the moon's orbit coincided with the plane of the ecliptic, the moon would eclipse the sun totally at every new moon (once a month) and once a month (at full moon) there would be a total eclipse of the moon. But these eclipses do not occur every month because of the 5°8′ tilt of the moon's orbit; an eclipse (either of the sun or of the moon) can occur only when the moon lies close to the ecliptic either at new moon or at full moon. This can occur only twice every 246.6 days (possibly three times in one calendar year), so eclipses are relatively rare. Two solar eclipses must occur every year, but there may be as many as five. There need be no lunar eclipses in any one year, but as many as three may occur. The total number of eclipses (solar and lunar) cannot exceed seven: either four solar and three lunar or five solar and two lunar.

Not only is the moon revolving in its own orbit around Earth but, like Earth, it is spinning around its own axis. But its rate of rotation is much smaller than our rate of rotation. Instead of spinning once a day, the moon spins once every 27.322 days; it rotates once on its axis every time it revolves in its orbit. Owing to this equality of the moon's period of rotation and period of revolution, the moon keeps the same face to Earth at all times. As the moon revolves around Earth, it oscillates a small amount, which allows an observer on Earth to see about 59% of the moon's surface instead of 50%. On the moon the sun is above the horizon (daytime) for about 14 days and below the horizon (nighttime) for 14 days. Owing to these long periods of light and darkness on the moon, the moon's temperature changes considerably from midday to midnight. At midday the lunar temperature is about 110°C and at midnight, −160°C.

From the known distance to the moon its size can be determined. Its diameter is 3475.8 km, approximately $\frac{1}{4}$ of that of Earth. From this we deduce that the surface area of the moon is about 1/16 of that of Earth and its volume, about 1/64.

Before the space age, determining the moon's mass was not an easy task. It could only be done by carefully measuring the moon's gravitational pull on Earth. We mentioned earlier that Earth's pull on the moon keeps the moon revolving in its orbit around Earth, but this is only part of the story. The moon pulls on us with a force that exactly equals (but is opposite in direction to) our pull on the moon. But the response of Earth to this pull is quite different from that of the moon, because Earth's mass is considerably larger than that of the moon. Galileo and Newton discovered that the more massive a body, the more sluggishly it responds to a given force acting on it; if the same force acts on two bodies, we can determine how much more (or less) massive one of the bodies is than the other by determining (measuring) how much less (or more) responsive this one body is to the force than the other.

This technique was applied to the Earth–moon system, and careful observations of nearby celestial bodies showed that Earth's response to the moon's gravitational pull is 81 times smaller than the moon's response to our pull. We find that the center of Earth and the center of the moon both revolve with different speeds in orbits of identical shapes, but of different sizes, around a common point that lies on the line connecting the two centers. This point, the center of mass of the Earth–moon system, lies 81.33 times closer to the center of Earth than to the center of the moon. This means that, owing to the moon's gravitational pull on us, our center moves in an elliptical orbit that is 81.33 times smaller than the moon's elliptical orbit and our speed in this orbit is 81.33 times smaller than the moon's speed in its orbit. From this we see that Earth must be 81.33 times more massive than the moon. Since we know the mass of Earth, a simple calculation shows the mass of the moon to be 73.5 trillion trillion grams.

Today the mass of the moon can be determined more easily by applying Newton's law of gravity to an artificial satellite that is in a parking orbit around the moon. If we carefully measure the period of such a satellite (the time required to complete one orbit around the moon) and also its mean distance from the center of the moon, we can calculate the mass of the moon. The mass just equals $4\pi^2$ times the cube of this mean distance divided by the product of the square of the period and the universal gravitational constant.

The mass and size of the moon determine its surface gravity. Because the moon's mass is 81 times smaller than our's, the moon's surface gravity would be 81 times smaller than our's if the moon were as large as Earth. But the moon's radius is about $\frac{1}{4}$ that of Earth so that the moon's surface is about four times closer to its own center than Earth's surface is to its own center. This, to some extent, offsets the smallness of the moon's mass and increases the force of gravity on the moon's surface: it introduces an additional factor of about 14 for the moon's surface gravity so that the net result is that the moon's surface gravity is not 1/81 of that of Earth but rather about 1/6. This means that an object on the surface of the moon weighs only about 1/6 of what it does on Earth, and a freely falling body on the moon falls much more slowly than it does on Earth. All bodies in a vacuum on Earth fall with a speed that steadily increases at a constant rate (the acceleration of gravity). At the end of each second the speed of every such falling body increases by 32.2 feet per second or 981 cm per second. On the moon the speed of a freely falling body increases by 5.3 feet per second or 162.0 cm per second every second.

Because the moon's surface gravity is smaller than that of Earth, the speed of escape from the moon is smaller than that from Earth by a factor of slightly more than four. The speed of escape from the moon is 2.38 km per second; because of this small speed of escape the moon has no atmosphere. If the moon were supplied with an atmosphere like our's, it would be lost in about a week because the atoms and molecules in such an atmosphere, heated by the sun's rays, would, on the average, be moving at speeds well in excess of the speed of escape from the moon. This would be true not only of hydrogen and helium but also of such heavier molecules as oxygen, ozone, nitrogen, and carbon dioxide.

That the moon has no atmosphere was known long before astronauts first landed there, for it was observed that the moon's rim is extremely sharp instead of fuzzy, as it would be if the moon had an atmosphere. Stars disappear suddenly and sharply when they are occluded by the moon. Moreover, telescopic observations of the moon (as shown also on telescopic photographs of the moon's surface) show that shadows cast in the sunlight by objects on the bright side of the moon are almost pitch-black. If the moon had an atmosphere, the molecules of this atmosphere would scatter the light into the shadows and the latter would not appear black at all. All of this, of course, has been verified by the direct observations by the lunar astronauts. Since there is no atmosphere, there can be no free water on the surface of the moon: any such water would quickly evaporate, and the free water vapor molecules would then escape from the moon.

From the mass and volume of the moon, we calculate its mean density (the amount of mass on the average in a unit volume, e.g., the number of grams per cubic centimeter) to be 3.34 grams per cm^3. This is to be compared with Earth's mean density of 5.517 grams per cm^3. Since the mean density of rock is about the same as that of the moon, we believe that the moon consists almost entirely of rock. There are no measurable quantities of such heavy elements as iron and nickel inside the moon so that the moon's center is quite different from the Earth's.

At the present time the moon appears to be almost entirely inactive with no volcanic outbursts or lunar quakes (of any appreciable magnitude) occurring. The seismometers left on the lunar surface by the astronauts of the Apollo space missions do record slight tremors on the moon, but these are hardly significant as far as the evolution of the moon is concerned. With no wind or running water to erode the lunar surface and its rocks, no volcanism to melt the rocks or cover the surface with layers of lava, and no moonquakes to shift lunar masses about, conditions on the moon have remained essentially unchanged for billions of years. This has been verified by the data gathered by lunar astronauts and the various unmanned missions to the moon, and is correct from an overall point of view, but the data from the Apollo mission show also that lunar erosion has been occurring due to the continual bombardment of the lunar surface by micrometeorites. Such bombardments have not affected the surface greatly, but they have fragmented and pulverized some of the rocks so that substantial layers of rock dust are present. This moon dust consists of substantial quantities of small glass beads, prompting Neil Armstrong to say that the "rocks are rather slippery." These beads are formed when the intense heat generated by the impact of a meteorite (traveling at about 20 miles per second) melts the lunar material that it strikes.

In addition to reducing some of the rocks to dust, the micrometeorites leave pits and, in some cases, fairly large craters. They also tend to wear down the walls of the small old craters and to shift around the upper layers (the top 50 feet) of the moon's surface. Their overall effect, however, is negligible compared to erosion on Earth, and so we may say that the moon's surface and rocks have remained very nearly the same for the last 4 billion years. This finding is one of the reasons that the Apollo mission was so important and promising. If the lunar rocks are essentially the same now as they were billions of years ago, these rocks might reveal the secret of the origin of the solar system.

This hope was too optimistic, for the rocks brought back by the astronauts can be traced back only about 4 billion years and not to the time, some $4\frac{1}{2}$ billion years ago, when we believe the solar system and the moon were formed. The first half billion years of the moon's history was obliterated from the rocks by intense volcanic activity and heavy meteoric bombardment, which greatly transformed the initial rocks. The extensive rills, cracks, rows of craters, vast beds of lava flows, and volcanic rock attest to the intense volcanic activity on the moon during its early stages. We now believe that the moon passed through six stages: its formation, the separation and solidification of its crust, a period of volcanism, its bombardment by massive meteorites, a second period of volcanism, and its present quiescent stage. The first five stages lasted about $\frac{1}{2}$ billion years and the last stage lasted 4 billion years. The moon was probably formed by accretion from the same reservoir of solar debris from which Earth was formed. This accretion occurred in no more than a few thousand years from material that was circling Earth. This accretion of high-speed particles in a fairly short time led to high temperatures and to the melting of the entire outermost layer of the moon, down to a depth of a few hundred miles. As the temperature of this initial shell of molten material dropped, the lighter rocks now found in the crust of the moon crystallized out and rose to the top to form the lunar crust. The basaltic nature of these rocks and much of the present lunar crust indicates that it solidified from a molten stage.

These two stages of accretion and crust formation were followed by secondary melting of some types of rock in the solid crust, which resulted in volcanism with the eruption and deposition of extensive layers of lava. The intense bombardment of the lunar surface by massive meteorites in the next stage produced most of the large craters and vast peaks on the moon. During or shortly after this bombardment, another period of volcanism set in, generating huge floods of lava that flowed into the basins previously gouged out by the impacts of the meteorites. Creation of these dark regions ended about 3 or 4 billion years ago. Today these regions are referred to as lunar seas, a designation that stems from Galileo, who, on observing the moon through the telescope he built, called them maria because he thought they were bodies of water. Today these lunar "seas" look as they did some 4 billion years ago.

The moon has an important gravitational effect on Earth that was first correctly explained by Sir Isaac Newton. The ebb and flow of the ocean tides along the continental shores have been studied for thousands of years and are quite important for navigation, fishing, and other marine and coastal activities. The ancients already knew that the tides were related to the rising of the moon in some manner because the high tides, occurring on successive days, were delayed by about 52 minutes each successive day, just as the rising of the moon was. Just why the tides are related to the moon was not understood, however, until Newton showed that they are caused by the moon's (and, to a lesser extent, by the sun's) gravitational pull on the oceans relative to its pull on Earth as a whole.

The moon pulls all of Earth toward it as though the entire mass of the planet were concentrated at its center. The oceans on the surface of Earth directly below the moon (when the moon is up) are 4000 miles closer to the moon than is the center of Earth. Hence, gram for gram, they experience a larger gravitational pull toward the moon than does Earth as a whole and are thus pulled away from Earth into a bulge. As the moon moves across the sky, this bulge of water (a high tide) follows it. Since the oceans on the side of Earth away from the moon are 4000 miles farther from the moon than is the center of Earth, they experience a smaller gravitational pull toward the moon gram for gram than does Earth as a whole. Hence, Earth as a whole is pulled away from such oceans, causing them to bulge (a high tide) away from Earth and moon. There are thus two high tides each day—one on the side of Earth directly under the moon and one on the opposite side of Earth; they occur 12 hours apart in any given region of Earth.

Tides are produced not only in water but also in the atmosphere and in the crust and mantle of Earth. One may reasonably assume therefore that the tides affect the weather, the rotation, and the structure of Earth. Just how the tides affect such things as the motions of continental plates (tectonics), earthquakes, and volcanoes is not well understood, but such effects unquestionably exist.

Even though the gravitational pull of the sun on Earth is about 180 times as large as that of the moon on Earth, the tidal action of the sun on Earth is somewhat less than half as large as the moon's tidal action. The reason for this is that the sun is much farther from Earth than the moon is. In calculating and estimating the tides, one must combine the moon's and the sun's tidal actions. These two actions reinforce each other when the sun, Earth, and moon lie along one line as in new moon and full moon. We then refer to the tides as spring tides. When the moon is in quadrature (the line to the moon and the line to the sun form a 90° angle) or at half phase, the tidal action of the sun opposes that of the moon and we have neap tides.

The tides have an important effect on Earth's rotation and the dynamics of the Earth–moon system. As the tidal waters are dragged up by the moon and held in place momentarily, these waters press against the continental shorelines, exerting a frictional force opposite in direction to that of Earth's rotation. Owing to this friction, Earth's rotation is slowed and the length of the day is increased (by 1/1000 of a second per century). At the same time, the distance between Earth and moon increases as the reaction of the tidal bulges on the moon accelerates the moon causing it to move in an ever larger orbit. This interplay between tidal action, Earth's rotation, and the lunar orbit will continue for some billions of years until the length of the day and the length of the month become equal, and the moon will then no longer cause the tides to rise and fall. There will be just two fixed lunar high tides on opposite sides of Earth along the line from Earth's center to the moon's center. However, the sun will still cause some tides to rise and fall, with the result that Earth will be slowed still further and the day will become longer than the month. The moon will then rise in the west and set in the east. The tidal bulges will act to pull the moon closer to Earth. This will go on until the moon comes to within 10,000 miles of Earth and is torn apart by the tidal action of Earth on it.

CHAPTER 3

The Planets and Their Motions

The snow had fallen many nights and days;
The sky was come upon the world at last,
Sifting thinly down as endlessly
As though within the system of blind planets
Something had been forgot or overdriven.

—GORDON BOTTOMLEY, *The End of the World*

Ancient astronomy probably began when the earliest observers of the sky recognized the difference between what the Greeks called the *fixed stars* and the *wandering stars* or planets. It was probably easy enough to discover that the sun and the moon shift their apparent positions with respect to the constellations, but rather difficult to find that five other celestial bodies—Mercury, Venus, Mars, Jupiter, and Saturn, the naked eye planets—do so as well. This discovery probably came with the observation that the two brightest planets, Venus and Jupiter, shift their positions with respect to the sun quite differently from the way the stars do. It probably was also noted that the light from Venus and Jupiter is quite steady whereas the light from the stars varies continuously (the twinkling of the stars). These phenomena must have aroused considerable curiosity and wonder among the ancient observers and led to careful observations of the apparent motions of the naked eye planets and to the development of the earliest models of the solar system.

The apparent motions of the planets, as seen from Earth, are not simple. Instead of drifting to the east each day relative to the constellations as the sun and moon do, the planets move eastward at certain times and then reverse their paths in a looping motion and move westwardly. The apparent motion of each planet consists of a series of connected loops that change their motion from east

to west and then back to east again. Since the ancients [except for
Aristarchus of Samos (ca. 300 B.C.) who believed that the sun is at
the center of the solar system] considered Earth to be fixed at the
center of the universe, this kind of planetary motion was difficult
to understand or to describe by a simple model. This difficulty was
compounded by the observations that the apparent motions of Mercury and Venus are not the same as those of Mars, Jupiter, and
Saturn. Whereas the loops in the motions of the three latter planets
are in no way connected with the apparent motion or position of
the sun, the loops of Venus and Mercury move right along with the
sun so that the sun always appears to be near the center of these
loops. Mercury's loop is about half the size of that of Venus so that
Mercury appears to oscillate from east of the sun to west of it and
back quite rapidly whereas Venus does so more slowly. Mercury
is never found more than 28° to the east or west of the sun whereas
Venus can be as far as 48° on either side of the sun. When Venus
is to the west of the sun, it is visible in the early morning before
the sun rises, and when it is to the east of the sun, it is visible above
the western horizon in the late afternoon or early evening.

Hipparchus, who lived about 130 B.C., was the first philosopher
after Aristarchus to propose a mathematical model of the solar system to explain all the observed motions. Placing Earth at the center
of the solar system, Hipparchus introduced a series of epicycles
(circles on top of circles, the largest of which for all the planets had
Earth as their common center) for each planet. This system of epicycles was improved and enlarged by Ptolemy who used a special
set of epicycles for Mercury and Venus that were different from
those for the other three known planets. Ptolemy's geocentric solar
system as described and expounded in his book *The Almagest* was
accepted by Western civilization for almost 1500 years and was
hardly questioned until Nicolaus Copernicus in 1543 published his
great book *De revolutionibus orbium coelestium* and proposed (actually, revived) the heliocentric model of the solar system.

Like many of his fellow scientists, Copernicus's background
and upbringing were not calculated to instill that spark of creative
insight that would inspire him to offer such a radical view of Earth's
proper place in the universe. He was born on February 19, 1473,
in Thorn, a town that had recently been taken over by the Polish
monarch. Copernicus's father was a merchant who had lived in
Krakow until moving to Thorn some 15 years before Nicolaus's

birth. Nicolaus was raised in a conservative household and received his university education at Krakow where he studied mathematics and astronomy, becoming quite familiar with the geocentric theories of the ancient Greeks. It was probably at Krakow that Nicolaus first heard of Aristarchus's sun-centered theory but, given the heavy hand of religious orthodoxy prevailing at that time, it is unlikely that Aristarchus's work was discussed in great detail in the classroom. Copernicus returned home in 1493 with no definite career in mind, but his maternal uncle, who happened to be the Bishop of Ermland, convinced Nicolaus that the greatest opportunities for him lay with the Church. He offered to pay for Nicolaus's studies abroad and the offer was gratefully accepted. Nicolaus enrolled at the University of Bologna in 1497, where he studied the classics and struck up an enduring friendship with the astronomer Domenico Navara, who encouraged Copernicus to study the heavens although Copernicus was more interested in understanding the actual structure of the solar system than he was in making celestial observations. This disinterest in observational astronomy continued throughout his life as he recorded only 27 observations in his *De revolutionibus*, which he wrote during the last three decades of his life.

Copernicus stayed in Bologna for three years before traveling to Rome in 1500, where he gave a series of well-received mathematics lectures. The following year he returned to Poland and accepted the canonry at Frauenburg, which had been kept open for him owing to the influence of his uncle. Despite the years spent abroad, Copernicus decided to pursue a doctoral program before taking up his official duties. He took a leave of absence and went back to Italy to study medicine and law at Padua. In 1503 he received his doctorate in canonical law, but he remained in Italy for three more years. "At latest in the beginning of 1506 Copernicus left Italy, where he had spent nine years, and although we know next to nothing about the people he may have associated with or the manner in which he prosecuted his studies, we cannot doubt that his long residence at the two most renowned Italian universities had put him in full possession of all the knowledge accessible at that time, in classics, mathematics and astronomy as well as theology."[1] Copernicus returned home to take up his duties as canon at Frauenburg. "As this cathedral town was situated in the disputed territory between the Teutonic knights and the kingdom of Poland, he had much to do with war and administration; but his main interest was always astronomical, and he devoted the whole of his private life to the effort to find a more rational picture of the heavens, which he set out in final form in his book [*De revolutionibus*] which was printed only in the very year of his death in 1543."[2] The timing of the publication was not accidental as Copernicus was quite aware that his admittedly heretical views would bring down on him the wrath of the Church dogmatists so it was not until he was on his deathbed that he was given a copy of the book. Its central thesis was that the planets move in circular orbits around the sun. Copernicus argued that by setting Earth loose along a circular path, many puzzling observations about the movements of the planets could more easily be explained. He was convinced that nature could not be so complicated as to create the epicycle-laden Ptolemaic universe although the difficulties Copernicus encountered in explaining the deviations of the planets from their predicted paths forced him to keep some of Ptolemy's epicycles. He did not realize that the planets actually move in elliptical orbits; this discovery remained to be made by Kepler.

Copernicus was not an observer of the heavens but he was a bold and imaginative thinker. To him the whole system of epicycles seemed a needlessly involved and complex description of the motions of the planets. He saw very clearly that placing the sun rather than Earth at the center sweeps away most of Ptolemy's epicycles and greatly simplifies the picture of the solar system. Thus prompted by his love of symmetry and simplicity rather than by any hard observational data, Copernicus proposed his heliocentric theory for which he will always be revered. His system still required epicycles because he assigned circular orbits to the planets, which he considered as the most suitable for celestial bodies. The circle held a special appeal for the ancient Greeks and the early European astronomers because they considered circles to be the "most perfect" figures.

Having adopted circular planetary orbits, Copernicus saw at once that the sun could not be at the center of such orbits, since the distances of all the planets from the sun vary from moment to moment. From the instant a planet is closest to the sun (its perihelion), its solar distance increases steadily to a maximum value (the planet's aphelion). This distance then decreases to its perihelion value again, and this pattern of motion repeats itself over and over again. Even with the sun off center, Copernicus found it necessary to introduce epicycles (many fewer than Ptolemy) to account for all the observed apparent motions of the planets. This flaw in his model was a great disappointment to him, and all his adult life, he tried, unsuccessfully, to find orbits that do not require epicycles.

The great astronomer Johannes Kepler (1571–1630) succeeded where Copernicus had failed; he discovered the nature of the planetary orbits and laid down the three basic laws of planetary motion. Owing to these discoveries, Kepler is often called the "law-giver" of the solar system. This is not to say that Kepler enjoyed a peaceful life that provided him with the leisure to contemplate the motions of the planets, for his was one of the most difficult and impoverished lives ever endured by one who made such important contributions to the advancement of our knowledge about the universe.

Like Newton, young Johannes was a sickly child who was tormented by violent illnesses that began in his earliest years. His parents often despaired as to whether he would live to reach the age of ten. Because his persistent sickness prevented him from obtaining much formal education during his youth, he was taken out of school. His father ran a tavern and put his son to work as a potboy after the boy reached his ninth birthday. Despite his lack of formal education, young Kepler intended to enter the Church. He enrolled at a monastic school but soon found the Lutheran orthodoxy to be too stilted for his tastes. Kepler's time in the monastic school was not totally wasted, however, as it gave him a taste of a hitherto unknown world of knowledge that he entered with a passion after he left there to enroll at the University of Tübingen. It was at Tübingen that Kepler finally found his calling as a scholar, graduating second in his class. Although he knew little of Copernican astronomy except for occasional references in university lectures, Kepler's scholastic achievements led to an appointment as a lecturer in astronomy at Graz, which he accepted after some deliberation. His new status instilled in him an added sense of enthusiasm for his subject and "he speedily became a thorough Copernican, and as he had a most singularly restless and inquisitive mind, full of appreciation of everything relating to number and magnitude—was a born speculator and thinker just as Mozart was a born musician, or Bidder a born calculator—he was agitated by questions such as these: Why are there exactly six planets? Is there any connection between their orbital distances, or between their orbits and the times of describing them? These things tormented him, and he thought about them day and night."[3]

In 1596 Kepler published his first work, *Mysterium Cosmographicum*, which "deals with the problem of finding the law connecting the relative distances of the planets."[4] In the book Kepler described how he had experimented unsuccessfully with a number of methods for solving the distance problem before he was inspired to insert the five regular solids between the six planetary spheres so that the "sphere of Saturn is circumscribed to a cube in which the sphere of Jupiter is inscribed; the latter is circumscribed to the tetrahedron, and so on."[5] Unfortunately, this agreement was actually quite coincidental; we now know that at least nine planets exist (so that there is no one-to-one correspondence between the five regular solids and the number of known planets), and their distances agree only very roughly with Kepler's geometric scheme.

Kepler was not an observational astronomer but rather a theoretician who worked with mathematics and depended on the observations of others for the data he needed to deduce his laws. Fortunately the greatest of all naked eye observers in the history of astronomy, Tycho Brahe, who lived a short but turbulent and extremely productive life from 1546 to 1601, invited Kepler to become his assistant in 1599. Tycho, who had been expelled by the new king of Denmark from his great observatory on the island of Hveen in 1597, had been invited by Emperor Rudolf to continue his astronomical work in Prague as imperial mathematician, and it was there that he invited Kepler to come. Kepler eagerly accepted the invitation, for he knew that if he was ever to discover the correct planetary orbits, Brahe's data would show him the way. He stated this observation in his usual poetic way: "Tycho possesses the best observations and consequently, as it were, the material for the erection of a new structure; he has also workers and everything else which one might desire. He lacks only the architect who uses all this according to a plan. For, even though he also possesses a rather happy talent and true architectural ability, still he was hindered by the diversity of the phenomena as well as by the fact that the truth lies hidden exceedingly deep within them. Now old age steals upon him, weakening his intellect and other faculties or, after a few years, will so weaken them that it will be difficult to accomplish everything alone."

Kepler was sure that with Tycho's data he would accomplish

his task in a week or two, but things did not go smoothly at all. During the year 1600 Kepler found it very difficult to work with Brahe who supplied data to him only very sparingly and rarely of the kind he wanted. Working with the orbit of Mars, Kepler kept pressing for Brahe's Martian observations but Brahe was so very grudging and miserly about them that Kepler was driven almost to distraction. He despaired of ever completing his work, but a year later Brahe died and Kepler was appointed to Brahe's post as imperial mathematician, inheriting all of his predecessor's data. Even so his task was a long and tedious one: it was not until a few years before his death that he completed his monumental work and set mankind moving along the correct scientific path.

Despite Kepler's post in the royal court, he was valued primarily as an astrologer, for Emperor Rudolph was a superstitious man who sought to find clues to his nation's destiny in the stars. While Brahe was still alive, the emperor had begun to compile tables of planetary motion, which are now known as the Rudolphine tables. Because Brahe died before finishing the calculations, the emperor entrusted them to Kepler to finish; it was an onerous task that Kepler loyally undertook. "The Imperial funds were by this time, however, so taxed by wars and other difficulties that the tables could only be proceeded with very slowly, a staff of calculators being out of the question. In fact, Kepler could not get even his own salary paid: he got orders, and promises, and drafts on estates for it; but when the time came for them to be honoured they were worthless, and he had no power to enforce his claims."[6]

Eventually Kepler had to abandon his efforts as he found his personal circumstances becoming more dire. Despite the distractions caused by his financial difficulties, however, Kepler began to question seriously Aristotle's teaching that all of the planets move in harmonious circles. He concentrated on Mars, which he had studied throughout his career at Prague, but despite his best efforts he was unable to obtain the true positions of the planets using circular orbits. After numerous tedious calculations that yielded few useful results, he tried various oval orbits and found the calculations to be much improved but still not perfect. Fortunately, he had finally convinced himself that Aristotle's perfect circles were not the answer but merely an outdated philosophical obstacle to a deeper scientific truth. However, the mathematical calculations became progressively more time-consuming and difficult and Kepler approached the point of exhaustion. When he had almost given up after years of unremitting labor, he had a brilliant insight and substituted an ellipse in place of the oval. "But did it satisfy the law of speed? Could the rate of description of areas be uniform with it? Well, he tried the ellipse, and to his inexpressible delight he found that it did satisfy the condition of equable description of areas, if the sun was in one focus. So, moving the planet in a selected ellipse, with the sun in one focus, at a speed given by the equable area description, its position agreed with Tycho's observations within the limits of the error of experiment. Mars was finally conquered, and remains in his prison-house to this day. The orbit was found."[7]

It would be difficult to exaggerate the joy that Kepler felt at this great discovery, which forever ended the Aristotelian obstacles to the development of modern astronomy. However, his joy was short-lived as his patron Emperor Rudolph died in 1612, further complicating his efforts to get paid for his past labors. While Kepler debated whether he should leave Prague for good, his wife and one of their sons died within 11 days of each other. Perhaps taking this double tragedy as an omen that things were unlikely to improve in Prague and seeing little chance that he would ever obtain the salary due him, Kepler accepted a professorship at Linz.

Kepler supplemented his meager income at Linz by making astrological tables. However, he continued to think about the long-standing problem of deriving "some possible connection between the distances of the planets from the sun and their times of revolution, i.e. the length of their years."[8] These considerations ultimately led him to his third law, which shows that the cubes of the respective distances of the planets are proportional to the squares of the corresponding times.[9] His last great achievement, however, was to complete the Rudolphine tables and—in spite of his poverty—to secure somehow the funds to pay for its publication, which made available the first truly accurate tables that navigators ever possessed.[10] Sadly, Kepler spent his final days in Prague trying to get his salary paid; but it was to no avail, and he became ill while traveling home and died a short time later of a fever at the age of 59.

Before we state Kepler's three laws of planetary motion, we introduce a few concepts associated with planetary motions that will help us understand these laws. We have, to begin with, the mean distance of each planet from the sun. The larger the mean distance of a planet from the sun, the larger is its orbit. The mean distances of the planets, expressed in terms of Earth's mean distance, which is called one astronomical unit (93,000,000 miles), are as follows: Mercury, 0.387; Venus, 0.723, Earth, 1; Mars, 1.524; Jupiter, 5.203; Saturn, 9.540; Uranus, 19.18; Neptune, 30.07; Pluto, 39.44. A planet's mean distance is obtained by taking half the sum of its closest (perihelion) and farthest (aphelion) distances from the sun. We assign to the motion of each planet an interval of time called the sidereal period of the planet. This is the time it takes the planet to circle the sun once. As seen from the sun, it is the time the planet takes to return to the same point in the sky at which it was first observed. Taking the planets in the same order as above, these sidereal periods (expressed in years) are: 0.241, 0.615, 1.000, 1.881, 11.862, 29.458, 84.013, 164.79, 248.4.

Another period associated with the motion of a planet is the time it takes the planet to reappear in the same relative position with respect to the sun and Earth. This is called the synodic period; it tells us how long we have to wait before we see the planet again in the same direction relative to the sun's direction. The synodic periods of the planets (expressed in days in the same order as above, skipping Earth), are: 115.88, 583.92, 779.94, 398.88, 378.09, 369.66, 367.49, 366.74. The reason the synodic period of a planet is different from its sidereal period is that Earth is revolving around the sun just as the planet in question is. The synodic period is the time it takes Earth to overtake the planet (if Earth is moving faster than the planet) or the time it takes the planet to overtake Earth (if the planet is moving faster than Earth) from the moment Earth and the planet lie on the same line from the sun. The apparent complex motion of a planet as seen from Earth stems from the combination of Earth's true motion and the planet's true motion around the sun. The failure of the ancient and medieval astronomers to understand this complex motion led them into error regarding the structure and dynamics of the solar system. Copernicus *did* understand this motion and saw that in the heliocentric theory of the solar system the motions of the planets are greatly simplified.

Another important quantity associated with the motion of a planet is its total energy. A body acquires energy when we do work on it by pushing or pulling it (setting it in motion by exerting a force on it); work and energy are thus equivalent, and we may consider energy as stored work done on a body. Since energy can never be destroyed or created (the principle of the conservation of energy), the body that has acquired this energy can (under appropriate conditions) give it back to us in the form of work again. The energy that a body acquires when we do work on it by pushing or pulling it can appear in the form of the motion of the body (kinetic energy) or in the form of its height above the ground (potential energy). Just as we have to do work to impart motion to a body, we have to do work to lift it off the ground. The faster we make a body move, the more work we have to do on it and the greater is its kinetic energy. The higher above the ground we lift a body, the more work we have to do on it and the greater is its potential energy.

Scientists have discovered that the kinetic energy of a body can be calculated very easily if we know its mass and speed. We simply multiply half its mass by the square of its speed ($\frac{1}{2}mv^2$, where m stands for mass and v its velocity). The potential energy of a body relative to Earth at any height above the ground is calculated by multiplying the weight the body would have at that height if it were at rest on a plateau at that height (the weight of a body decreases as its height increases) by its distance from the center of Earth. The potential energy of a body is always taken as a negative quantity since, as a matter of convenience, the potential energy of a body is taken as zero when the body is infinitely far away. The sum of the potential energy of a body and its kinetic energy is its total energy, which may be negative, zero or positive depending on how fast it is moving. If the body is moving at a speed less than the speed of escape, its total energy is negative; if its speed equals the speed of escape, its total energy is zero; and if its speed exceeds the speed of escape, its total energy is positive.

The unit of energy, the erg, is an extremely small amount of energy; one calorie, which is the heat required to raise the temperature of 1 gram of water by 1°C, equals about 42 million ergs. A 2-gram mass, moving at a speed of 1 cm per second, has 1 erg of kinetic energy. If the height above the ground of a 1-gram mass is increased by 1 cm, its potential energy is increased by 980 ergs.

A second important dynamical quantity associated with the motion of a planet is its rotational motion or its angular momentum, as scientists call it. This is somewhat more difficult to comprehend than energy, but it is something we meet in our lives all the time whether we know it or not. Like energy, it is conserved—it can neither be created nor destroyed. If one body loses angular momentum, some other body or system of bodies gains it. Thus, angular momentum, like energy, is passed from body to body. The easiest way to get at the concept of angular momentum is to consider a particle moving with a fixed speed in a circle of a definite size (a definite radius). Associated with this circular motion is a certain quantity that does not change as the body moves from point to point. This quantity, which we call the magnitude of the angular momentum of the body, is the product of the body's mass m, its speed v, and the radius r of the circle (mvr). Not only does this quantity remain constant, but the plane in which the orbit lies remains the same. This is another way of saying that the direction of the axis of revolution or rotation of the body remains unaltered.

Although we have introduced the concept of angular momentum by considering a circular orbit, it applies to the motion of a body regardless of the kind of orbit it is traversing. This tells us something important immediately. Suppose we contemplate the angular momentum of a planet when it is farthest from the sun (the distance r is a maximum). If we know the mass of the planet and its speed at that distance, we can calculate its angular momentum as the product we wrote down above. This quantity must not change as the planet gets closer to the sun. But since r gets smaller as the planet gets closer to the sun and m remains the same, v must increase to compensate for the decrease in r. We thus see that a planet, moving in a noncircular orbit, must speed up as it approaches the sun and slow down as it recedes from the sun. The planet thus moves fastest in its orbit at perihelion and slowest at aphelion.

We are now in a position to consider Kepler's three laws of planetary motion as he deduced them from Brahe's data. The first law is a statement about the shape of the orbit of a planet: Every planet moves around the sun in a fixed ellipse with the sun at a focus of this ellipse.

To explain this law we first define an ellipse. We do this most easily by explaining how it can be constructed. Draw a straight line on a piece of paper and mark the endpoints A and A'. Take a piece of string whose length exactly equals the length of the line AA' and nail the two ends of the string to two points F and F' on the line that are equidistant from A and A' (the length AF equals the length F'A'). Place the point of a pencil in the loop of the string and, holding the string taut, move the pencil point along the paper tracing out a curve. This curve is an ellipse and the two points F and F' are called the foci of the ellipse. If this curve represented the orbit of any planet, the sun would be at either F or F'. We note that the point of the pencil must pass through A and A' because the distance from A to A' just equals the length of the string. These two points are thus points on the orbit of the planet. Having constructed the ellipse as described, we see that we can define the ellipse as follows: It is the collection of all points the sums of whose distances from two fixed points (the points F and F') are equal (equal to the length of the string or to AA').

If the sun is at focus F (which then becomes a point that the orbits of all the planets have in common), point A becomes the perihelion of the planet and point A' the aphelion. The mean distance of the planet from the sun is just half the length of the string or half the distance from A to A'. This distance, which gives the size of the ellipse, is called the semimajor axis of the ellipse; it is the largest radius of the ellipse (the radius from A or A' to the center of the ellipse). The shortest radius of the ellipse (the radius from the point on the ellipse, one quarter of the way around from A, to the center) is called the semiminor axis. The total energy of a planet in its orbit depends only on the semimajor axis (or size) of the ellipse. The larger the semimajor axis (the bigger the ellipse), the greater is the planet's energy.

What we have just said about the orbit of a planet around the sun holds for the orbit of an artificial satellite and the moon around Earth. When a satellite is launched into orbit around Earth, it moves in an elliptical orbit with Earth's center at one focus of this ellipse. The major axis is obtained by taking half the sum of the satellite's shortest and longest distances from Earth's center (actually 4000 miles plus half the sum of the satellite's closest and farthest distance from Earth's surface). The larger this orbit (or the larger the mean distance of the satellite from Earth), the more fuel we have to use to place the satellite in orbit and the more energy it has in its orbit.

We come now to another aspect of an elliptical orbit—its shape. It is clear that we can have an infinite variety of elliptical shapes ranging from practically a straight line to a circle for the same size or the same semimajor axis. To see this we go back to our string and straight line AA', and note that if the ends of the string (or the two foci F and F') are close to A and A' the ellipse is very flat and narrow. The semiminor axis is small compared to the semimajor axis and the area enclosed by the ellipse is small. If F and F' coincided with A and A' the ellipse would be a straight line and the enclosed area would be zero. On the other hand, if the ends of the string are very close to each other (the foci F and F' are close to the center of the line AA'), the ellipse is very round; if the two ends of the string touch (the two foci coincide with the center), the figure becomes a circle. We thus see that the circle is a special case of an ellipse.

Just as the size of the ellipse (its semimajor axis) determines the total energy of the planet or satellite moving in that ellipse, its shape (for the given size) determines its angular momentum. As the shape of the orbit changes from very nearly a straight line to a circle, the total energy of the body moving in the orbit (planet or satellite) remains constant, but its angular momentum increases from very nearly zero to a maximum value—the rounder the orbit or the more area it encloses (for a given size), the larger the angular momentum. We note that since the size and shape of the orbit of a planet do not change in the course of time (actually there are changes but these are so minute as to be negligible except over long periods of time), both the total energy and the angular momentum of the planet remain constant. These two unchanging features of the motion of any planet demonstrate two important principles in nature: the principle of the conservation of energy and the principle of the conservation of angular momentum.

We have seen that an infinite range of sizes and of elliptical shapes, from the very narrowest to very round ones, is possible for planets or satellites moving under the gravitational pull of the sun or a planet. What then determines the size and shape of an orbit? The size of a planet's orbit is determined by the speed of the planet at any given point in its orbit and its distance from the center of the sun at that point. The orbit's shape is determined by the direction of its motion and its speed. If the perpendicular distance from the sun's center to the line along which the planet is moving, at any point in its orbit, is smaller than the distance from the sun to the planet, the orbit is highly elliptical (narrow) and the angular momentum is small. If this perpendicular distance is not much smaller than the sun-to-planet distance, the orbit is round and the angular momentum is large. For the orbit to be a circle, two conditions must be fulfilled exactly. The direction of the planet's motion at any point must be *exactly* at right angles to the line from the center of the sun to the planet; the speed of the planet must have exactly one and only one precise value (called the circular speed) at all points in its orbit. Since only this particular velocity out of an infinitude of possible velocities at a given point leads to a circular orbit, the probability for a circular orbit is zero so that circular orbits do not exist.

We have discussed the nature of the planetary orbits in some detail because without such a discussion the significance of Kepler's first law cannot be fully grasped, and the second law, which we now state and discuss, would not be as meaningful: The radius vector (the line) from the center of the sun to a planet sweeps out equal areas in equal intervals of time as the planet revolves in its orbit.

We may illustrate this law by considering the motion of Earth in two given 24-hour periods at two different times of the year, say from January 1 noon to January 2 noon and from June 1 10 A.M. to June 2 10 A.M. If two lines are drawn from the center of the sun to Earth's positions on January 1 and January 2, these two lines and the arc of Earth's orbit during that 24-hour period enclose an area (like a segment of a pie). In the same way we obtain an area enclosed by the two lines and the arc of the orbit for the June 1 to June 2 24-hour interval. The second law states that these two areas are equal and that this equality is true for all such areas cut out during any 24-hour period at any time of the year. Since Earth is closer to the sun in January than in June, the January area can equal the June area only if Earth moves faster in January than in June. We thus see that Kepler's second law tells us that the planets move faster as they get closer to the sun.

If a planet's orbit were a circle, there would be nothing surprising about the law of equal areas because then the planet's distance from the sun and its speed would always be the same. But since a planet's orbit is an ellipse, it must have been somewhat surprising to Kepler that the radius vector sweeps out equal areas in equal times. Not knowing the law of gravity, Kepler could not understand why equal areas should be swept out in equal times, even though he accepted this law without question.

Having discussed the relationship of the shape of a planet's orbit and the area it encloses to the angular momentum of the planet, we can now understand Kepler's second law because it, too, deals with areas. Since the second law states that some characteristic or property of a planet's motion remains constant as the planet moves from point to point, this law must refer to one of the only two quantities, energy and angular momentum, associated with a planet's motion that remain constant. Since the energy of a planet depends only on the size of its orbit and not on its shape, whereas the area for a given size orbit does depend on the shape of the orbit, the law of areas cannot refer to the energy; it must therefore refer to the fact that the angular momentum of a planet is constant throughout the orbit and so is a restatement of the principle of conservation of angular momentum. Without going into details, we note that the angular momentum of a planet in its orbit (orbital angular momentum) equals twice the planet's mass times the rate at which the radius vector from sun to planet sweeps out area. Since the mass of a planet remains constant and the rate at which area is swept out remains constant (Kepler's second law), twice the product of these two quantities (the angular momentum) remains constant.

By studying Brahe's data for the motion of Mars, Kepler discovered the first two of his three laws within the 6-year period after Brahe's death. He described these laws in his famous book *The New Astronomy*, published in 1609. This book is outranked in importance only by Copernicus's *The Revolution of Celestial Bodies* and Newton's *Principia*. More than 20 years elapsed after this book appeared before Kepler discovered and stated his third law, which he considered the greatest achievement of his life and which expresses the great harmony in the motions of the planets that he had been seeking all his adult life. Various fanciful schemes for relating the speeds of the planets in their orbits to their distances from the sun had occurred to him, but none gave the correct relationship, which he found only after the most laborious, time-consuming, and boring calculations: The squares of the sidereal periods of the planets are proportional to the cubes of their mean distances from the sun.

We may state this law somewhat differently. Take the time a planet requires to revolve around the sun once and square it (multiply the time by itself). Take the mean distance of the planet from the sun (the semimajor axis of the planet's orbit) and cube it (multiply it by itself twice). Divide this number into the previous number (the number obtained by squaring the period) and the result is the same for all planets, that is, a constant. In doing this we must measure periods in the same units (e.g., years) for all of the planets, and all the mean distances from the sun in the same units (e.g., in terms of Earth's mean distance A, the astronomical unit). To find the constant in this proportionality of the squares of periods to the cubes of the mean distances when time is measured in years and mean distances in astronomical units, we consider Earth. Since its period is exactly 1 year and its mean distance is exactly 1 astronomical unit, we obtain 1/1 or 1 for the constant of proportionality. Hence, Kepler's third law tells us that the square of any planet's period, expressed in years, divided by the cube of its mean distance, expressed in astronomical units, equals 1.

Kepler deduced this law from Brahe's data so it is only as good as Brahe's observations, which were not accurate enough to show the slight departures from Kepler's third law that we now know exist. As one goes from Mercury to the massive outer planets, the ratio of the square of the period to the cube of the mean distance does not remain constant but decreases very slightly for the massive planets. It is smallest for Jupiter, the most massive planet. Brahe's observations were not accurate enough to reveal this difference from planet to planet so that, based on Brahe's data, Kepler could not formulate his third law correctly. We shall see that we can deduce the correct third law from Newton's law of gravity.

The importance of Kepler's third law of planetary motion is that with it we can obtain a picture of the solar system to the correct scale, because the sidereal periods of the planets can be measured quite accurately. Since the third law tells us that the cubes of the mean distances of the planets measured in astronomical units equal the squares of their periods measured in years, knowing the periods permits us to calculate the mean distances, expressed in terms of Earth's mean distance. We can thus construct a correctly scaled picture of the solar system using Earth's mean distance (the astronomical unit) as our unit of distance.

Even before Kepler discovered his three laws of planetary motion, he knew that the planets move the way they do because of the pull of the sun, but he was unable to formulate the law of gravity correctly. Throughout his writings one finds references to a force exerted over a distance by one body on another. For example, at one point he notes that the tides on Earth are probably caused by some kind of force stemming from the moon. In discussing the motions of the planets he correctly states that the sun is the prime mover, directing the planets by exerting a force on them that becomes weaker as their distances from the sun increase. His error in describing this lay in his assumption that the force does not spread out in all directions from the sun like light, but rather spreads out only in the plane close to which the planes of the planetary orbits lie. He thus concluded that this force diminishes as the distance increases rather than as the increase in the square of the distance, which Newton discovered. Kepler further erred in picturing the force from the sun as not only pulling the planets but also acting laterally like a broom and sweeping them along in their orbits. With no correct law of force to guide him, Kepler could not deduce the orbits of the planets mathematically from basic principles. This was left for Newton after he had discovered the correct law of gravity.

Although Isaac Newton (1642–1727) is best remembered for developing what is now known as Newtonian mechanics and proposing his universal law of gravitation, his birth to the wife of a yeoman farmer on Christmas Day at Woolsthorpe was notable for how close he came to perishing. Isaac was a sickly baby who was said to be so small that he could fit in a quart jar. Despite this unpromising beginning, Isaac managed to survive the year and, in time, became an inquisitive boy who was interested in mechanical devices and puzzles but somewhat shy and awkward when in the company of others. He had no real role model during his early childhood as his father died before he was born. His widowed mother married the pastor of a nearby parish, leaving her son in the care of his maternal grandmother at Woolsthorpe. Because Isaac did not see his mother regularly during the decade she spent with her second husband, he was often on his own and, despite an uneven primary education, found a welcome refuge in the few books that he could lay his hands on. Although his mother returned to Woolsthorpe after the death of her second husband, the damage done to Isaac in the interim appears to have been substantial. His personality was characterized by widely swinging moods and tantrums; traits that were manifested repeatedly in later years when he engaged in acrimonious disputes about the priority of various inventions and discoveries with other scientists such as Robert Hooke and Gottfried Wilhelm Leibniz. In spite of the emotional turmoil he suffered, Isaac was a good student whose desire to succeed academically seems to have originated with the taunts of a schoolmate who had received a better grade on an assignment. Although Isaac's mother had thought that her son would eventually run the family farm, his lack of interest and utter incompetence in farm management soon became apparent and his mother despaired about what to do with him.

Fortunately, a maternal uncle was perceptive enough to see that Isaac was simply bored with a farmer's life and that his neglect of his duties did not stem from lack of intelligence. Isaac's uncle had seen the tenacity with which his nephew had approached his subjects in school and suggested that Isaac would be better suited for a calling that could make more use of his considerable intellectual abilities such as the clergy. Although Isaac was never overly religious, he saw the higher education required to join the clergy as the key to getting off the farm, so he enrolled at Cambridge University in 1665, where he passed his undergraduate career—if the university's records are to be believed—without distinction. Indeed, the only man whom he seems to have made any impression on was Professor Isaac Barrow, the Lucasian Professor of Mathematics at Cambridge. Barrow recognized that Newton had some potential as a scientist and took him under his wing.

While Newton was a student at Cambridge, he was exposed to the ideas of Galileo and Descartes as well as the classical philosophers who had done so much to ensure that science would be virtually moribund for more than a millennium following the sack of Rome. Given the deference toward the classical works that still existed at that time, it is likely that Newton began to formulate his more important ideas based on readings he did on his own rather than the officially sanctioned classroom texts. Because he was blessed with the ability to concentrate on a problem with such single-minded devotion that he could turn it about in his mind for months on end until it yielded an answer to him, it is likely he was first seriously motivated to exercise his mental muscles when he wondered about the hodgepodge of theories purporting to explain the dynamic forces guiding both the universe and Earth.

Toward the end of his undergraduate career, the bubonic plague reappeared and Cambridge University was closed. Newton returned to the family farm where he spent the next two years thinking about the nature of space and time and light and formulating a branch of mathematics called calculus that eventually became an invaluable tool in all branches of science. Although various explanations have been offered by science historians about how Newton could formulate his laws of motion, his law of gravity, his corpuscular theory of light, and the calculus at such a young age and in such a brief time, perhaps the best answer is to be found in the solitude he enjoyed at Woolsthorpe, which gave him the leisure to concentrate, without distraction, on these matters. It is interesting that both Newton and Einstein were about the same age when they made their monumental discoveries—a similarity that suggests that scientific intuition is most acute in the early stages of a career when the mind has not yet become rigid with prejudices and paradigms.

Newton returned to Cambridge in 1667 at the age of 24; by that time he had developed his important ideas about optics, mechanics, and gravitation. However, he was reluctant to share his ideas with others. Throughout his career he was never eager to publish his scientific findings and only produced his greatest work, the *Principia*, after his friend Edmond Halley, the noted astronomer, offered to underwrite its costs and personally oversee its production.

Of his great discoveries, the only one that he committed to paper during his early years was an account of his development of the binomial theorem and the differential and integral calculus; he presented the handwritten manuscript to Barrow in 1669. Astounded by the evident brilliance of his former student, Barrow resigned his chair at Cambridge, recommending that Newton be appointed as his successor. Despite the reluctance of some of the university administrators to promote such a young man to one of the most coveted faculty positions in England, the sheer weight of Newton's mathematical contributions overrode all objections to his appointment to Barrow's chair.

In 1669 Newton began giving lectures in optics, which included his own theories about the nature of light and its propagation; his papers on the subject were written in Latin and deposited in the university archives. His lectures were poorly attended but his interest in the behavior of light led him to devote much of his spare time to the grinding of lenses and the construction of a magnificent reflecting telescope that attracted the attention of the Royal Society. In 1671 the Royal Society invited Newton to submit one of his telescopes for examination. Based upon the Royal Society's favorable assessment of its scientific value, the membership invited Newton to join their ranks. As a result, Newton became a member of that august body based not on the work he did that was to dominate scientific thinking for more than 200 years (much of which remains an essential cornerstone of the physical sciences) but for his less important work as a craftsman (albeit a very able one) in improving the telescope.

In 1672 Newton submitted his first paper to the Royal Society;

it contained the results of his experiments refracting a beam of sunlight using a prism. Newton thus demonstrated that light is not a single homogeneous color as had previously been supposed but instead a spectrum of different colors that are mixed together into a polychromatic whole and are separated from each other only by virtue of their differing degrees of refrangibility. This conclusion was so startling and at odds with the views of the ancient philosophers such as Aristotle and Galen that Newton was severely criticized by some of his contemporaries, most notably Christian Huygens and Robert Hooke, who wanted Newton to explain how the continuity of the spectrum could be accounted for by purely mechanical principles. In addition, Huygens believed that light consists of waves—not particles as believed by Newton—and offered several criticisms of Newton's theory, although the two theories shared some significant similarities and each man admired the work of the other greatly. Newton brought his work in optics to its full maturity some 30 years later with the publication of his book *Opticks* in 1704.

Despite the ease with which Newton probed the mysteries of the universe, he never tried to offer answers for the great "why" questions that philosophers have grappled with over the centuries—thus giving rise to his famous statement "I do not form hypotheses." Instead he was concerned primarily with developing theories that can somehow explain recurring patterns of physical phenomena, such as the falling of objects to the ground or the orbit of the moon around Earth in terms of the same attractive force, which he attributed to matter itself.

Newton's scientific career was characterized by great bursts of creativity followed by long periods of inactivity during which time he completely withdrew his attention from scientific matters. He often complained of the tedious labor required to prepare his papers for publication and the controversies that invariably arose between him and his colleagues. By 1675 he had done most of his work in mechanics and formulated his universal theory of gravitation but had never documented his ideas in print. It took Robert Hooke's unproven assertion that the motions of the planets could be explained by the inverse square law of attraction to rouse Newton from his apathy toward his profession. It took a visit by his friend Halley to Cambridge to prod Newton to reveal that he had calculated that a body subject to the inverse square law of attraction travels in an ellipse. Newton later sent a proof of this work to Halley; Halley was so impressed with Newton's ingenuity that he pleaded with Newton to expand his treatment of gravitation and mechanics into a treatise. The *Principia* took about 18 months for Newton to write and contained the basic tenets of Newtonian mechanics, which brought an unprecedented sense of order and coherence to the apparent chaos of nature by showing how the motions of the planets can be explained in terms of a single law of gravitational attraction.

Newton's masterwork originally appeared in Latin, as was true of most learned books of that time, and was laden with detailed propositions and extensive mathematical proofs. Although he could have simplified the text greatly by using his calculus, he chose the traditional classical geometry to illustrate his arguments. In the first book he deduced the laws of simple orbits from the inverse square law and hypothesized that every particle in space is attracted to every other particle—a brilliant and audacious idea that enabled him to solve problems involving the mutual gravitational interactions of two bodies. The second book dealt with the motions of fluids and offered the first mathematical treatment of wave motion; it constitutes what E. N. Da C. Andrade called "the first text-book of mathematical physics, and of hydrodynamics in particular, and is embellished by an account of many experiments, carried out with the highest degree of precision and skill." However, it was the third and final book of the *Principia* in which Newton used his universal law of gravitation to demonstrate how satellites revolve around planets and planets around the sun. He also calculated the masses of several planets and the density of Earth and explained most of the variations in the movement of the moon around Earth. His use of the gravitational force to explain the orbit of the moon also laid the groundwork for the modern study of tides. Throughout the *Principia* Newton offered his own careful scientific experiments to support his conclusions, illustrating his often overlooked skill as an experimental physicist.

Despite Newton's hardheaded approach to the physical sciences and his unflagging efforts to treat rigorously a wide variety of natural phenomena ranging from the propagation of light to the motions of the planets around the sun, he also had a curiously mystical side that manifested itself in a lifelong interest in alchemy. He wrote nearly a half million words on the subject, which, aside from their historical value, have been regarded by most scholars as largely worthless. Although he made a number of extensive chemical investigations, he published only a single note on chemistry; he also appended a series of queries to the later edition of the *Opticks*, one of which suggests that atoms are gravitationally attracted to each other and do not hook together as had been previously supposed. Newton also engaged in theological speculations and produced two massive books, *The Chronology of Ancient Kingdoms Amended* and *Observations upon the Prophesies of Daniel and the Apocalypse of St. John*, which contain little of the critical analysis which Newton applied to the physical world.

The great effort required to produce the *Principia* left Newton with little appetite for further scientific work and for the next few years he devoted himself to his duties at Cambridge. In 1688 he was elected a member of Parliament in recognition not only of his scientific achievements but also his courageous defense of Cambridge against the attacks of the recently deposed King James II. His political career was undistinguished as he preferred to observe the parliamentary debates in silence instead of offering his own views on particular subjects. Over the next few years he tried unsuccessfully to obtain an administrative post in London and suffered from long bouts of melancholy and sleeplessness. In 1793 he was said to have suffered a mental breakdown, but its permanent effects were overexaggerated by some of his colleagues as Newton made a full recovery and his mental powers remained unimpaired. He did little scientific work during this time because he was bored with science although he did continue to engage in some scientific correspondence. He was jolted from his complacent state when he was appointed Warden of the Mint in 1696, a job that required that he leave the academic life forever and take up residence in London. Three years later he was promoted to Master of the Mint.

While at the Mint, Newton actively participated in the implementation of a new recoinage scheme. What time he did devote to science consisted largely in the preparation of his *Opticks* manuscript for publication; Newton appears to have been quite content with this situation and found the cosmopolitan bustle of London preferable to the scholastic solitude of Cambridge. He was elected president of the Royal Society in 1703 and knighted by Queen Anne in 1705. For the next 22 years until his death in 1727, he enjoyed the acclaim and honors that were accorded to the man rightly declared to be the most important scientist of all time. He was the toast of fashionable London society and received a steady stream of distinguished visitors. After his death, he was buried in Westminster Abbey.

To understand Newton's universal law of gravity and see how it permits us to deduce Kepler's three laws (the third law comes out correctly), we must define the concepts of velocity, acceleration, force, and mass precisely. We have used these concepts rather loosely in our discussion so far, but now we must be more precise. When a body moves along a straight line, we define its average speed as the distance it travels in a given time divided by the time. The speed is thus expressed as distance per time (e.g., (cm/sec, mi/hr, mi/sec, km/sec). If we consider the motion of the body during an extremely short interval of time (e.g., a millionth of a second) we obtain what is very nearly the instantaneous speed. In principle, the instantaneous speed can be found only if we watch the body for an infinitesimal period of time, which, of course, is not possible in practice. In spite of this difficulty Newton devoted much time to the consideration and analysis of changes occurring in infinitesimal time intervals and was thus led to the discovery of the calculus—the single most powerful intellectual tool invented by man. In general, the instantaneous speed of a body changes from moment to moment as it moves along the given line. If we specify the direction in which the body is moving as well as its speed, we speak of the velocity of the body. The speed is thus the magnitude of the velocity of a body. Just giving the speed by itself does not give a complete description of the motion of the body. To say that the speed of Earth as it moves around the sun is $18\frac{1}{2}$ miles per second does not tell us very much about Earth's motion.

If either the speed of a body, its direction of motion, or both change, we say the body is accelerated. In other words, when a body is accelerated, its velocity changes. The acceleration of the body is not the change in the velocity itself but the rate of change (the amount of change per unit time) of the velocity. Thus, if the speed of a body changes quickly (e.g., a change of 100 miles per hour in a second) its acceleration is large, and if its direction of motion changes quickly (e.g., by 180° in a tenth of a second) its acceleration is large. If the speed of a body changes continuously but its direction of motion remains unchanged, the motion of the body can only be straight-line motion, and the acceleration at any moment is then either in the same direction as the velocity or in the opposite direction. For a body that is falling freely (nothing resists its motion) toward the center of Earth (its direction of motion is along the vertical), the acceleration (the acceleration of gravity as it is called) is in the same direction as the velocity of the body so that the speed of the freely falling body increases steadily. Near the surface of Earth the speed of such a body increases by 980 cm/sec every second or by 32.2 feet/sec every second so we say the acceleration of gravity on the surface of Earth is 980 cm/sec per sec or 32.2 feet/sec per sec and is always directed downwardly along the vertical (always points to Earth's center). If a body is thrown directly upwardly in a vacuum (this is considered to be free-fall also even though the motion is upward), its speed decreases continuously at a rate of 980 cm/sec or 32.2 feet/sec every second. The same acceleration of gravity acts as for a falling body but its direction is still always downward and thus opposite to the velocity of the rising body. Here, too, the direction of motion is always along the same line.

If the speed of a body always remains the same but its direction of motion changes continuously and by the same amount in equal intervals of time, the motion of the body *must be circular*. The body is accelerated in this case too (remember that acceleration means change of speed, change of direction of motion, or both) but the direction of the acceleration in circular motion is exactly at right angles (perpendicular) to the velocity of the body. In fact, the acceleration in this case is always in the direction from the body to the center of the circle (always points to the center of the circle). Since the body is moving from point to point along the circumference of the circle, the direction of the acceleration thus changes continuously. This type of motion is of particular importance in astronomy because the orbits of the planets and satellites are almost circular.

To determine the acceleration of a body moving with constant speed in a circle is somewhat more difficult than for straight-line motion, but it can be calculated with the aid of a fairly simple formula. To understand this formula, we recall our experiences rounding a curve in an automobile. If the road curved gently, we had no trouble following it even at a high speed, but if the road curved sharply, we always had to slow down considerably before negotiating the curve. This taught us that the higher the speed of a car, the greater is its acceleration as it moves along a curve and the sharper the curve, the greater is its acceleration.

It is clear that the acceleration of a body moving rapidly in a circle is larger than that of a body moving more slowly in the same circle if we note that the acceleration measures the rate at which the directions of the velocities of the two bodies change as they move along the circle. Since the faster body is changing its direction of motion more rapidly than the slower body, the acceleration of the faster body is greater than that of the slower one. If two bodies are moving at the same speed in two circles of different size, the body moving in the smaller circle suffers a larger acceleration than the body moving in the larger circle. To see why this is so, we note that each body moving in its own circle changes the directions of its motion by 180° when it traverses half the circumference. Since the two bodies are moving equally fast, the one in the smaller circle covers half its circumference, and hence changes its direction of motion, faster than the body in the larger circle does, and that proves our contention. We conclude, then, that the acceleration of a body moving in a circle is greater the greater its speed and the smaller the radius of the circle. The actual numerical value of the acceleration of the body is obtained by dividing the square of its speed by the radius of the circle.

We have spent some time discussing the concept of acceleration because it plays a very important role in the way bodies move when they are pulled or pushed, that is, acted on by a force. The relationship between the force acting on a body and that body's acceleration is the content of Newton's laws of motion, which stem from Galileo's fundamental work on dynamics. We generally think of a force as acting between two bodies when they are in contact with each other, as when we take hold of a spring and stretch it (our hand is in contact with the spring). The amount by which the spring is stretched as we pull it is then a measure of the force we are exerting on the spring, which exactly equals the pull of the spring on our hand. But bodies can exert forces on each other without direct contact between them as in the case of the gravitational force. (All forces in nature are in fact acting over a distance; there is no real contact between any molecule or atom in our hand and any molecule in the spring when we pull it.) Newton's laws of motion tell us when a body is being subjected to a force of this sort, for according to these laws a body must either remain at rest if it was initially at rest or, if moving, continue to move with constant speed in the same straight line if no net force acts on it. If, then, a body is at rest or *moving with constant speed along the same straight line*, no *net* (or unbalanced) force of any kind is acting on it no matter what individual forces it may be subjected to. All the separate forces acting on it must cancel out or balance each other to give a net force of zero. Thus, a book resting on a table has no net force acting on it because the downward pull of Earth on it is just balanced by the upward push of the table. In the same way if you push a heavy object across the floor at a constant speed, the net force acting on the object must be zero; the force you exert to push it equals the force of friction exerted by the floor on the object in the opposite direction. Note that in considering a combination of forces we must take into account that each force is exerted in a definite direction because a force is a directed quantity. The rule for combining (adding) such quantities is not the same as that for adding numbers.

If a body is accelerated (if its velocity—its direction of motion or its speed—is changing), a force must be acting on it whether any other body is touching it or not. Thus, acceleration—*not velocity*—is the sign of the action of a force. The force and acceleration imparted to the body by the force are related to each other in a simple way; the acceleration is in the same direction as the force, and the larger the force the larger the acceleration.

One other element enters into the relationship between the force acting on a body and the acceleration it imparts to the body: it is a characteristic or quality of the body called the inertia or mass of the body. We use the force–acceleration relationship to define the mass of a body. Although the acceleration imparted to a body in a given direction by a force is doubled when the force is doubled, tripled when the force is tripled, and so on, it is incorrect to say that the force equals the acceleration since force and acceleration are two entirely different things. We may, however, say that the force is proportional to the acceleration, which means that the force equals the acceleration multiplied by some quantity. We call this quantity the mass of the body. If we call F the force acting on the body, a the acceleration imparted to it, and m the mass of the body, we express this relationship in the form $F = ma$.

This expression is Newton's famous second law of motion. It states that the smaller the acceleration imparted by a given force to a body, the larger is the mass of that body. This law enables us to compare the masses of any two bodies in a very simple way: we apply the same force to each and compare the resulting accelerations. That body that acquires the smaller acceleration has the larger mass, and its mass is larger than that of the other body by the same factor that its acceleration is smaller than the other acceleration. The unit of mass is called the gram; it is that amount of mass (matter) that experiences an increase of speed of 1 cm/sec every second (an acceleration of 1 cm/sec^2) when a unit force is applied to it. The unit of force is called the dyne; a force of about 981 dynes is required to lift 1 g of matter off the ground, and to lift 1 pound a force of 445,000 dynes must be applied. The dyne is thus a tiny unit of force. Since the more massive a body is, the less readily it responds to a force, the more inertia it has. The inertia of a body is thus measured by its mass.

We note one more important characteristic of the force or interaction between two bodies. If body A exerts a force on body B, then B exerts an exactly equal but opposite force on A. This relation is known as Newton's law of action and reaction (third law of motion), which states that forces always come in equal and opposite pairs.

Having introduced the concepts of force, acceleration, and mass and having noted how they are related to each other through Newton's law of motion, we now introduce Newton's law of gravity. We shall then see how Kepler's three laws of planetary motion can be deduced from Newton's law of gravity when it is properly combined with his law of motion.

Of the four known forces in nature, the gravitational force is the most pervasive, the most obvious, and the most easily recognizable one. If there were no gravity, there would be no stars, no solar system, no planets, and no advanced forms of life. If the force of gravity were suddenly to vanish, our atmosphere and water would quickly disappear, all objects (however massive) not attached to Earth's crust would gradually float off into space, and the concepts of up and down would have no meaning. It is remarkable that, in spite of the unescapable influence of gravity on the motions of all earthly objects, no one, until Newton, recognized gravity as a force: quite simply, people thought that objects fall not because Earth pulls them but because being on the ground is their natural state, which they seek to achieve when they are off the ground.

Once Newton's second law of motion was accepted, it became clear that since a freely falling body is accelerated as it falls, a force must be acting on it. This force, which we call the weight of the body, is the gravitational pull of Earth on the body. The remarkable thing about this force is that it imparts the same acceleration to all bodies in a vacuum, at the same height above Earth regardless of their weight as was first discovered by Galileo. Stated differently, we may say that however hard or however gently Earth pulls different bodies toward it, their accelerations toward Earth are all exactly equal if these bodies are at the same height.

Since the weight of a body is just the gravitational force of Earth on it, and this force imparts to the body the acceleration of free fall (the acceleration of gravity), we see that the weight of a body (according to Newton's law of motion) equals its mass multiplied by the acceleration of gravity. We write this as follows: $W = mg$, where W is the weight, m the mass, and g the acceleration of gravity. If m is expressed in grams and g in cm/sec per sec, the weight W is obtained in dynes. Thus, 1 g weighs 981 dynes. We note, by the way, that a body in free fall (i.e., moving freely in response to Earth's gravitational pull) is weightless since any scale to which it is attached falls right along with it so that the body cannot push against the scale.

Newton, who was well aware that the weight of a body at rest on Earth is just Earth's gravitational pull on it, could not, from that knowledge alone, deduce the nature of the gravitational force. He felt instinctively that the shapes and sizes of the bodies as well as their masses, enter into the formula for the gravitational force, and, not knowing how the shapes or sizes would influence this force, he decided to formulate the force for particles that have no size or shape, namely points of matter, or mass points. Since mass is the only thing associated with a mass point, only the masses of two such gravitationally interacting particles and the distance between them can play any role in their mutual gravitational attraction.

Newton assumed, quite reasonably, that the gravitational force exerted by a particle spreads out equally in all directions, like light from a point source. This means that this gravitational force falls off as the square of the distance between the particles. If the distance is doubled the force is one-fourth as large; if the distance is tripled the force is one-ninth as large, and so on. We thus speak of Newton's inverse square law of force. With the influence of the distance on the gravitational force taken care of, Newton had to take account of the influence of the masses of the two particles on this force. He proposed the remarkable idea that the force depends on the product of the masses of the two particles. Combining these two ideas, Newton stated his law of gravity as follows: The gravitational force between two particles separated by a given distance is along the line connecting the two particles, is proportional to the product of the masses of the two particles, and is inversely proportional to the square of the distance between them; the two particles exert exactly equal but oppositely directed forces on each other.

If F is the gravitational force between two particles of masses m_1 and m_2, separated by the distance r, Newton's law is stated as

$$F \propto \frac{m_1 m_2}{r^2}$$

This expression, as it stands, is not an equality because the right-hand side is not a force, whereas F is, and two different kinds of things cannot equal one another. Newton was well aware of this and therefore multiplied the right-hand side by a quantity G, called the Newtonian universal constant of gravity because it is part of the law of force under all possible conditions (universally). Taking G into account, Newton wrote his law in the form

$$F = G \frac{m_1 m_2}{r^2}$$

Newton did not know the numerical value of G; not until some time after his death was it first measured in a very ingenious way by another great British physicist, Henry Cavendish. If length is measured in centimeters, time in seconds, and mass in grams, the numerical value of G is 6.78/100,000,000 (the units are cm^3/sec^2 g). The gravitational force between ordinary bodies, even when they are very close to each other, is negligible precisely because of the minute value of G. The gravitational force plays no role at all in the structure of molecules or atoms.

The gravitational interaction between two extended bodies (bodies other than mass points) cannot be represented by Newton's simple formula because the way the mass is distributed within each body (its shape) plays an important role. For spheres, however, in which the mass is distributed in uniform layers or the density is the same throughout, Newton's formula does apply because such spheres behave gravitationally as though their masses were concentrated at their centers. This is very fortunate because most of the bodies we deal with in astronomy are spherical. Thus, the gravitational force between Earth and the moon is obtained by multiplying the product of Earth's mass and the moon's mass by G and dividing the quantity thus obtained by the square of the distance between the centers of Earth and the moon.

Using this formula, the weight of a body can be found by multiplying the gravitational constant by the product of the mass of the body and the mass of Earth, and then dividing this quantity by the square of Earth's radius (distance of the body from Earth's center). If we equate this to the product of the mass of the body and the acceleration of gravity g, we can calculate the mass of Earth. Earth's mass is just the product of the acceleration of gravity and the square of Earth's radius divided by the gravitational constant G. (It is to be noted that an object weighs more at the poles than it does at the equator, because it is closer to Earth's center at the poles than it is at the equator. Thus, a pendulum beats slightly faster at the poles because the acceleration of gravity there is slightly larger than at the equator.)

Having obtained Newton's law of gravity we can now apply it to the motions of a planet around the sun and see how it leads to Kepler's laws. We do this in a general way without using mathematics, even though little more than elementary algebra is required to demonstrate the relationship between the law of gravity and the three laws of Kepler. We picture a sphere of mass m (the planet) revolving around a very massive body (the sun) whose mass M is many times larger than m. Under these conditions the motion of M is negligible compared to that of m, and we may disregard it. The pull of m on M is exactly equal to that of M on m, but M's response to this pull is as many times smaller than m's as M's mass or inertia is larger than m's.

To understand why a planet moves in one particular orbit rather than another, we note that its orbit is the combination of its tendency to move in a straight line with constant speed (its inertia) and its response to the sun's gravitational pull. If the sun exerted no pull on the planet, the planet (completely dominated by its inertia) would move off into space along a straight line at constant speed. But the sun constantly pulls the planet along the line from the sun's center to the planet's center, altering the planet's motion from that of a straight line at constant speed to a curved orbit of fixed size and shape in which the planet's speed, in general, is variable. To see just how the size and shape of a planet's orbit are determined by the planet's velocity at any given point in its orbit, we first consider a perfectly circular orbit. In such an orbit the speed of the planet must *always* have the same *constant* value and its direction of motion must *always* be exactly at right angles to the line from the center of the sun to the position of the planet. The slightest departure from these two exact conditions will lead to a noncircular orbit. If now, the planet were moving in a circle of radius r with constant speed v, this speed would have to have just the right value to make the outward centrifugal force acting on the planet owing to this speed, actually its inward radial acceleration, exactly equal to the inward gravitational pull of the sun. The centrifugal force on each gram of the planet's mass is just v^2/r and the sun's pull or inward acceleration of each gram of the mass is just GM/r^2. If we equate these two quantities (which we must if the orbit is to be a circle), we obtain $v^2 = GM/r$ for the square of the speed of a planet moving in a circle of radius r around the sun. Only if the planet has exactly the speed $\sqrt{GM/r}$ at every point of its orbit is the orbit a circle. From this we see that circular orbits cannot exist since the chance that the planet will have exactly one particular speed out of an infinitude of possible speeds is zero.

Consider what happens if the speed of the planet is not $\sqrt{GM/r}$ when it is at the distance r from the sun and moving exactly at right angles to the line from the sun's center. It is certainly not moving in a circle; the size and shape of the orbit then depend on how much slower or faster the speed of the planet is than the speed required for a circular orbit (the circular speed). If the speed of the planet is smaller than the circular speed, the planet's orbit will lie within the circular orbit since the arc of its orbit will then fall away from the circular arc toward the sun. This is clear since its speed will be too small to keep its distance from the sun equal to r. It will thus get closer and closer to the sun, traveling faster all the time as it does so (as though it were rolling down a hill) until it is diametrically opposite its initial position on the other side of the sun. It will then be at its smallest distance from the sun and traveling at its maximum speed. At that point it will again be moving exactly at right angles to the line from the sun's center to its own center, and the planet will then begin its return trip to its starting point ascending along an arc identical in shape to the arc along which it descended. The orbit is thus an ellipse. Under these conditions the planet's initial point is its aphelion (farthest distance from the sun) and the diametrically opposite point its perihelion. The smaller the initial speed at the distance r, the smaller and narrower is the elliptical orbit.

If the speed at the initial distance r is larger than the circular speed, the arc of the orbit will extend beyond the circular arc, and the planet will move farther and farther away from the sun, slowing down continuously as though it were climbing a hill. Its speed will continue to decrease until it reaches the diametrically opposite point on the other side of the sun. It will then have its smallest speed and be at its maximum distance from the sun. This point is the planet's aphelion and its initial point is its perihelion. When it passes its aphelion, the planet will move back toward its initial position along an arc exactly like the one it traversed when it came to its aphelion. Its speed will then increase steadily as it gets closer to the sun.

The larger the speed of the planet at its initial position r, the larger and more elliptical is its orbit. As the initial speed increases beyond the circular speed, the aphelion recedes farther and farther from the sun until at a certain speed it recedes to infinity. The orbit is then a parabola and the speed is called the parabolic speed or the speed of escape at the distance r. The value of this speed is given by $\sqrt{2GM/r}$, which is larger than the circular speed by the factor $\sqrt{2}$. If any object at the distance r from the sun has this speed, it will move off to infinity along a parabola regardless of the direction of its motion. The direction will merely determine whether the parabola is wide or narrow. If the initial speed of the body is greater than the speed of escape, the orbit is a hyperbola. Since one can obtain all these curves by taking a cut through a cone, we say that they are all conic sections.

We have spent some time discussing the relationship between the velocity of a body at a given distance from the sun and the nature of its orbit around the sun because the initial speed for a given distance determines the size and shape of the orbit. But there is another way of describing the orbit if we introduce the energy of the body. Each planet has a total amount of energy E in its orbit, which is the sum of its kinetic and potential energies and which is a constant (the same at every point in its orbit so that the energy of each planet is conserved). Consider now a body of mass m at a certain distance r from the sun and moving with speed v. What is the total energy E of the body and how does this energy depend on the nature of the body's orbit? The kinetic energy of the body, which depends only on its speed and mass, is $(\frac{1}{2})mv^2$ (we neglect the motion of the sun), and its potential energy, which depends on its mass, its distance from the sun, and the sun's mass, is $-GMm/r$. The potential energy, as already noted, is always taken as negative because we adopt the convention that the mutual potential energy of the sun and a body is zero when the body is infinitely far from the sun. Adding the kinetic and potential energies we obtain the total energy:

$$E = \frac{1}{2}mv^2 - \frac{GMm}{r}$$

Since E must always be the same, as the body moves, r and v must change together to keep E constant: as r decreases (the body approaches the sun so that its potential energy decreases) v must increase (the body moves faster as it gets closer to the sun). When v increases, the reverse is true.

Considering E we see that three possibilities exist: (1) $(\frac{1}{2})mv^2$ is always smaller than GMm/r, and the total energy E is negative; (2) $(\frac{1}{2})mv^2 = GMm/r$, and the total energy E is zero; (3) $(\frac{1}{2})mv^2$ is always larger than GMm/r, and the total energy is positive.

What do these three cases imply about the orbit of the body? Case (1) means that the orbit is a closed curve and the body returns to each point in its orbit over and over again. This case applies to each planet. To see that case (1) leads to a closed orbit, note that GMm/r can remain larger than $(\frac{1}{2})mv^2$ only if r does not increase forever; if it did, a point would be reached where GMm/r would become equal to or smaller than $(\frac{1}{2})mv^2$. But if r cannot increase indefinitely, it must begin to decrease, which means that the orbit will curve around and turn back on itself, and we obtain an ellipse. One finds in this case that the total energy E equals $-GMm/2a$, where a is the semimajor axis of the ellipse. We thus have

$$-\frac{GMm}{2a} = \frac{1}{2}mv^2 - \frac{GMm}{r}$$

or

$$-\frac{GM}{a} = v^2 - 2\frac{GM}{r}$$

This enables us to calculate the speed v of a planet at any distance r from the sun if we know the size of its orbit (its semimajor axis), or it permits us to find the semimajor axis of a planet's orbit (the mean distance of the planet from the sun) if we know its speed v when it is at a given distance r from the sun.

In case (2), the total energy of the body in its orbit is always zero, and this means that the energy at each point of the orbit must be the same as the energy at infinity. Thus, the orbit must be infinitely large, and we have a parabola. Since for case (2) $(\frac{1}{2})mv^2 = GMm/r$, we see that $v^2 = 2GM/r$ and, as we have already noted, this is the square of the speed of escape.

In case (3), the total energy of the body is positive. The orbit is then a hyperbola.

We have indulged in a lengthy discussion of the relationship between Newton's law of gravity and Kepler's three laws of planetary motion to show how orbits can be deduced (though not rigorously) from the laws of motion and Newton's law of gravity. As we have seen, if only two bodies are involved (the sun and a planet) we obtain a closed elliptical orbit. This ideal situation, however, does not occur in practice. The real situation is much more complex than this since we are then dealing with many bodies revolving around the sun. The complexity arises because the planets all pull on each other (and also on the sun) gravitationally, thus distorting the various elliptical orbits that would exist if there were no gravitational interactions among the planets. One has here what mathematicians call the n-body gravitational problem, which may be stated as follows: Given n bodies moving under their mutual gravitational attractions, find the orbits of all the bodies; that is, find a set of mathematical formulas that predict for all time exactly how each of these bodies moves.

Mathematicians have been trying to solve this problem for almost two centuries without success. We now know that this problem is so complex that neither the mathematics now available nor the mathematicians are powerful enough to obtain a general solution. Even the three-body problem (the sun and two planets) is so complex that we cannot obtain a general solution. When we go from two to three bodies, so many possibilities arise, ranging from the very simplest to the most complex motion, that it is impossible to find a simple solution. And, of course, the larger the number of bodies, the greater the complexities become and the more remote the chance of obtaining a general solution. How, then, does one go about predicting the motions of the bodies in our solar system that are interacting with each other?

Here we are aided by the fact that dominating the entire dynamics is one central massive body, the sun, so that to a first approximation we may solve the problem of the motion of any planet as though we were dealing with only two bodies, namely, the sun and the planet. The effects of the other planets are of secondary importance, and may be introduced as small corrections to the first approximation. This is known as the method of perturbations and can be carried out to any desired accuracy and quite rapidly with high-speed electronic computers.

In this way we can calculate the orbits of all the planets, taking into account not only the attraction of the sun but also the effects of all the other planets and satellites. We find, when this is done, that all the observed motions, with few exceptions, agree very well with Newtonian theory. Most of the exceptions are rather trivial and will be eliminated as computing techniques improve, but one is extremely striking and indicates a flaw in Newtonian theory.

When the theory of perturbations had been developed to a point where the effects of the sun together with those of the other nearby bodies on the motion of a planet could be computed, the theory predicted that the planets should not move in closed elliptical orbits but almost in such orbits. A slight correction had to be introduced such that the effect of the other planets on any one planet causes the entire orbit of the planets to move around slowly.

We can illustrate this roughly in the following way. Imagine an ellipse drawn on a tabletop that can rotate. If we now set a particle moving around this ellipse according to Kepler's laws and then allow the table to rotate slowly in the same direction as the motion of the particle, we have an approximation of the effect of the other nearby planets on the motion of the planet being considered. This additional motion is generally referred to as the advance of the perihelion of the planet.

This perturbing action means that a planet really does not move in a closed orbit because by the time it has come back to the perihelion point of its orbit (the point nearest the sun), that point itself has shifted its position in an eastwardly direction. This effect would be absent if the orbit of the planet were a circle because then there would be no perihelion point and one could not detect the slow rotation of the orbit. The effect is the greater the more pronounced the ellipticity of the orbit, so that the effect should be greatest for Mercury. This was indeed observed, so that the Newtonian theory of gravitation appeared to be more firmly established than ever. However, it soon became evident that all was not perfect in the Newtonian kingdom of gravity.

The major axis of Mercury's orbit advances eastwardly at a rate of a little more than $5\frac{1}{2}$ sec of arc every year, but this is not quite in agreement with the calculations based on Newtonian theory. In 1845 Leverrier found that the perihelion of Mercury advances a further 43 sec of arc every 100 years than the theory predicts. To one not acquainted with the great precision in astronomical measurements, this might seem to be a discrepancy so small as to be hardly worth worrying about, but it is precisely these small deviations that are of greatest interest to the scientist.

At first, the discrepancy in Mercury's motion was not of too much concern, for it was thought that there might be some unknown planet between Mercury and the sun that was producing the perturbation. But when all endeavors to find this planet failed, it became clear that science was confronted with a fundamental flaw in the Newtonian law of gravity. Although this effect is quite small for the other planets, it adds up from year to year and it has already become measurable for Earth and Mars.

But even while difficulties were piling up for Newton's theory of gravity, there were other branches of classical physics in which theory was running into contradictions with observation and experiment. In the 19th century, such rapid advances were made in electricity, magnetism, and radiation that it became possible with the aid of very accurate optical instruments, such as spectroscopes and interferometers, to subject the basic Newtonian concepts to the most precise kinds of tests. Two investigations in particular were of extreme importance in precipitating the downfall of the Newtonian universe, even though they seemed of little significance at the time they were conducted. One was an attempt to determine the absolute speed of Earth through space by measuring the speed of light in different directions with respect to Earth's motion; the other was concerned with determining the properties of the radiation emitted by a hot body.

Both of these investigations gave results that cannot be explained in terms of the Newtonian concepts of space and time, and it soon became clear (at the beginning of the 20th century) that revolutionary changes would have to be introduced into the fundamental concepts of physics. At first it appeared that these new concepts would be of theoretical importance only and have little effect on the practical applications of physics in such fields as astronomy. And this, indeed, was true as long as the astronomer was concerned with the study of the motions of the planets and with the visible properties of the stars. But very early in the 20th century the astronomer found himself in the very midst of the new ideas that were flowing so rapidly from the pens of the physicists, for the astronomer was beginning to penetrate into the stars themselves and to wander far out in the universe to the distant galaxies. In these new realms, beckoning with such exciting promises, the astronomer cannot find his way without using the new ideas of space and time and of atomic and nuclear structure.

The revolutionary changes introduced into physics cannot be ignored by the astronomer if he wants to understand the structure and evolution of stars, or such things as pulsars, quasars, neutron stars, and black holes. Since we propose to discuss precisely these topics, it is necessary for us to learn something about modern physics. We are going on some very exciting trips in this book, and we shall require every bit of our skill as travelers and explorers if we are not to get lost in a maze of bypaths, or are not to be misled by the many signposts that will beset our way. We shall, for example, travel into the deep interiors of some typical stars to see if we can understand why a star like the sun behaves the way it does. The sun has been with us for billions of years, pouring out into space vast quantities of energy every second. What keeps it going and how long will it continue shining so brightly? We can answer these questions only by probing its interior, but when we do so, we find ourselves surrounded by atoms, electrons, and bare nuclei moving about at tremendous speeds and interacting with each other in very complex ways. And all of this activity takes place in a sea of intense radiation flowing steadily outward from the central region of the sun where it is produced by thermonuclear reactions. We therefore have to know something about the nature of fundamental particles such as electrons, protons, and neutrons, of which atoms are composed, if we are to understand stellar interiors; we have to acquaint ourselves with the most recent theories about the structure of atoms and their nuclei.

Once we have learned about the structure of individual stars, we shall see how these stars group themselves together to form galaxies such as our Milky Way. We shall journey far out into space among these galaxies to get a picture of the universe itself, and here again we shall have to call upon the most recent developments in physics if we are to understand the picture that unfolds. Two theories, in particular, are of great importance to us: the *theory of relativity* and the *quantum theory*. The first of these shows us how the Newtonian absolute concepts of space and time were replaced by a single space–time concept in which distances and time intervals between the same two events are different for different observers; the second of these theories enables us to understand how atoms emit and absorb radiation.

We consider the theory of relativity first and see how it explains the discrepancies between the observed motions of planets like Mercury and the orbits predicted by Newton's law. Before leaving this chapter, however, we show how Kepler's third law of planetary motion can be deduced from Newton's law of gravity with very little effort.

To do this in the simplest possible manner, we prove it for a planet moving in a circular orbit, which establishes it for any kind of orbit since the third law does not depend on the shape of the orbit but only on its size. From Newtonian gravitational theory we deduced that the square of the speed of a planet moving in a circular orbit of radius r around the sun is GM/r, where M is the mass of the sun. If we divide this into the square of the circumference of the orbit, i.e., into $4\pi^2 r^2$, we obtain the square of the time it takes the planet to revolve around the sun. This is just the square of the planet's period P so that we have

$$p^2 = 4\pi^2 r^2/(GM/r)$$
$$= 4\pi^2 r^3/GM$$

which is just Kepler's third law, since r is the mean distance as well as the instantaneous distance of the planet from the sun.

This formula, and Kepler's third law, would be exact if the mass of the sun were infinite so that the sun could not move. But this is not so; the sun does move under the gravitational attraction of the planet so that its period must be governed by the same law that governs the period of the planet. Indeed, we are dealing here with a single period that must be given by a single formula containing both the mass M of the sun and the mass m of the planet, and this formula must be symmetric in the two masses since the law of gravity does not distinguish between them. The formula for the square of the period must therefore be $P^2 = [4\pi^2/G(M + m)]r^3$. The square of the period of the planet divided by the cube of its mean distance from the sun is thus not the same for all of the planets because the mass m varies from planet to planet.

CHAPTER 4

The Special Theory of Relativity

Nothing puzzles me more than time and
space; and yet nothing troubles me less, as I
never think about them.

—CHARLES LAMB, *Letter to T. Manning*

If Newton's theory was incapable of correctly accounting for planetary motions down to their smallest measurable details, how was the theory to be altered? It did not take astronomers and physicists long to see that nothing could be gained by just making a slight change in some of the numbers that appear in Newton's formula, for example, changing the exponent from 2 to 2.0000001 in his law of gravity. This does bring Mercury's motion into step with the theory, but then the rest of the army is out of step; the motions of all of the other planets are thrown into disagreement with the theory.

Some fundamental change had to be made, and the hint as to what this change was to be came from a quarter that seems to have no relation whatsoever with the problem of planetary motion. It was concerned, of all things, with the behavior of light. In a later chapter we consider what changes we have to make in our notions about the nature of radiation to account properly for the way a hot body sends out light; but now we concern ourselves with the speed of light and how different observers measure this speed and the results they get.

Before we can answer the question of how fast light travels as measured by different observers, we must define the phrase "different observers" and also agree upon some scheme for comparing the results that these observers obtain. There is no sense in speaking of the differences between two observers unless we have a consistent way of comparing these differences. For the time being, our "different observers" will be those moving with uniform speed in the same straight line (no acceleration) with respect to one another. Thus, for our purpose, all observers in empty space that are not moving with respect to each other are to be considered as equivalent as far as the laws and phenomena of nature are concerned.

It is quite clear that if two different observers carry out experiments or make observations on the same set of phenomena, the comparison of their results must involve definite concepts of space and time. There must be some precise scheme involving these concepts of space and time for translating the results of one of the observers into the language of the other observer. Mathematicians call this scheme a transformation from one system to the other system, and they express this by means of a set of transformation equations.

We may look upon these transformation equations as a sort of mathematical dictionary that enables us to translate from the mathematical language of one observer to that of the other. We all know that a good dictionary must meet certain elementary requirements if it is to give us a correct translation from one language to another: it must leave the meanings of equivalent words in both languages unchanged. If a Frenchman gives a precise description of some thing or event, and this description is then translated into English with the aid of a French–English dictionary, we expect it to correspond in every detail with a precise description of the same thing or event given by an Englishman. If this were not so, the dictionary would be of no value. In other words a good dictionary must leave the essential details of an event or thing unchanged, or *invariant*, in the language of the mathematician. It must neither add to nor detract from the intrinsic features.

We now consider the nature of the "dictionary" that is used for transforming an observation from one observer to another in Newtonian mechanics. To make things specific, let one of the observers be sitting in a railroad station and the other in a train that is moving with constant speed along the track, which we take as a straight line in a fixed direction. Our events are the explosions of two bombs at two different points along the track and at two different times. The description that is to be given by each observer is the distance of each bomb from his own position at the moment the bomb explodes and the time when each explosion occurs, as determined by his own clock. Each observer is to measure the distance in his own way and with his own instruments (e.g., identical rulers). The clocks also are identical and exactly synchronized. We further assume that the clocks read zero at the moment when the man in the train is opposite the man in the station, and that both observers start computing time on their clocks from that moment on.

The observer in the station, after carrying out his distance measurements, records the numbers X_1 and X_2 giving the distances (say, in feet) of the two bombs from him and he also records the numbers T_1 and T_2 as given by his clock to indicate the times when the bombs exploded.

What about the man in the train? He also records four numbers describing the two explosions, but because he is moving, these numbers are not the same as the four given by the fixed observer. To differentiate between the two sets of four numbers we add primes to those of the moving observer (X'_1, X'_2 for the distances, T'_1, T'_2 for the times). To compare these descriptions to see whether the two observers are really talking about the same events, we must have a transformation "dictionary" that has the property of neither adding to nor subtracting from the actual events. Our transformation must leave the intrinsic features unchanged.

But how are we to know what the intrinsic features of any particular set of events are? If there is a discrepancy between the two descriptions, are we to blame the dictionary, the observers, or both? To be on the safe side, the first thing to do is ensure that the transformation that we adopt satisfies the most general conditions that are consistent with our fundamental axioms and self-evident truths about space and time.

There are two general conditions that must be met if our dictionary is to be a useful one. First, it must be consistent with our concepts of space and time; and second, it must not change the content of a law of nature while translating it from one system to another. The latter is called the *principle of invariance*, and it has played an important role in the development of modern physics.

What, if any, are the self-evident truths or axioms about space and time that are to guide us in our search for the right transformations? Up until 1905 it seemed obvious that space and time are absolute concepts, unchanging and the same for all observers. They were also assumed to be completely independent of each other. In terms of our two bombs, this means the following: if the man in the railroad station finds that the distance between the two bombs is 5 miles, then the man in the moving train must find it to be the same; and if the time interval between the two explosions is found to be 1 hour by the moving observer, then it must also be 1 hour for the fixed observer. In other words, distances and time intervals between events must be the same for all observers regardless of how they are moving with respect to each other. These were the "self-evident truths" that were accepted about space and time and imposed as a requirement on the transformation "dictionary." These are the absolute space–time concepts of Newton.

It is clear that if we adopt this picture of space and time, we can then get the description of any event in the moving system from that in the fixed system by means of the following prescription. First, place all the T values equal to the T' values. In other words, time in the moving system is always the same as that in the fixed system. Second, to get the X' value of any event, that is, its distance from the moving observer, multiply the velocity of the moving observer by the time of the occurrence of the event and subtract this from the X value of the event, that is, the distance of the event from the man in the railroad station. The mathematician writes this transformation or "dictionary" as follows:

$$X' = X - vT$$
$$T' = T$$

where v is the speed of the man in the train.

These equations say only that the moving clock and the fixed clock keep identical time, and that the distance from the man in the train of any point on the railroad track is smaller than the distance of the same point from the man in the station by the distance from the station that the man in the train has moved.

What we have said above about space and time appears so obvious and in accord with common sense that it hardly seems worth talking about, and yet we have to see whether or not our second condition is fulfilled. Do the above transformations leave the laws of nature unchanged when we translate from one system to another? This is certainly true for the laws of mechanics, like Newton's three laws of motion and the principles of conservation of momentum and energy. For a long time, therefore, no one had any reason to suspect that something might be wrong, and yet new developments were occurring in science that would soon cast doubt about all the "self-evident truths" concerning space and time that had been so readily accepted.

In the middle of the 19th century, Michael Faraday, one of the greatest experimental physicists, laid the foundation upon which, a decade later, Clerk Maxwell built his beautiful electromagnetic theory of light. Maxwell discovered that he could write a fairly simple set of equations showing that all radiation (e.g., light, radio waves, X rays) is the propagation of an electromagnetic disturbance in the form of a wave. An important mathematical consequence of these equations is that the speed of light is a constant, independent of the motion of the source and the motion of the observer. This means that if a sudden flash of light started at the railroad station and both the man in the station and the man in the moving train were to measure the speed of this light as it moved along the track parallel to the motion of the train, then both of them should get the same result. The same thing should be true if the beam of light originated from any star in the heavens instead of from a point on Earth. In other words, the speed of light measured by each observer relative to his own frame of reference must be the same.

According to the Maxwell theory this should be a law of nature and therefore, if we use the "dictionary" described above, we should get the same answer when we transform from the fixed observer to the moving observer. However, this does not happen. When we use the transformation equations given above, the speed of light as measured by the man in the train is different from that as measured by the man in the station. In other words, according to the concept of space and time upon which these transformation equations are based, the speed of light should be different for different observers.

Now this conclusion seems quite reasonable, for if the man in the train is moving in the same direction as the light, why should he not find the speed of the light somewhat less than that found by the stationary observer? Is this not exactly what we find for moving bodies? If you are moving in a car at 40 miles an hour and another car is moving in the same direction at 50 miles an hour, then does not this other car draw away from you at 10 miles an hour? Why should light be different? There must be something wrong with Maxwell's theory.

This, at any rate, is the way the 19th century physicists looked at it, and that is how it stood until about 1880, when an important experiment produced a very puzzling observation. To see just how this experiment arose, we note that if the speed of light is different for a moving observer and a stationary observer, it should be possible, by measuring the speed of light and comparing the result with the value given by Maxwell's theory, to determine our speed through space. In other words, it should be possible to detect absolute motion. To see how this applies to Earth, we need merely consider the speed of light as measured for a beam moving in the same direction as Earth's motion and a beam moving at right angles to this direction. If the absolute concepts of space and time were right, these speeds would be different (they would differ by about 19 miles per second) and Maxwell's theory would be wrong.

In 1880 Michelson and Morley devised a very ingenious experiment to try to measure Earth's motion based upon the theory that there should be a difference in the speed of light in different directions. This experiment was carried out with great care and was done with sufficient accuracy to detect a result 1/20 as large as the one that was expected, and yet the results were negative, indicating either that Earth is not moving at all, which, of course, is contrary to the known facts, or that Maxwell is right and our "dictionary" wrong.

This unexpected experimental result remained one of the great unsolved puzzles until 1905, when there appeared in the *Annalen der Physik* a paper on "The Propagation of Electromagnetic Phenomena through Rapidly Moving Media" by Albert Einstein, who was then just 26 years old and an examiner in the Swiss patent office. In this paper the solution to this puzzle was given, but it introduced an entirely new and revolutionary concept of space and time and led to a new transformation "dictionary."

Einstein started out with two fundamental assumptions. The first is that all the laws of nature must remain unchanged when one transforms from a given system to another system that is moving with uniform velocity with respect to the first system (the principle of invariance). If our transformation equations are the correct ones (based on a correct concept of space and time) and a so-called law does not conform to this principle, then the law must be altered so that it does conform. We see that this is a very powerful analytical gauge that can guide the physicist in formulating the correct laws that govern our universe.

The second assumption made by Einstein is that the speed of light is a constant for all observers. This invariance is really more than an assumption because the Michelson–Morley experiment had already established it as an observed fact.

These two assumptions led to a new set of transformation equations (H. A. Lorentz had previously used these equations to describe certain properties of the electron, but they were derived anew by Einstein, entirely on the basis of his new concepts of space and time), which eradicated forever the concept of absolute space and absolute time and replaced it by a single space–time concept in which space and time are welded into a single entity.

According to the new "dictionary" derived by Einstein, time is no longer the same for all observers, and the X and T values of the man in the station are related to the X' and T' values of the moving observer by a set of equations that are somewhat more complicated than those above:

$$X' = (X - vT)/\sqrt{1 - (v/c)^2}$$

$$T' = \left(T - \frac{v}{c^2} X\right) \Big/ \sqrt{1 - (v/c)^2}$$

These formulas are almost like the ones above and, in fact, reduce to those if the velocity, v, of the moving observer is small compared to the speed of light, c. The difference between the old way of looking at time and space and the new way becomes important only if one is considering bodies that are moving with speeds that get close to the speed of light. At these speeds, time and space become intermixed in a complicated way and the only thing that still remains the same for all observers is a certain space–time interval.

Like Newton, Albert Einstein's humble beginnings as the only son of Hermann Einstein and the former Pauline Koch did not hint at the greatness he would one day attain. Albert was born on March 14, 1879, in Ulm, Germany, a town nestled in the foothills of the Swabian Alps and bounded by the Danube River. Albert's father was a good-natured engineer who preferred taking the family on Sunday excursions into the mountains to the daily routine of overseeing the family workshop. His mother, the daughter of a well-to-do grain merchant, was a cultured woman who enjoyed both German literature and music. Their marriage appears to have been a happy one although it was doubtless strained by Hermann's business difficulties. Within a year of Albert's birth, Hermann's business went bankrupt, prompting the family to leave Ulm for Munich where Hermann and his brother Jakob set up a small electrochemical facility.

As a child, Albert was slow to learn to speak and his propensity to daydream and ignore the world around him caused his parents to fear that he might be retarded. He attended a Catholic school until the age of 10, a placement not altogether unusual given the lack of religiousness in the Einstein family and their belief that a proper education was more important than the particular religious orientation of the institution. The atmosphere of his primary school was tolerant and, despite its overwhelmingly Catholic enrollment, Albert was not made to feel self-conscious about his Jewish heritage. He was then enrolled in the Luitpold Gymnasium, a regimented institution of a more distinctly German character that stressed the merits of obedience to authority and conformity of thought and deed. For young Einstein, who had grown up in a family that had largely rejected most of the tenets and rituals of Judaism, the heavy-handed discipline of the Gymnasium instilled in him a lifelong distrust of authority. He later saw the supposed virtue of respect for authority as contributing to the strangling of free thought and fostering an unhealthy deference to the opinions of others. The Gymnasium's Prussian character also fostered in Einstein a distinctly anti-German perspective, which led him less than a decade later to renounce his German citizenship.

Because Einstein found little in the curriculum to interest him, he neglected his studies of the classics (which formed the core of the Gymnasium education), preferring to devote his spare time to mastering mathematics and the philosophical arguments of Spinoza and Kant. In any event, he had few friends and did not impress his teachers; his academic performance was so poor that he was expelled from the Gymnasium in 1894. He journeyed to Italy to rejoin his family, who had moved to Milan 6 months earlier following the failure of the family business in Munich. Unable to enter a university without a Gymnasium certificate, Einstein eventually enrolled in the Federal Polytechnic Academy in Zurich in 1896 with the intent of becoming a teacher. After his graduation in 1900, he tried to secure an academic post but was unsuccessful as his rebellious attitude toward authority in general and his former professors in particular caused many potential employers to shun him. Moreover, Einstein had never been a diligent student and this evident unwillingness to devote himself fully to academics did not bode well for his career prospects.

He obtained his first regular job when he accepted an appointment as a patent examiner in the Swiss Patent Office in Bern in 1902. Although his training was more theoretical than mechanical, he soon made up for his educational deficiencies and became a valued member of the staff. During the early years of his 7-year stint with the Office, Einstein completed his dissertation and received his doctorate from the University of Zurich in 1905. However, 1905 was to prove even more memorable for the young Einstein as he published three papers in the *Annalen der Physik* on Brownian motion, the photoelectric effect, and the special theory of relativity, any one of which would have sufficed to establish his reputation as a world-class physicist. Although the cumulative effect of these papers constituted a full-scale frontal assault on classical physics, the impact was not immediate as news of Einstein's work traveled slowly and his conclusions were not uniformly accepted. Einstein's postulate that the speed of light is the ultimate speed limit in the universe and that space and time are really complementary aspects of a four-dimensional entity dubbed "space–time" seemed an affront to commonsense perspectives rooted in Newtonian mechanics. However, the ensuing years saw Einstein's views steadily win over the majority of physicists until by the 1920s his theory of relativity had become accepted as a cornerstone of modern physics.

Among Einstein's earliest admirers was Max Planck, the German physicist who had offered his quantum theory of radiation in 1900 and thus unknowingly laid the foundation for the other revolution that would upset the tidy Newtonian–Maxwellian view of the physical world that prevailed until the end of the 19th century. The aristocratic Planck lent his considerable prestige to popularizing Einstein's views among his more skeptical colleagues. Planck also devoted several of his university lectures to the still-obscure patent examiner's ideas and thereby gave them an additional aura of scholarly respectability.

Although Einstein gradually became recognized as a brilliant theoretician before he reached the age of 30—a remarkable accomplishment considering his isolation from the academic world—he continued to work at the Swiss Patent Office until 1909 when he accepted an appointment as associate professor at the University of Zurich. Although his salary was about the same as that of his patent office post, he was glad to be in a university environment where he hoped he would be able to carry out his research without distractions and complete his work on his general theory of relativity, which had already occupied his attention for several years.

Although Einstein enjoyed the relative serenity of Zurich and was a popular lecturer, he was not destined to remain there for very long. In 1911 he accepted an appointment to the German University in Prague. While in Prague he devoted much of his theoretical work to the question of whether the path of a beam of light would be bent by the gravitational field of a massive object such as the sun. He also attended the Solvay Conference in Brussels where he first met many of the outstanding physicists of his day such as Ernest Rutherford, Marie Curie, Hendrik Lorentz, and Max Planck. However, Einstein found that his lecture duties interfered with his theoretical work, so, soon after his arrival in Prague, he began to search for an acceptable alternative, which appeared in 1912 in the form of a professorship at the Zurich Polytechnic. Einstein returned to Switzerland that year as a celebrity because his special theory of relativity was now the subject of many newspaper and magazine articles; his classroom lectures were usually filled to overflowing. While Einstein enjoyed his life in Switzerland and had been a Swiss citizen for more than a decade, he was still intellectually restless and seemed to be vaguely unsatisfied with his professional life. Max Planck and Walther Nernst, two of the outstanding German physicists at the University of Berlin, got wind of Einstein's reluctance to settle down into the predictable and tidy routine of Swiss life. After much discussion, they persuaded Einstein to come to Germany and accept a post at the University of Berlin, a position that would permit him to devote himself fully to research with none of the administrative duties most professors must perform. He was also offered the directorship of the soon-to-be-created Physics Institute at Berlin and a vacant chair in the Prussian Academy of Sciences. The financial aspects of the offer were far in excess of what he was earning at Zurich, and Einstein, who was interested in securing a sufficient income to support his wife, Mileva, and the couple's two children, decided in favor of Berlin.

While the financial considerations figured greatly in Einstein's move from Zurich to Berlin, he was also attracted by the prospect of joining the faculty of the greatest center for theoretical physics in the world. The University of Berlin at that time boasted a constellation of scientists worthy to be colleagues of Einstein. The thought that he would be able to test his ideas on this esteemed group of individuals and benefit from his daily contacts with them greatly influenced Einstein's decision to leave Switzerland. In any event, Einstein moved to Berlin in the spring of 1914.

The outbreak of the Great War in August 1914 fractured the bonds of the international scientific community. At the same time, the austere life of Berlin helped to bring to an end what had been a steadily deteriorating marriage. Mileva soon returned to Switzerland with their two children and Einstein—never much of a family man—took advantage of the solitude to wrap up his work on the general theory of relativity. As the war progressed, many of his colleagues joined various government organizations to do their part to aid Germany, but Einstein's Swiss citizenship permitted him to avoid tasks that would contribute to the war effort. He had never forgotten his dislike for all things Prussian; the speed with which his colleagues rushed to contribute their talents to the defense of the fatherland only intensified these feelings. Einstein's personal and social isolation coupled with his humanism attracted him to pacifism, socialism, and internationalism. However, he was noted not for any influence he exerted in the political arena but for the way in which he served as a lightning rod for criticisms by persons who believed that Einstein's pacifist sentiments were inappropriate when the nation was locked in a vicious two-front war and those who wished to vent their anti-Semitism against the most famous Jew in Germany.

Einstein spent many months perfecting his general theory of relativity before it was finally published in 1916 in the *Annalen der Physik*. The general theory, which deals with bodies in accelerated relative motion, went beyond the special theory, which applies only to bodies in uniform relative motion. Einstein also showed how mass curves space and thereby bends the path of light, thus hinting at the possibility of a relativistic basis for cosmology—a field that was initiated the following year with the publication of a paper by Einstein on the structure of the universe. The strain of years of uninterrupted concentration and toil seems to have taken a heavy toll on him, and he fell ill in 1917 and was only slowly nursed back to health by his second cousin Elsa, whom he later married in 1919, after the divorce from his first wife became final.

Einstein's work on general relativity marked the peak of his career and he spent much of the rest of his life clarifying subtle aspects of the theory. For his theory to become generally accepted some sort of experimental confirmation was needed. An opportunity for this arose in 1919 when the British astrophysicist Arthur Eddington led an expedition to Principe Island in the Gulf of Guinea and took a series of photographs during an eclipse, which showed that the light path from stars grazing the sun was deflected by the sun's gravitational field by the amount predicted by Einstein's general theory.

Although news of the confirmation of Einstein's theory became public only with a series of articles by the *Times* of London, it was evident that the Newtonian vision of the universe had been fundamentally altered. The grandness of this theoretical achievement coupled with Einstein's own amused indifference to the media attention and his eccentric dress and personality brought him worldwide fame in 1919. He was the subject of countless interviews and articles and was invited to give lectures and accept honorary degrees throughout the world. For a man such as Einstein who had labored in relative obscurity and isolation for much of his career, his newfound fame came with all the impact of a tidal wave and thrust him into the hitherto unimagined public eye. He soon learned how to control the many requests for his time but he did take advantage of the access afforded him by the press to call for renewed cooperation among the European powers. He also became acquainted with some of the more prominent Zionists and allowed his name to be used to promote their fund-raising efforts. In 1920 he again became a German citizen, ostensibly to give his support to the struggling Weimar Republic and help create a more positive image of the German nation. However, he was personally attacked by a number of anti-Semitic writers who criticized his work not for any proven flaws but because it was the handiwork of a Jewish pacifist. Many of his colleagues rushed to defend in print the integrity of his work, fearing that the personal attacks would cause Einstein to leave Germany for good. Planck, whom Einstein idolized, visited him on several occasions to dissuade him from abandoning the German scientific establishment owing to the baseless attacks of a few extremists. The effort apparently succeeded, and Einstein declared in 1920 that he would remain in Berlin.

Einstein traveled throughout Europe and to the United States in the early 1920s, lecturing on his theory and trying to obtain money for the Hebrew University then being built in Jerusalem. He received the Nobel Prize in Physics in 1921 and, as part of his 1919 divorce settlement with his first wife Mileva, had the cash award forwarded to her in Switzerland. Ironically, Einstein received the award not for his theory of relativity, which was then being profitably used by physicists probing the constitution of the atom, but for his discovery of the law of the photoelectric effect. The reason for this designation was a clause in the will of Alfred Nobel, which stipulated that the Nobel Prize was to be awarded only for practical discoveries that had improved the lot of humanity. Because the theory of relativity, despite its contributions to fields as diverse as astronomy and philosophy, did not appear to meet this criterion, the reference to the photoelectric effect was used. Einstein was not present in Stockholm at the time of the award as he had previously accepted an invitation to lecture in Japan.

Einstein spent much of the 1920s writing articles and books about his theory of relativity and working on a unified field theory to unite electromagnetism and gravity into a single mathematical framework that would account for the motions of the planets as well as the propagation of light. Although he believed, on more than one occasion, that he had found the mathematical solutions that would support such a theory, he was destined to suffer repeated disappointment because the two forces were simply too different in nature to be reconciled using existing physical concepts. Despite the slowness with which his unified field theory progressed, Einstein continued to develop it, believing that he was the only one who could successfully unite gravity and electromagnetism. He also felt that his previous scientific accomplishments had provided him with enough of a mantle so that he could justify to himself what became an uninterrupted effort lasting some 30-odd years to deduce the forces of the universe from a single series of equations.

At the same time modern physics was being revolutionized with most of the younger physicists, such as de Broglie, Schrödinger, Heisenberg, Dirac, Niels Bohr, and Max Born ushering in quantum mechanics, with its indeterministic statistical probabilities, holding the key to understanding the subatomic world, and a few of the older physicists, such as Einstein, insisting that nature did in fact have a causal direction and that "God does not play dice." The result was that Einstein gradually became isolated from most of his contemporaries even though his work was largely responsible for the initial ventures into what became known as the quantum mechanics. Although many historians have pointed to Einstein's increasing involvement with political issues and a growing unwillingness to question freely his own views about the nature of physical laws and even a declining intuition for physics as causing Einstein to become less productive as a physicist insofar as original ideas were concerned, it appears that his difficulty stemmed from the absence of a mathematical extension of the Riemannian geometry that he had put to such good use in formulating his general theory of relativity from which he could develop a unified field theory. In short, Einstein found himself having to devote more and more time to the mathematical details of his work leaving less time for actually considering the physical foundations of his ideas.

The rise in the political fortunes of Adolf Hitler and the Nazis in the early 1930s made Germany an increasingly uncomfortable home for Einstein. He visited the California Institute of Technology at Pasadena several times and considered a tempting offer to join the faculty there. In 1932 when the Nazis were on the verge of taking over the German government, Einstein decided that he would have to leave Germany or face the very real prospect of being killed. He accepted a generous offer from the newly established Institute for Advanced Study in Princeton, New Jersey. After he left Germany for good in 1933, he renounced his German citizenship for the second and final time. He also resigned from the Prussian Academy of Sciences after concluding that its administrators were more interested in supporting the policies of the new government than in remaining an impartial body that would act as a forum for differing scientific opinions.

In 1936 Einstein filed the necessary papers to become a naturalized American citizen and moved into a two-story house at 112 Mercer Street in Princeton, an address that became famous, entertaining from time to time an impressive assortment of scientists and statesmen. The death of his wife Elsa two years later simply encouraged Einstein to devote himself more fully to the vexing problems of the universe, particularly his unified field theory. By the mid-1930s, however, many of his cherished beliefs about determinism and causality and his dislike for the philosophical implications of quantum theory had caused him to be regarded as something of a historical curiosity by many of his colleagues. Einstein continued to enjoy his comfortable routine at Princeton as the clouds of war darkened over Europe. However, his once absolute views on pacifism had undergone a metamorphosis in the preceding decade as the need to stop Hitler by force had become apparent to him. When his friend Leo Szilard approached him about sending a letter to President Franklin Roosevelt warning of the possibility that extremely destructive bombs could be built using uranium, Einstein was quite willing to comply with the request. The now-famous letter, which Einstein later called "the greatest mistake of my life," prompted Roosevelt to order the establishment of a committee to investigate whether such atomic weapons were feasible and to monitor progress in atomic research. The letter convinced the government to increase its funding of atomic energy research programs in the nation's universities.

In 1943 Einstein agreed to serve as a consultant to the U.S. Navy's Bureau of Ordnance where he dealt with problems such as the dynamics of explosions. Einstein was generally unaware of the progress then being made by the physicists affiliated with the Manhattan Project and, as he had not bothered to keep up with the scientific literature on atomic research, he was generally ignorant of the specifics of the American program. He was doubtless aware of the tremendous efforts being made by the American and British governments to develop the atomic bomb and, having decided in the early 1930s that force was necessary to stop the scourge of Nazism, he did not oppose the goals of the program. It is also possible that Einstein was curious to see whether a fission bomb could actually be made because his own intuition told him otherwise.

The ease with which two atomic bombs destroyed Hiroshima and Nagasaki in 1945 drove home to Einstein the need for a world government to control atomic weapons and prevent a third world war. Like Bohr, Einstein knew that it was only a matter of time before the Soviet Union acquired atomic weapons, so he urged that the bomb be brought under the control of an international agency while it was still feasible to do so. However, his assumption that the Soviets would find it in their interest to cooperate with the Americans to promote the peaceful use of the atom was wrong.

Einstein continued to speak out against the dangers of atomic weapons and stressed the need to internationalize the production of atomic energy. His poor health prevented him from traveling far from Princeton but he maintained a steady work schedule at the Institute. Following the death of Chaim Weizmann in 1952, Einstein was invited to become the president of Israel, a largely ceremonial post, but he modestly declined the invitation. His last act shortly before his death in the spring of 1955 was to add his name to the so-called Russell–Einstein Declaration, which urged that all nations resort to peaceful means to resolve their disputes.

To see the difference between the old absolute way of looking at space and time and the relativity space-time concept, we again consider the two observers and the two bombs exploding somewhere along the track. According to prerelativity ideas the distance between the bombs should be the same for the moving and the fixed observer, and the same thing should be true for the time interval between the two explosions. In the theory of relativity, however, this is not so because distances and time intervals vary with the speed of the observer. But there is a certain space-time interval between any two events that remains the same for all observers even in relativity theory.

This space-time interval is obtained in the following way. Take the distance between the two events (the exploding bombs) and square this number. Now take the time interval between the two events and square this quantity and multiply this number by the square of the speed of light. Subtract this last product from the square of the distance and take the square root of the number found. This is called the space-time interval between the events and is the same for all uniformly moving observers. From the theory of relativity it follows, then, that each observer cuts up the space-time fabric of the universe into a space part and a time part in his own way depending upon how fast he is moving, but the space-time fabric is the same for all observers.

The concepts of simultaneity and distance are no longer absolute. If the events A and B are observed as having occurred simultaneously by one observer, it may appear that A happened before B to a second observer, and to a third observer it may be that B took place before A. Since this order depends on how the observers are moving relative to the events, and there is no way of telling what the absolute motion of any of the observers is, each observer is correct in insisting that his version is the right one. If one observer measures a definite distance between the events, and another finds that the events coincide in space, both observers are to be considered right.

To see just how space and time change as we go from one observer to another, we suppose that the man in the station and the one on the train are holding yardsticks that were exactly the same length before they were given to the two men. If the man in the train holds up the stick at the moment he is passing the station so that it is parallel to the tracks, and the man in the station measures it at that same moment (he has to use some kind of optical method to perform this measurement), then this stationary observer finds that the length of the moving yardstick is smaller than the length of his own stationary yardstick. In fact, the moving yardstick appears to shrink by the factor $[1 - (v^2/c^2)]^{1/2}$.

But if this procedure is reversed, so that the moving observer compares the yardstick in the station (held parallel to the track) with his own yardstick, he finds the former yardstick to be smaller than his yardstick and by the same fraction as in the previous case. In other words, lengths in the direction of motion appear to contract whether the observer is moving or the length is moving; this is a completely reciprocal effect. The amount by which the length appears to shrink increases with the relative velocity of the observer to the length until the length shrinks to zero for the speed of light. This shows us that no object that can be brought to rest can move at the speed of light.

If there were a series of clocks at regular intervals along the railroad bank, all synchronized with the clock in the station, and an observer were stationed at each of these clocks and each observer compared his clock with the clock in the train at the moment that it passed by, the clock on the train would lag further and further behind the clocks along the track. In other words, the moving clock would appear to be running slow by the factor $[1 - (v^2/c^2)]^{1/2}$ compared to the clocks along the track. Of course, to the man in the train it would appear that the clocks along the track were all running slower than his clock, and by exactly the same amount. This effect becomes increasingly more important until, at the speed of light, the moving clock would appear to have stopped.

From this we obtain a remarkable result if we consider two clocks that move away from each other and then meet again. We suppose that one clock remains here on Earth and that the other clock moves off with a speed v to some distant point and then returns. How much time has elapsed for each clock? Or, if we picture an observer associated with each clock, how much has the observer who remains here on Earth aged as compared with the observer who goes out and returns with the moving clock? If the time T has elapsed as measured by the clock here on Earth, and if T' is the time measured by the traveling clock, it is clear that we must have

$$T' = \sqrt{1 - \frac{v^2}{c^2}}\, T$$

since T' is just the proper time of the moving clock, which runs slower than the fixed clock. This shows us that the traveling observer ages more slowly than the observer remaining back on Earth by the factor $[1 - (v^2/c^2)]^{1/2}$. Thus, if an observer moves off with a speed $(3/5)c$ (where c is the speed of light) to a distant point and returns after 20 years as measured by a clock here on Earth, then the traveling observer will have aged by an amount $T' = 16$ years.

The faster the traveling observer moves, the more slowly he ages with respect to the fixed observer. We can look upon this in a different way to get the same result. If the observer moves out into space with a speed v, the distance he has to travel shrinks by the factor $[1 - (v^2/c^2)]^{1/2}$ by the Einstein–Lorentz contraction of space, so that the time required to cover this distance is smaller by this factor. Therefore, to this moving observer less time has elapsed on the trip than that measured by a fixed clock.

Of course, to the moving observer it appears that the observer left back on Earth is moving away with a speed v, and that therefore the clock on Earth has slowed by the factor $[1 - (v^2/c^2)]^{1/2}$. As long as the two observers continue to recede from each other, they can never compare the effect of the journey on their clocks or on their aging, and each will think that the other is aging more slowly. But if the traveling observer returns to Earth, he must have experienced a series of accelerations and decelerations that the other observer has not, and this series of changes in velocity makes the difference between the aging of the two observers. Because accelerations are involved in this problem, a rigorous analysis cannot be given in the framework of the special theory of relativity (i.e., by means of the Lorentz transformations) but must be treated by general relativity theory. The result obtained, however, is that given above. One can also obtain the above result by using the relativistic Doppler effect to analyze signals sent from one observer to the other.

Another important effect associated with a moving body is that its mass increases with its velocity by the same fraction that its length in the direction of its motion diminishes. As the speed of the body approaches the speed of light, its mass becomes infinite. In fact, if m_0 is the mass of a body when it is at rest, its mass is $m_0/[1 - (v^2/c^2)]^{1/2}$ when it is moving at the speed v. This effect has been experimentally verified for rapidly moving cosmic ray particles. We can derive from this Einstein's famous relationship between mass and energy, which states that a body has energy even when it is at rest and this energy equals m_0c^2.

A further consequence of the space-time "dictionary" is that regardless of how fast two observers are moving, their relative speed can never exceed that of light. Thus, if two observers were on trains that were moving in opposite directions, and each train had a speed very close to that of light (say, 99/100 of the speed), each observer would still find that the speed of the passing train is smaller than that of light.

However queer and contrary to common sense these results appear, we must accept them, for without the theory of relativity, we could explain none of the discoveries of modern physics; the events in the atom and in its nucleus would be entirely incomprehensible. We can no more go back to the prerelativity idea of absolute space and time than we can to the Ptolemaic picture of the universe or the philosopher's stone of the Middle Ages.

CHAPTER 5

The General Theory of Relativity

*A theory is the more impressive the greater
the simplicity of its premises is, the more
different kinds of things it relates, and the
more extended is its area of applicability.*

—ALBERT EINSTEIN, *Autobiographical Notes*

From what was said in the previous chapter, it is clear that the entire field of mechanics had to be revised because Newton's three laws of motion are not valid according to relativity theory. If we consider the law that tells us that the force acting on a body is obtained by multiplying the mass of the body by its acceleration, we see that this has no precise meaning because the mass of a body changes as it moves, and we have no way of knowing which value of the mass to use in expressing Newton's law. Moreover, the idea of acceleration is not what it was in prerelativity days, so that we must discard the notion of force. As a matter of fact, a whole new body of mechanics has been developed, called relativistic mechanics.

What of the validity of Newton's law of gravity in the light of the theory of relativity? It obviously cannot be considered a law because the principle of invariance imposes the condition that laws must be left unaltered by the Einstein–Lorentz space-time "dictionary." This is not true of Newton's law of gravity; this law involves the masses of bodies and the distances between them, and both of these quantities vary from observer to observer. In other words, the "law" changes from observer to observer so that we are at a loss to know what mass to use for the interacting bodies, or what distance to take for their separation.

Einstein was well aware of this failure of the Newtonian gravitational principle and soon became involved in the problem of reconciling relativity and gravitation. But this was no easy task; it took the greatest mind of modern times 10 years to find a solution in the form of what is now known as the general theory of relativity. But how did Einstein know what road to take in solving this problem? Fortunately, there was a very important hint lying around, which was almost crying out to be recognized, but nobody before Einstein saw the importance of it.

Since the gravitational problem is concerned with the mass of a body, Einstein quite naturally turned to the consideration of this concept and was at once struck by the appearance of mass in two different places in mechanics. To begin with, it is the quantity that is obtained when one divides the force exerted on a body by the acceleration that the force imparts to the body (Newton's second law of motion). This quantity is referred to as the inertial mass of the body. On the other hand, the mass also appears in Newton's formula for the force of gravity between two bodies. This quantity is often referred to as the gravitational mass of a body. Now it is an experimental fact that has been known for a long time that the inertial mass of a body and its gravitational mass are numerically equal. This greatly intrigued Einstein, for, as far as he could see, this equality between two apparently unrelated physical quantities associated with a body is no mere accident of nature. One has to attach a deep significance to this equality, according to Einstein, and its proper interpretation must certainly lead to the discovery of profound and hitherto unsuspected laws of nature.

How does one use this important fact to unearth the secrets of nature? Here Einstein was aided by his desire to generalize his theory of relativity so that it would apply not only to different observers moving with uniform speed, but to all observers regardless of their motion. Actually, if we wish to be exact, there are no systems in the universe, as far as we know, that are moving with uniform speed so that the special theory of relativity really has no application to nature as we know it. We apply it because most of the systems we deal with approximate quite closely those for which the special theory holds. If we could say that the laws of nature must be the same for all observers even if some of them may be moving with an accelerated motion (observers on Earth, for example), we would then have a general principle of relativity. This general principle would free us from the fear that some of the laws we have discovered here on Earth are really not general because they depend on the assumption that Earth is moving uniformly through space when in fact it is being accelerated. This was the drive that motivated Einstein's search for a general theory.

But this presents a serious difficulty: if we accept the principle that the laws of nature are the same for all observers, we must also agree that all systems, regardless of how they are moving, are equivalent. This means, however, that there can be no way of differentiating between a system that is moving with uniform speed and one that is being accelerated through space. And here we must momentarily object, for we know, from experience (sudden starting or stopping of a train or elevator) that there is quite a difference between uniform motion and accelerated motion. Accelerated motion makes itself felt by jerkiness and the pushing and pulling that we experience as our velocity changes. How is it permissible, then, for us and Einstein to assert the equivalence of all motion?

It is just at this point that Einstein used the equality of the inertial and the gravitational mass of the body to justify this equivalence. He showed that because of this equality, all accelerated motion is equivalent to the presence of a gravitational field. In other words, if you are in a train, and the train suddenly speeds up, you know what is happening to the train because you suddenly find yourself being pulled toward the back of the train. This is your inertial mass asserting itself. If you watch carefully, you observe that all bodies in the train, regardless of their masses, start going back at exactly the same speed as you do. But this is precisely what would happen if, instead of the train's speeding up, there were suddenly created a huge mass behind the train that caused all the bodies in the train to fall toward it. Thus, in principle, there is no way for you to know whether your backward motion is due to your being in an accelerated system or whether you are really being pulled backward by a gravitational field. Einstein called this the principle of equivalence; it is the basis of the general theory of relativity. This principle states essentially that inertial forces (forces arising from accelerations) are equivalent to and cannot be differentiated from gravitational forces.

We can best see the importance of the principle of equivalence if we consider one of Einstein's famous ideal or thought experiments (carried out in the mind). We consider two observers, one sitting on Earth under an apple tree and the other in an elevator out in empty space. We suppose that there is an apple tree in the elevator and that the elevator observer is sitting under it. We finally imagine that attached to the roof of the elevator is a rope and that a giant is pulling the elevator toward him so that it has an acceleration of 32.2 feet per second per second.

The observer on Earth watches an apple fall to the ground and is suddenly inspired with a profound idea: a force, produced by Earth, is pulling the apple and, indeed, all bodies, causing them to fall toward the center of Earth with the same acceleration regardless of their mass if they are at the same height. He calls this force "gravity."

The observer in the elevator also notes that an apple detaches itself from the tree and falls with increasing speed to the floor of the elevator. He also is inspired with a profound idea, but his reasoning is somewhat more complex than that of the other observer. At first, he is inclined to ascribe the behavior of the apple to some force that emanates from the floor of the elevator, for he observes that not only the apple but all objects fall to the floor with the same acceleration. He also feels his feet pushing against the floor as though this same force were pulling him down. But on second thought, knowing about the equality of inertial and gravitational mass, he reasons that all the effects he observes may really arise because he is on a system that is undergoing uniform acceleration. In fact, he notes that there is no experiment or observation that can tell him whether he is at rest in a gravitational field or moving in an accelerated system. All the effects that he observes can be due to the fact that the floor of the elevator is really moving up with increasing speed to meet the objects that seem to fall toward it. And this upward acceleration of the elevator causes its floor to push against his feet as though he had weight.

This reasoning led Einstein to one inescapable conclusion: Just as there is no way of determining the absolute velocity of a body, so, too, there is no way in which the absolute acceleration can be found. From this it must follow that no meaning can be given to the notion of a gravitational force, since one can never distinguish in an absolute sense between the effects of being under the action of a gravitational force and those of being in an accelerated system. But this leads us to a dilemma, for we know that when we throw a body into a gravitational field (a stone thrown through the air) it moves in a curved path, as though a force were acting, and not in a straight line as demanded by Newton's first law of motion when no force is acting. We must therefore either discard the accepted notion of a straight line or retain the concept of a gravitational force. But since the latter choice is already excluded by the principle of equivalence, we are forced to reject the usual notion of a straight line.

But how can we change our concept of a straight line, and how will that help us understand the nature of gravity? We might begin by asking what one usually considers a straight line to be. We soon discover that it is the curve one obtains by connecting two points on a flat surface in such a way that the distance between the two points along this curve is shorter than that along any other such curve. In other words, a straight line is the shortest distance between two points on a flat surface. But this definition is not very helpful in practice because we are then left with the problem of defining what is meant by a flat surface.

To illustrate the difficulty, we note that if we mark two widely separated points on the surface of Earth (a sphere), we can draw only one curve (a great circle) through these points that is entirely on the sphere and is the shortest distance between them. But if we were flat creatures without any height perception, we would have no way of knowing by direct observation that the sphere is curved and therefore we would call this a straight line. However, we know from our three-dimensional experience that Earth has a curved surface and that this "straight line" is really the arc of a great circle.

Actually, even if we were two-dimensional flat creatures living on Earth, we could still determine whether the surface of Earth is flat or not by means of certain simple mathematical tests that were developed more than 100 years ago by the German mathematician Karl Gauss. These tests enable one, just by measuring the distance between two neighboring points of a surface, to determine the curvature of the surface. If the curvature is zero, the surface is flat and the shortest distance between two points is the usual straight line, but if the curvature is positive (greater than zero) or negative (less than zero) the surface is either elliptical or hyperbolic and the shortest distance is a different kind of "straight line."

We see then that there are many kinds of "straight lines" depending on the curvature of our space-time surface (three dimensions of space and one of time); we must therefore allow for the possibility that the space-time of our universe is not flat so that there may be no straight lines in the ordinary sense. What then happens to Newton's first law of motion? We must replace it by the following statement. *All freely moving bodies move in space-time along a path that is the shortest distance between any two points on that curve.*

This second feature of Einstein's general theory of relativity was the starting point of the mathematical formulation of the theory. The approach Einstein took was that a freely moving body (one that is not subjected to a push or a pull by some other physical object) moves in the ordinary Euclidean straight-line path unless a gravitational field is present, in which case it moves in a curved orbit that is still the shortest four-dimensional or space-time distance between any two points. In other words, the existence of a gravitational field acting on a body is no longer represented as a force, but rather as a curvature in our space-time manifold. This means that one of the fundamental phenomena of nature (gravitation) is really a manifestation of the departure of the geometry that governs our universe from Euclidean geometry (plane geometry or geometry of a flat surface).

Starting from this point, Einstein first extended to four-dimensional space-time the Gaussian method of determining whether a surface has a nonzero curvature; second, he formulated a law (his field equations) revealing how the curvature of space-time is determined by the amount of mass that is present, or rather how the deviation of the geometry of a given region of space-time from flat Euclidean geometry defines the gravitational field. Once this was accomplished, one could determine the shortest distance between two points in this space (mathematicians call such a curve a *geodesic*) and this curve would then be the orbit of a freely moving particle—a planet, for example.

Fortunately for Einstein, in the latter part of the 19th century, the German geometer Georg Riemann had extended the work of Gauss to curved spaces of any number of dimensions; all Einstein had to do was apply Riemann's technique to the special case of space-time. It was also fortunate that in the development of this geometry Riemann had used a special kind of mathematics known as *tensor calculus*, which has the very desirable property of leaving intrinsic characteristics (e.g., curvature) unchanged when one passes from one frame of reference to any other. This is just the kind of mathematics that is needed to express the laws of nature, for if a law can be written in tensor form it is invariant under any kind of transformation and therefore has one of the properties required of a law. The first part of Einstein's task had thus almost been completed for him by Riemann, and he had but to turn his mind to the second part.

The second part was considerably more difficult, for it involved finding some characteristic of space-time geometry that could be expressed in tensor form and that contains within it the correct law of gravity. Moreover, this tensor would have to be built up of the mathematical quantities (they are called the metric potentials of space-time) that enter into the expression for the distance between two neighboring points in our space-time. The reason for this is that the nature of the geometry of a given region of space-time is determined by the distance between two points separated by a very small space. Such a distance can be expressed as a sum of terms, and the metric potentials (which depend on the masses present) are the coefficients of these terms.

Since there are ten different metric potentials, it appears at first that this is a most formidable task with a vast array of possible combinations. However, most of these combinations do not fulfill the condition of invariance.

Einstein did not go about this job blindly, but was guided by the work of Gauss and Riemann. He argued that since the presence of a gravitational field manifests itself as a departure of space-time from flatness and this departure is exhibited as a curvature of space-time, then the tensor that measures this curvature determines the gravitational properties of space. Such a tensor had already been introduced into the geometry of many dimensions by Riemann and Christoffel, and all Einstein had to do was to adapt it to his needs by showing that such a tensor depends on the matter present in space. This task was comparatively easy and Einstein was then able to write the law of gravity for a point in empty space in the form of a curvature tensor set equal to zero. This is Einstein's law of gravity.

One immediate check of this law is to see whether it contains Newton's gravitational force as a first approximation. This is indeed true, so Einstein knew he was on the right track. However, he could not be sure his law was correct until it was verified observationally. Before this could be done, the mathematical equations that expressed this new law had to be solved. This was done for the first time for a particle (a planet) moving in a gravitational field emanating from a single mass point by Karl Schwarzschild of the University of Göttingen in 1917. This derivation is equivalent in general relativity to Newton's derivation of Kepler's three laws of planetary motion. To understand all the consequences of Einstein's theory requires the mastery of advanced mathematical techniques. All we do here is indicate briefly how some of the results of the theory follow quite simply from the principle of equivalence.

A theory as general and as profound as this one developed by Einstein must have a great deal to say about many fundamental facets of nature, and even today all its ramifications have not been investigated. We consider here only three conclusions that have been drawn from the theory and that have important applications to astronomy. We cannot derive these results here, but we shall see how they may be justified by an appeal to the principle of equivalence. To see how this principle works, we first consider a special kind of space in which an artificial gravitational field is induced and see how the geometry of that space is affected by such a field.

We consider a large rotating disk on which an observer is stationed, and the conclusions he draws about the space in which he lives. The first thing he notes is that stationary bodies left to themselves do not obey Newton's first law (they are acted on by centrifugal forces); he might then ascribe their strange behavior to a gravitational field that acts radially outward from the center. However, if he measures the distances between neighboring points on his disk, he soon discovers that his geometry is non-Euclidean.

Since different parts of the disk are moving at different speeds, rulers contract differently at different places as a result of the special theory of relativity. In particular, rulers laid along the radius of the disk show no change in length whereas rulers laid along the circumference of the disk shrink. From this it follows that the relationship between the radius and the circumference of a circle is not the same on the rotating disk as it is on the stationary disk. This kind of geometry is a special kind of non-Euclidean geometry called hyperbolic geometry. The shortest distance between two points in this case is not a straight line but a curved line, and if one forms a triangle with these "straight" lines, the sum of the angles is not 180 degrees but less.

But all of these effects are the result of the observer's being in an accelerated system, so that one can describe the situation either in terms of a fictitious gravitational field (i.e., a non-Euclidean geometry) or in terms of accelerations induced by the rotation. It may also be noted that in this system the farther away a clock is from the center of the disk, the greater is its velocity and therefore the more slowly it runs, so that time, too, is affected by a gravitational field.

The importance of this example of the rotating disk is that it enables us to see just how the principle of equivalence allows us to determine the nature of the non-Euclidean geometry in a gravitational field. The gravitational field on the rotating disk is, of course, fictitious, but the procedure we used to investigate the geometry can be applied to all gravitational fields.

What about the geometry in the sun's neighborhood where a real gravitational field is present? This differs in one respect from the rotating disk. In the latter it is possible by means of a simple transformation to eliminate the fictitious gravitational field (just step off the spinning disk), whereas in the case of the sun or Earth no single accelerated system (transformation to some moving system) can be set up that eliminates the gravitational field everywhere. Nevertheless, in small enough regions of space we can duplicate the sun's gravitational field by means of an accelerated system.

We intend here to duplicate the gravitational field of the sun at any point on its surface (or at a given height above its surface) by means of an elevator at that point that has an outward radial (i.e., directed away from the center of the sun) acceleration exactly equal in magnitude to the acceleration that a freely falling body would have at that point. If we picture elevators of this sort, all with different accelerations, placed around the sun, whose gravitational field we imagine as having vanished, they reproduce all the gravitational effects of the sun, and a person in any one of these elevators experiences all the gravitational phenomena that he would if he were in the actual gravitational field of the sun. The elevators that are closer to the surface of the sun have, of course, larger accelerations than those farther away from the sun. As the first example of an important application of the general theory of relativity to astronomy, we consider the deflection of light in the gravitational field of the sun.

If, during a total eclipse of the sun, we photograph the stars around the rim of the sun, we find that the images of two stars on diametrically opposite sides of the sun are farther away from each other than the same images are on a photographic plate taken of the same star field 6 months later when the sun is in another part of the sky. This can be understood if we accept the conclusion of the general theory of relativity that the path of a ray of light is bent by a gravitational field. To see why this is so, we use our elevators and the principle of equivalence.

We consider a beam of light that enters on one side of an elevator and leaves through a window on the other side of the elevator. Since the elevator is in motion, the apparent direction of the light when it enters, as seen by the observer is different from its true direction of motion because of the aberration effect. We understand this aberration effect if we recall our experience on a boat with the wind blowing. If the true direction of the wind is at right angles to the direction of the motion of the boat, it appears to us on the boat that the wind is coming from a direction that lies somewhere between the direction of the boat and the true direction of the wind. The same thing holds for the direction of a beam of light as seen by a moving observer.

Since change in the direction of the beam of light depends on how fast the observer is moving, the direction of the light when it leaves the elevator is different from its direction when it enters the elevator. By the time the light has moved across the elevator to the other side, the speed of the elevator will have increased because of its acceleration so that the aberration effect is larger, and it appears to the observer in the elevator that the direction of the light when it leaves the elevator is different from that at which it entered. In fact, the light appears to bend toward the floor of the elevator just as though it were a material particle in a gravitational field. We may thus conclude from the principle of equivalence that a light ray is bent when it passes through a gravitational field.

This prediction of the general theory, as we have already noted, was verified by observing the positions of stars in the neighborhood of the sun during a total solar eclipse. Of course, since a massive body like the sun introduces a curvature of space in its neighborhood, we should expect the path of light to be bent since light must move in accordance with this curvature. In fact, the behavior of light as it passes the sun can be taken as a proof that space-time is non-Euclidean in the neighborhood of the sun. From an analysis of the path of a light beam, we find that the curvature of space caused by a massive body is positive so that the geometry is elliptical.

This is the kind of geometry that we have on the surface of a sphere and it is characterized by the following properties: the sum of the angles of a triangle is greater than 180 degrees; the circumference of a circle is smaller than π times its diameter, not equal to it, as for a circle on a flat surface; there are no parallel lines in such a geometry since all "straight" lines (i.e., great circles) intersect. This is quite different from the geometry on the rotating disk (hyperbolic geometry), since there it is possible to draw through a point an infinite number of "straight" lines that do not intersect a given "straight" line not passing through the point.

We now consider how a clock behaves in a gravitational field as compared to a clock far removed from any gravitational influence. We, in fact, take a whole array of clocks stationed radially at increasingly larger distances from the surface of the sun. Again we go to our collection of elevators, which we may assume to be equivalent to the sun's gravitational field, and suppose that each elevator is supplied with a clock that is to be compared with the clock at a great distance (infinitely far) from the sun. It is clear, from what we know from the special theory of relativity, that the clocks inside the elevators will not keep time with the distant clock because they are moving with respect to the distant clock.

Another way of analyzing the behavior of clocks in the gravitational field of a body like the sun is to set up an array of clocks at varying distances from the surface of the sun along a straight line from its center. These clocks are not to be in elevators but are to be fixed with respect to the sun. We must now compare these clocks with a clock that is not affected by the gravitational field of the sun. We may do this as follows: we consider an elevator with a clock in it that starts from rest at a great distance from the sun and falls freely toward the sun's center, passing the stationary clocks on the way. Since this elevator is falling freely, an observer inside it experiences no gravitational field, so that by the principle of equivalence his clock is free of the gravitational influence of the sun. If the observer in the elevator now compares his clock with each of the stationary clocks as he passes them, he finds that these stationary clocks are running slower than his clock. This is so because the stationary clocks appear to the man in the elevator to be moving past him, and by the special theory of relativity, lag behind his clock.

The closer the falling elevator gets to the surface of the sun, the faster it moves, and therefore the slower the fixed clocks in the gravitational field of the sun run as compared to the clock in the elevator. This means, by the principle of equivalence, that the stronger the gravitational field (the gravitational field of the sun gets stronger as one gets closer to its surface), the slower a clock in it runs. We can apply the same kind of analysis to the behavior of measuring rods. If a ruler is placed in the gravitational field of the sun so that it points toward the center of the sun, it shrinks in size, and this shrinkage is larger the closer the ruler is to the surface of the sun. If the ruler is placed horizontally on the sun's surface instead of vertically, its length remains unchanged. This shows us that the geometry in the gravitational field of the sun is of a non-Euclidean nature, and is of the elliptical kind because the circumference of a circle as measured by this ruler is smaller than 2π times its radius as measured by the same ruler. Since the freely falling elevator, starting from an infinite distance, has the speed $\sqrt{2GM/r}$ (the speed of escape) when it is at the distance r from the center of the sun whose mass is M, the fixed clocks and radially aligned rods appear to be moving with the speed $\sqrt{2GM/r}$ past the falling elevator. Hence, these fixed clocks and rods slow down and shrink by the factor $[1 - (2GM/c^2r)]^{1/2}$, where c is the speed of light.

The slowing of clocks in gravitational fields is of considerable importance in astronomy because we can apply it to measure the intensity of gravitational fields on the surface of certain stars. To do this we must find the equivalent of a clock on the surface of the star, and since any vibrating structure behaves like a clock, we can apply this to an atom that is emitting radiation. We take the reciprocal of the frequency of the radiation that it is emitting (given by the color of the light it sends out) as its unit time interval. If we now compare the radiation emitted by such an atom when it is in an intense gravitational field with the radiation emitted by the same atom when it is on the surface of Earth, we see that the frequency in the former case is smaller (the period between two beats of the atom is longer) than in the latter case.

This is known as the Einstein redshift and simply states that the light emitted by atoms on small massive stars (large gravitational fields) is redder than light emitted by the same atoms on Earth. This effect is used to determine the sizes of certain small stars (the white dwarfs).

We come now to the application of general relativity to celestial mechanics and the explanation of the discrepancy in the motion of Mercury. We have noted that the geometry in the neighborhood of the sun is not Euclidean but rather is of an elliptical kind. But this means that the relationship between the circumference of a circle (say, the orbit of Mercury) and its radius (the distance of Mercury from the sun) is not the same as it would be for Euclidean geometry. Actually, the circumference is smaller, so that during the planet's period as determined by Euclidean geometry the planet overshoots its non-Euclidean orbit. This exhibits itself as a precession of the perihelion. This effect has been studied very carefully and the most recent astronomical data, not only for Mercury but also for Earth and Mars, are in very good agreement with the theory of relativity.

In later chapters we shall come back to the general theory of relativity and see how profoundly it has influenced our entire concept of the structure and geometry of the universe, enabling us to develop a rational cosmology to account for the observed behavior of the distant galaxies. We shall also see that some of the most exciting discoveries dealing with quasars, pulsars (neutron stars), and black holes can be understood only if the general theory of relativity is applied.

We note that in a gravitational field the square of the space-time interval between two events is not the same as in the special theory. One obtains the correct expression in a gravitational field by dividing the square of the radial separation of the two events by $1 - (2GM/c^2r)$ and multiplying the square of the time interval between these two events by this same factor, where r is the distance of the events from the center of the gravitational field. Doing this takes account of the shrinking of rulers held radially and the slowing of clocks in gravitational fields. These factors in the space-time interval slow light in a gravitational field and account for phenomena in the neighborhood of a black hole.

CHAPTER 6

The Properties of the Planets

Our Souls, whose faculties can comprehend
The wondrous Architecture of the world;
And measure every wand'ring planet's course,
Still climbing after infinite knowledge.

—CHRISTOPHER MARLOWE,
Conquests of Tamburlaine

Owing to the vast growth of the American and Soviet space programs since the launching of Sputnik, our knowledge of the conditions on the surfaces of the planets that have been surveyed with space probes has increased tremendously. It is fair to say that through space probes we have learned more in the last few years about the bodies in our solar system than we did in all previous times with our Earth-based telescopes. Indeed, so much information has reached us and is still being sent that we can hardly keep up with or interpret the facts as they accumulate. Remarkable photographs have been obtained of Jupiter and of the surfaces of Mercury, Venus, and Mars; and the two landers of the Martian Viking missions have supplied some exciting and puzzling data about the Martian soil. Much remains to be done, both observationally and theoretically, before we can obtain a self-consistent picture of the chemistry and structures of the planets and of how the solar system originated, but we are well on the way to such a goal. Even before the modern space era, astronomers had learned a great deal about the properties of the planets through optical and radio telescopes. The most striking thing about the planets as a group is that the four inner planets—Mercury, Venus, Earth, and Mars—differ markedly from the four outer planets (we exclude Pluto because there are strong indications that it is an escaped satellite of Neptune)—Jupiter, Saturn, Uranus, and Neptune. This difference reveals itself in mass, size, density, rate of rotation, chemistry, internal structure, and satellites. Moreover, the large distance between Mars and Jupiter emphasizes the sudden change in planetary characteristics as one goes from the inner to the outer group.

If a planet has a satellite, the planet's mass can be found quite easily by measuring the satellite's period in its orbit around the planet and its mean distance from the planet, and then inserting these values into Kepler's third law. If the planet has no natural satellite, we can obtain its mass by observing an artificial satellite in orbit around it or by measuring its gravitational pull on a comet that comes close to it. All these methods have been used and we now have accurate values for the planetary masses. Expressed in terms of Earth's mass, which we take as 1, the masses of Mercury, Venus, Mars, Jupiter, Saturn, Uranus, and Neptune, in that order, are 0.054, 0.815, 0.108, 317.8, 95.2, 14.5, and 17.2. The masses of the four inner planets are all of the same order of magnitude (Earth is the most massive and Mercury the least), whereas the outer planets are considerably more massive than Earth.

Since the apparent size of each planet's disk can be measured on a photographic plate, and its distance can also be measured, the true diameter of the planet can be calculated. The procedure is simple and direct. One divides the size of the image on the photographic plate by the length of the tube of the telescope (its focal length) that was used to take the picture, and multiplies this by the planet's distance when the photograph was taken. The length of the tube must, of course, be expressed in the same units of length as is the size of the photographic image; the diameter of the planet is then obtained in the same units as those used to express its distance. Such procedures have given us the following diameters for the planets (in the same order as above and expressed in terms of Earth's diameter): 0.380, 0.950, 0.532, 11.18, 9.42, 3.84, 3.93. Again we see the big jump when we go from the inner to the outer planets.

The mean densities (masses divided by volumes) of the inner planets are considerably larger than those of the outer planets, and, more than anything else, indicate the chemical differences between the two groups of planets. Starting with Mercury, the densities (in g/cm^3) are 5.4, 5.2, 5.518 (Earth), 3.95, 1.34, 0.70, 1.58, 2.30. The differences between these two groups of densities can be accounted for by the differences in chemical composition: the inner planets consist of rocks and such heavy metals as iron and nickel whereas the outer planets consist of hydrogen, helium, water, carbon dioxide, methane, and ammonia.

As we go from the inner to the outer planets, we note a marked increase in the rate of spin. Mercury rotates once every 59 days, Venus once every 244.3 days (in the opposite direction, i.e., retrograde), Earth once every 23 hours, 56 minutes, 4.1 seconds, and Mars once every 24 hours, 37 minutes, 22.6 seconds. The very slow rotations of Mercury and Venus are due to the sun's tidal action, which slowed their spins when these planets were still in a molten state. Earth and Mars were too far away from the sun for the solar tides to have had as much of an effect as on the two inner planets and so they escaped the fate of Mercury and Venus. The periods of rotation of the four outer planets are 9 hours, 50 minutes, 30 seconds (Jupiter); 10 hours, 14 minutes (Saturn); 10 hours, 49 minutes (Uranus); 15 hours, 48 minutes (Neptune).

We finally note that the four inner planets have only three satellites among them, two of which (the Martian satellites) are little more than giant boulders. The four outer planets, however, have some 30-odd satellites, five of which are considerably larger than our moon. The difference in the number of satellites associated with the inner and outer planets undoubtedly stems from the difference in the mean distances of the inner and outer planets from the sun. Most of the material surrounding the inner planets, when they were formed, was probably swept to the outer regions of the solar system by the intense solar wind and solar radiation so that little remained for the formation of satellites. At the distances from the sun of the outer planets, the solar wind was probably too weak to carry off much interplanetary material. Thus, enough material was left to form the satellites that we now observe.

Since Mercury, Venus, and Mars have been studied in greater detail, we can say much more about their surface and atmospheric features than about those of the four outer planets. In 1974 the unmanned spacecraft Mariner 10 was steered into an orbit around Mercury that brought it to within a few hundred thousand kilometers of that planet's surface every 176 days. Before using up the fuel that kept its orientation in space stabilized, Mariner 10 had sent three sets of high-resolution pictures back to Earth. These photographs reveal a heavily cratered surface hardly different from that of the moon. The resemblance between the moon's and Mercury's surfaces is further enhanced by the way they reflect sunlight and radar pulses and the way they emit infrared rays and radio waves.

Although Mercury looks like the moon on the outside, it is quite different on the inside. It is slightly larger than the moon but about four times as massive, which means that, unlike the moon, its interior consists of large amounts of iron and possibly nickel. Because of Mercury's high density but small size compared to that of Earth, it probably contains a much higher percentage of iron in its interior than does Earth. All the evidence about its interior indicates that Mercury's iron core extends three-quarters of the way out to its surface (the iron core is about as large as the moon) and that this core is surrounded by a thin silicate mantle and crust.

For a long time, astronomers were sure (based on the observation of Mercury with optical telescopes) that Mercury's period of rotation was exactly equal to its period of revolution around the sun. This led to the belief that Mercury always kept the same face toward the sun just the way the moon always keeps the same face toward Earth. It was argued that the sun and Mercury were locked into this synchronous state of rotation and revolution because of the sun's tidal action on early Mercury (molten state). With the introduction of radar astronomy after World War II, the rotation of Mercury was measured directly by bouncing radar beams off opposite edges of Mercury's visible disk, with the startling result that Mercury's period of rotation was found to be 58.65 days—just two-thirds of 88 days. This means that Mercury does not keep the same face toward the sun all the time, and explains why the temperature of the dark side of Mercury is not as low as it would be (not perpetually cold) if it never received any direct heat from the sun. The thermal emission from the dark side of Mercury is much more intense than could possibly be true for a perpetually cold surface. The total thermal radio flux (radio waves emitted owing to random motions of charged particles) from the bright and dark surfaces of Mercury indicates an average surface temperature of 170 to 200°F, exactly what one expects for the surface of an object rotating as Mercury does.

We now know that Mercury has no atmosphere whatsoever, contrary to what was thought to be before the Mariner 10 mission. An atmosphere on Mercury was thought to be required to account for the unexpected intensity of thermal emission from the dark side of Mercury when astronomers thought that Mercury's rotation was synchronous with its revolution. Under those conditions an atmosphere would keep the perpetually dark side warm by convective currents (winds). It should be noted that, owing to Mercury's small mass, the speed of escape from its surface is too small for it to retain an atmosphere. At the temperatures found on Mercury's illuminated face, molecules of any atmosphere would be moving fast enough to escape. Mariner 10 also confirmed the existence of a magnetic field on Mercury that has an intensity consistent with a 59-day period of rotation but not with an 88-day period. This strengthens the deduction that Mercury has an iron core.

Mercury's 58.65-day period of rotation means that it rotates on its axis three times while revolving around the sun twice. The rotation and revolution are thus tied together (so-called spin–orbit coupling) in a way that cannot be accidental, but rather was the result of the interaction of the gravitational force of the sun on Mercury and the rotation of Mercury. The only way this interaction could have occurred was through the tides raised on Mercury by the sun when Mercury was in a molten or plastic state. Since Mercury is about 3 times closer to the sun than Earth is, the tide-raising force of the sun on Mercury is about 27 times greater (it goes inversely as the cube of the distance) than it is on Earth.

Mercury's orbit around the sun is the most elongated (the most eccentric) of all the planets; its distance from the sun varies from 28 million miles at perihelion to 43.4 million miles at aphelion. At perihelion each square centimeter of Mercury's surface facing the sun receives ten times as much energy per second as does each square centimeter of the moon's lit surface. Owing to this intense solar radiation, which lasts for some 28 days on a given face of Mercury, Mercury's noon temperature reaches 700°K (absolute). At midnight the temperature drops to 100°K.

Although the surface of Mercury looks like that of the moon, there is a difference in the appearance of their craters. Whereas the moon's surface is densely populated with overlapping craters, many of which are very large, Mercury's surface is not completely covered with large craters and there is very little overlapping of craters. This difference undoubtedly stems from the difference in their surface gravity. The force of gravity on Mercury is about twice that of the moon, so that the stronger impact of a meteorite on Mercury has a smaller effect and ejects matter over a considerably smaller area. Thus, craters tend to be smaller and separated from each other on Mercury.

All the evidence obtained from Mariner 10 points to five major sequences of events in the formation of Mercury's surface. In the first, the large iron core and silicate crust were formed, and if any atmosphere was present it escaped shortly after the crust solidified. The second sequence of events probably resulted in the obliteration of any craters that were formed in the crust in the first phase. This may have been caused by volcanic activity. In the third stage the deep craters and the mountainous terrain that are shown on Mariner 10 plates were formed. This phase was followed by another volcanic period, which produced the broad plains that appear to consist of basalt and other volcanic rocks. The fifth phase in Mercury's history is its present quiescent period in which very little is happening. Mercury reappears in the sky in the same position relative to Earth and the sun once every 115.88 days (its synodic period).

Because the mass and diameter of Venus very nearly equal those of Earth, astronomers for a long time were convinced that the internal structure of Venus is quite similar to that of Earth and that surface conditions on Venus would, in time, also be found to be similar to those on Earth. Even though Venus receives about twice as much solar energy per second as Earth does, it was felt that the thick impenetrable cloud that surrounds Venus would allow just enough heat to trickle through to the surface to keep its mean temperature equal to that of Earth. Two other incorrect assumptions contributed to invalid conclusions about Venus's surface conditions. The first was that the atmosphere under the thick clouds is chemically similar to Earth's atmosphere (an oxygen–nitrogen mixture, with very little carbon dioxide), and the second was that Venus rotates on its axis at about the same rate as Earth does. Since the solid surface of Venus could not be observed, its period of rotation could not be measured by noting how rapidly fixed markings on the surface moved across one's line of sight. Only the upper layers of the clouds could be observed so that the period of rotation of the planet had to be surmised from the motion of the cloud cover. The various periods of rotation so obtained by different observers were quite contradictory but they did seem to indicate that the planet has a slow retrograde (from east to west) rotation.

The first breakthrough in determining the correct period of rotation of Venus came with the discovery that radar beams sent from

Earth to Venus penetrate the clouds quite easily and are reflected back to Earth with considerable information about the topography. In fact, the radar beams indicate that the surface of Venus looks the way Earth's surface would if all the oceans were absent. The radar beams show the presence of some craters, which one would expect to find if there were no water erosion on Venus. Thus, the absence of water on Venus is well established, which means that the clouds on Venus do not consist of water droplets. We know today that the clouds consist of droplets of some corrosive liquid— probably sulfuric acid.

When radar beams are reflected from the surface of Venus, the rotation of Venus alters their frequency (and wavelength) somewhat by an amount that depends on whether they are reflected from the edge of Venus that is approaching or receding from us as it rotates (the Doppler effect). From these changes in wavelength the period of rotation of Venus can be found quite accurately. Venus is rotating from east to west (retrograde rotation) with a period of 243 days. If there were no clouds on Venus, an observer there would see the sun rise in the west and set in the east. The 243-day period is remarkable in one respect. It means that every time Venus passes between Earth and the sun it presents the same face to us. This indicates a relationship between the motion of Venus and that of Earth, which might have been produced by some kind of tidal action of Earth on Venus.

The slow rotation of Venus seems to be at variance with the rapid motions of the clouds, which have a 4-day period of rotation and indicate winds with speeds of up to 100 meters per second (over 600 miles per hour). It is difficult to see how huge masses of the atmosphere can be set moving with such high speeds if the surface of the planet is moving so slowly. These high wind speeds are supported by the data obtained from the Soviet Venera series of spacecraft that dropped into Venus's atmosphere. In addition to streaming along the surface of Venus, the atmosphere appears to have a periodic vertical motion, so that the cloud tops move up and down every 4 to 6 days.

The most remarkable discovery about the surface was made with radio telescopes in the mid-1950s, when intense thermal radio emission from Venus's surface was detected. The laws of thermal radiation (whether in the visible, infrared, or thermal wavelengths) give a very precise relationship between the intensity of the emitted radiation and the temperature of the source. The intensity of the thermal radio waves emitted by Venus indicates surface temperature of 750°K, well above the melting point of lead. This amazing result has been completely verified by the data gathered by the Mariner 5 and Venera 4 probes of Venus's atmosphere. These show that the atmosphere is very deep, dense, and massive, as it must be to maintain the temperature difference between the tops of the clouds (250°K) and the surface (750°K).

As one descends through the clouds into the atmosphere, the temperature remains fairly constant down to a height of about 90 km. It then increases rapidly, reaching 500°K at a height of 30 km, 600°K at 15 km, and 750°K at the surface. The atmospheric pressure increases rapidly also, reaching a surface value of 90 times Earth's atmospheric pressure.

The high surface temperature on Venus can be accounted for by the great abundance of carbon dioxide in its atmosphere. Even before the Soviet and American probes were sent into Venus's atmosphere, carbon dioxide and carbon monoxide were detected there spectroscopically. Indeed, the first substance detected above the clouds on Venus was carbon dioxide, discovered spectroscopically in 1932 by W. S. Adams and T. Dunham of the Mt. Wilson Observatory. The absorption lines of carbon dioxide in the spectrum of the atmosphere above the clouds were so intense that one would need a 2-mile-long column of carbon dioxide at atmospheric pressure here on Earth to produce lines of the same intensity when light passed through such a column. This was a rather startling result because it was so different from the concentration of carbon dioxide in Earth's atmosphere and so different from what was expected. But it fits in neatly with Venus's high surface temperature. Gaseous carbon dioxide allows visible radiation (light) to pass through it quite readily but it blocks the passage of infrared rays (the heat rays). This means that carbon dioxide traps the solar radiation after it has passed through Venus's atmosphere and warmed the planet's surface (the greenhouse effect). The warmed surface reradiates this energy, not in the form of visible radiation, but as infrared radiation, which cannot return through the atmosphere because of the carbon dioxide. Thus, the surface gets hotter and hotter until a state of equilibrium is established in which the temperature of the surface reaches a value just high enough to maintain an outward flow of infrared radiation that just balances the visible radiation that trickles down through the atmosphere. The Soviet Venera experiment showed that only 1% of the solar radiation that strikes the top of Venus's cloud layer reaches the planet's surface. Small as this is, it is sufficient to keep the surface at 750°K if enough of the infrared rays are trapped.

To account for a surface temperature of 750°K, 97% of Venus's atmosphere must be carbon dioxide. This is such a large fraction of the atmosphere that few astronomers accepted it as correct. But this has been completely verified by actual measurements made by the Soviet Venera probes. Two other interesting features of the atmosphere were revealed by the Mariner and Venera probes: hardly any oxygen was found, and very little water vapor is present—certainly no more than 0.2%. The great abundances of carbon dioxide in Venus's rocks were too high to absorb the original carbon dioxide that was ejected in volcanic eruptions. This carbon dioxide then initiated the greenhouse effect that fed on itself until the present temperature was reached.

In two of its later Venera experiments the Soviets achieved a soft landing on Venus's surface and important information was radioed back for almost an hour in both landings. In the second of these experiments, photographs of Venus's terrain were obtained. These show a very arid, rock-strewn surface similar to desert regions on Earth. The rocks are fairly angular and sharp-edged, indicating very little wind erosion, contrary to what one would expect if winds of hundreds of miles per hour were the rule. These photographs seem to indicate breezes of a few miles an hour.

The formation and evolution of Venus were probably similar to those of Earth, and its internal structure is like ours even though no magnetic field has been detected on its surface. We have seen that the difference in Earth's and Venus's atmosphere can be accounted for by the fact that Earth's initial carbon dioxide was absorbed by the rocks in its crust whereas the rocks of Venus were too hot from the very beginning to form carbonates. When the sun becomes more luminous in a few billion years, the rocks on Earth may give up their carbon dioxide, and thus initiate a runaway greenhouse effect. Earth will then lose all its water and its atmosphere will become similar to Venus's present atmosphere. Venus revolves around the sun once every 224.701 days (its sidereal period) and reappears in the same position in the sky relative to Earth and the sun once every 583.92 days (its synodic period).

When Kepler began his great work on the planets as Tycho Brahe's assistant, he fervently hoped that Tycho would allow him to work on Mars because, unlike Venus, which moves around the sun in an almost circular orbit, the eccentricity (the flatness) of Mars's orbit is second to that of Mercury's orbit. The orbit of Mars was thus an excellent test of Copernicus's theory (Mercury was much too close to the sun for an observer even as excellent as Tycho to obtain any reliable data).

Kepler considered the correct determination of the orbit of Mars the greatest of all challenges in astronomy and wrote as he began on Mars, "I consider it a divine decree that I came [to Brahe] at exactly the time when Longomontanus [another of Brahe's assistants] was busy with Mars . . . that mighty victor over human curiosity . . . who kept the secret of his rule safe throughout all past centuries and followed his course in unrestrained freedom. Assuredly either through it [Mars] we arrive at the knowledge of the secrets of Astronomy or else they remain forever concealed from us."

Just as the orbit of Mars challenged Kepler to his greatest efforts, so the surface conditions on Mars have challenged the astronomers who followed Kepler. Since Mars, with a surface that changes periodically as the planet moves around the sun, as though responding to changes of seasons, seems to offer the greatest hope for extraterrestrial life in our solar system, it has been studied more thoroughly than any other planet. But even with the largest and best telescopes on Earth, very few details of the Martian surface could be discerned, and what could be seen still left the observers puzzled and uncertain. Indeed, it was not until the Mariner series of spacecraft, particularly Mariner 9, which went into orbit around Mars in 1971, and the soft landing on Mars of the two Viking landers that a detailed and definitive close-up picture of the Martian surface was obtained. The first three Mariner probes, which flew past the planet in 1965 and 1969, sent back over 200 photographs revealing a drab, lifeless, cratered surface something like the moon's in appearance. Since these photographs cover only a tiny fraction of the surface of Mars, they are rather inconclusive concerning geological or biological activity. In 1971 Mariner 9 went into orbit around Mars at a mean distance of a few thousand miles from its surface, and sent back excellent photographs of its entire surface, revealing details that had never been seen before by man, and proving beyond a doubt that Mars had a very active geological past. Although these photographs support, to some extent, the general conclusions that were drawn from the earlier Mariner probes, they show an incredibly diverse topography consisting of spectacular gorges and canyons, vast peaks and craters, huge volcanic cones, and vast windswept terrains. Other remarkable photographic features are huge sedimentary deposits, looking just like those that were laid down on Earth by running water during the past millions of years, long sinuous channels with tributary systems like dried-up riverbeds on Earth, and dark tails all pointing in the same direction, associated with the numerous craters on the surface.

The large volcanoes, canyons, and mountain peaks are considerably larger than those on Earth. The giant shield volcano, Olympus Mons (also called Nix Olympica), is more than 300 miles in diameter, and its cone rises more than 70,000 feet (some 25 km) above its base. At the foot of this volcanic cone a sharp circular cliff rises 2 km above the surface of Mars. The photographs taken by Mariner 9 reveal about a dozen large volcanic mountains, which resemble the peaks of solidified lava formed on Earth by successive flows of molten rock from the same source for millions of years. Thus, if Earth's water disappeared, photographs taken of its surface from a height of a few thousand miles would look like those of Mars.

The canyons on Mars are vast compared to those like the Grand Canyon, on Earth. They undoubtedly correspond more closely to the great rifts in our ocean beds and vast faults such as the San Andreas fault. The enormous Canyon Tithonius Chasma (also called the Mars Rift Valley or Coprates Canyon) has an overall length of 2500 km, an average width of about 150 km, and a depth of several kilometers. The many branching channels diverging from it resemble the channels of vast drainage systems here on Earth. Mars undoubtedly passed through a period of great volcanic activity that was followed by torrential rains and a great deal of water erosion. Its early history thus resembles that of Earth but subsequently it lost its water and much of its atmosphere and became the almost dead planet that it now is. The evidence from the Mariner photographs indicates that the past climate of Mars was very different from its present climate and much more favorable to life than it is now.

Mars is probably quite similar to Earth chemically, but since it was formed in a part of the solar system that is cooler than where

Earth was formed, the chemical compounds on the surface of Mars are probably quite different from those on Earth. Its rocks consist of compounds of aluminum, magnesium, iron, silicon, oxygen, and sulfur. The amount of free iron in metallic form is considerably smaller than that on Earth. The planet, owing to its small mass (one-tenth of Earth's mass), was formed quite rapidly—in less than a half million years—from the debris in the solar nebula. It was a molten homogeneous sphere initially devoid of an atmosphere. As it cooled rapidly a solid crust consisting of a variety of rocks (aluminum oxides and silicates) and some 50 km thick was formed. The heat generated by the heavy radioactive nuclei under the crust raised the temperature of the interior to a point at which mixtures of metallic iron and iron–sulfur compounds melted and sank to the center to become the molten core. Above this core a mantle consisting of ferromagnesium silicates was formed. The volatile gases were forced out through the crust to become the original atmosphere. Owing to the slight expansion of Mars that resulted from this separation of the various materials, the original surface was destroyed and a second one was formed.

During the next billion years the craters on the surface of Mars were formed, with lava flows that produced the lava plains now visible. The volcanic activity and the heating of the rocks stimulated the emission of water vapor, carbon dioxide, methane, nitrogen, and other gases. These formed the second atmosphere. The light gases escaped because the surface gravity on Mars is too weak to reduce (below the speed of escape) the random speeds of such atoms at the top of the atmosphere. Much of the carbon dioxide and water vapor condensed and became the ices that form the polar caps. The permanent caps consist of layers of ordinary ice (solid water); the temporary caps, which lie on top of the ice caps and which grow and decrease in size with the change of seasons, consist of solid carbon dioxide. Probably all the water on Mars, or most of it is trapped in the polar caps and in the granular material that lies above the bedrock.

The Martian atmosphere is extremely tenuous; its pressure on the surface is about 1/200 of Earth's atmospheric pressure at sea level. It consists mostly of carbon dioxide, some argon, some carbon monoxide, and oxygen, which make up about 0.1% of the atmosphere. The water vapor in the atmosphere varies, but rarely exceeds 0.01%. As much as 25% of the CO_2 probably condenses in each hemisphere during its winter season, forming the top layers of the polar caps, which may extend down to a latitude of 60°. The trace gases such as carbon monoxide, oxygen, ozone, and hydrogen in the atmosphere are probably formed photochemically from CO_2 as the latter is disintegrated by sunlight. The atmosphere contains clouds (condensations from gaseous CO_2 and water vapor), which are formed and disappear just as clouds do in our atmosphere. Vast dust storms occur quite frequently, obscuring large areas of the surface with winds of up to 100 miles per hour.

Although most of the craters on Mars were formed during the first billion years of its existence, cratering on a small scale is still occurring. The enormous shield volcanoes, hundreds of kilometers in diameter and taller than Mount Everest, are the most remarkable features on the Martian surface. The light and dark patches that were found to change seasonally as viewed through Earth-bound telescopes are caused by windblown dust. The Mariner photographs show that the streaks of sand that give the patchy appearance all point in the same direction as would be the case if caused by winds blowing in a fixed direction.

Our most recent information about surface conditions on Mars comes from the two Viking orbiters and their landers launched from the Kennedy Space Center on August 20 and September 9, 1975, respectively. Viking I went into orbit around Mars on June 19, 1976, and Viking II went into orbit on August 7. Lander I reached the surface of Mars in the Chryse Planitia basin on July 20, and Lander II settled safely in the northern regions of Mars in September. The photographs sent back to Earth from the orbiters tend on the whole to confirm much of what was learned about Mars from Mariner 9, but much more detailed information was obtained. For example, the existence of fresh craters was definitely established, and the presence of very little water vapor in the atmosphere was confirmed. The vapor, which lies very close to the surface and is probably in equilibrium with a thin surface fog, appears in the morning as the surface warms up, hugs the surface during the day, and condenses into ice at nighttime. The surface temperature in the equatorial regions rises rapidly after dawn from 142°K to 240°K. The general chemical composition of the atmosphere revealed by the Viking orbiters is similar to that revealed by Mariner 9: it is composed mainly of CO_2 with traces of nitrogen, argon, oxygen, and carbon monoxide.

The Viking I lander photographs (the first ever sent to Earth from the Martian surface in black-and-white and color) show the western slopes of Chryse Planitia as a boulder-strewn, deeply reddish desert with the rims of impact craters in the distance. A sky ranging in color from orange-cream to pink surmounts the entire panorama. The rocks are sharp-edged and their surfaces vary from smooth to extremely pitted, indicating heavy wind erosion. No gross biological activity was observed.

Since the presence of nitrogen, water vapor, and CO_2 in the atmosphere and sunlight favors some kind of biological activity, careful experiments involving the soil chemistry of Mars were conducted by Lander I. Although a very high rate of chemical activity involving CO_2 and oxygen was found to occur in the soil, no large organic molecules, such as one generally associates with biological activity, were discovered. The question of life on Mars is thus unanswered. Mars revolves around the sun once every 686.98 days and reappears in the same position in the sky relative to Earth and the sun once every 779.94 days.

In 1877, it was discovered that Mars has two tiny satellites. The inner satellite Phobos is only 5800 miles from the center of Mars and orbits the planet once every 7 hours 39 minutes. Its diameter is about 7 miles. The outer satellite Deimos is about 14,600 miles from the center of Mars and orbits once every 30 hours 18 minutes. Its diameter is about 4 miles.

The gap between Mars and Jupiter is about twice as large as one expects from the regularity of the spacings between the four inner planets and those of the four outer planets. It was therefore quite natural for astronomers to survey that region of the solar system with the hope of discovering a planet there. Although no large planet exists in that region, thousands of small ones (the minor planets or asteroids) do. The first and largest of these was discovered by the Italian astronomer G. Piazzi in 1801. The mean distance of this object, Ceres, from the sun is 2.8 astronomical units so that its orbit lies just where the orbit of a planet ought to lie in accordance with the regularity of the planetary spacings. The diameter of this asteroid is 350 km and its mass is about 1/10,000 of Earth's. Following quickly upon the discovery of Ceres, the smaller, less massive asteroids Pallas, Juno, and Vesta were discovered. To date, more than 3000 asteroids have been discovered and the orbits of 1780 have been determined.

The mean semimajor axis of the orbits of the asteroids is 2.7 astronomical units, and 99.8% of the semimajor axis lie between those of Mars (1.524) and Jupiter (5.203). The mean period of the asteroids is 4.5 years and there are conspicuous gaps in the sequence of periods at $\frac{1}{2}$, $\frac{2}{5}$, $\frac{1}{3}$ of Jupiter's period. On the whole the orbits of the asteroids are more eccentric (flatter) than those of the planets (with the exception of Mercury). The asteroid orbits are also more inclined to Earth's orbit than are the orbits of the other planets.

The total mass of all the known asteroids is less than 1/1000 of Earth's mass. Thus, all the asteroids have a negligible gravitational effect on the motions of the planets. The asteroids have been important in astronomy, because their orbits can be determined with great accuracy and checked against Newtonian gravitational theory. On the whole the agreement has been excellent. The asteroid Eros, whose orbit is closest to the sun, comes within 14,000,000 miles of the sun. Owing to this close approach its mean distance from the sun can be found in kilometers very accurately. From this datum and Kepler's third law applied to Eros, whose period is very well known, the astronomical unit can be calculated very accurately. Since the astronomical unit (the mean distance of Earth from the sun) is one of the most important constants in astronomy, Eros is an important object. The asteroids are probably the most primitive objects in the solar system (the debris of the original solar nebula), and they have hardly changed since the planets were formed. A close study of any one of these objects will therefore probably reveal a great deal about the origin of the planets. This will have to wait for suitable future space probes.

The largest, the most massive, and probably the most mysterious planet, Jupiter, moves in an orbit whose mean distance from the sun is 5.2 astronomical units. It revolves around the sun once every 11.862 years and its synodic period is 398.88 days. Until quite recently the only information about Jupiter came to us through Earth-based optical infrared and radio telescopes. In November 1973 the spacecraft Pioneer 10 flew past Jupiter at a distance of about 1,000,000 miles and in December 1974 Pioneer II came within 800,000 miles. Beautiful color photographs, with excellent resolution, and other important data were transmitted back to Earth. Because of its vast mass (it is $2\frac{1}{2}$ times as massive as all the other planets combined and 1/1000 the mass of the sun) it dominates the planets, having scooped up most of the mass in the solar system that did not fall into the sun. From the point of view of the distribution of mass we may neglect all the other planets and consider the solar system as a two-body gravitational structure—the sun and Jupiter. This, indeed, is the way it would appear to an observer approaching our solar system. If Jupiter had been three or four times more massive than it is, it would probably have evolved into a faint red dwarf star.

With a mass 318 times that of Earth and a radius of 11 Earth radii, Jupiter has a surface gravity that is more than $2\frac{1}{2}$ times greater than Earth's; so a person who weighs 100 pounds on Earth would weigh more than 250 pounds on Jupiter. Owing to Jupiter's large mass the escape velocity from its surface is 61 km/sec, which is more than five times larger than the escape velocity from Earth's surface. This has had a very important bearing on the chemical composition of Jupiter and its atmosphere. As already noted, the escape velocity from Earth is 11 km/sec, which means that all the light gaseous atoms (the hydrogen and helium atoms) escaped when Earth was formed; but with an escape velocity of 61 km/sec, Jupiter retained the light as well as the heavier elements so that its internal structure and atmosphere are quite different from those of Earth. Because Jupiter retained the same mixture of chemical elements that were present in the original solar nebula from which the sun was formed, the chemistry of Jupiter is very nearly the same as that of the sun.

The internal structure of Jupiter, however, is quite different from that of a star (the sun) or of a typical planet (Earth). At the center of Jupiter is a small solid hydrogen–helium core, which is surrounded by a thick liquid shell. Above the surface of the liquid shell is a thick dense atmosphere that extends for thousands of miles beyond the surface. That Jupiter consists almost entirely of the very lightest elements (hydrogen and helium) is evident from its overall density, which about equals that of the sun—1.33 g/cm³. This density is considerably less than that of Earth and can only mean that Jupiter consists almost entirely of hydrogen and helium. An interesting feature of the hydrogen–helium abundances in Jupiter's atmosphere is that the ratio of the hydrogen abundance to the helium abundance, as measured by special instruments on Pioneers 10 and 11, is very nearly the same as it is for the sun: 1 atom of helium for 20 atoms (10 molecules) of hydrogen.

A number of interesting features and properties of Jupiter that could not have been discovered with Earth-based optical telescopes alone, were discovered with radio telescopes, orbiting observatories, and the Pioneer series of Jupiter probes. In 1955, K. Franklin found that intense radio waves, ranging in wavelength from 3 cm to 70 cm, are emitted continuously from Jupiter. In addition, spasmodic bursts of very intense radio waves in the wavelength region of 15 m are emitted at random. The intensity of these long waves is great enough to make Jupiter the strongest radio source in the sky in this wavelength region. The 3-cm radio waves appear to be thermal in nature (resulting from the random thermal motions of ions) but the other waves are not. The latter can be accounted for by the interaction of streams of charged particles from the sun (the solar wind) with Jupiter's intense magnetic field. The Pioneer 10 and 11 missions measured Jupiter's magnetic field and found it to be ten times as intense as Earth's. Rapidly moving protons or other high-velocity ions from the sun, caught in such intense magnetic fields, produce the kind of radio emissions that we receive from Jupiter. The Pioneer missions also discovered that Jupiter radiates about twice as much energy per second than it receives from the sun. Thus, there must be some internal heat source in Jupiter.

If Jupiter were not rotating, its gravitational field would have shaped it into a perfect sphere, but it is much less spherical than Earth. It spins on its axis once every 9 hours 50 minutes at its equator and once every 9 hours 56 minutes at its poles. This difference in the period of rotation indicates very strong Jovian winds. Owing to its rapid rotation, Jupiter is quite distorted; its equatorial radius is 71,400 km whereas its polar radius (from pole to pole) is 67,000 km. Thus, the poles are flattened by 4400 km, which is almost two-thirds Earth's radius.

The gaseous envelope (atmosphere) surrounding the liquid surface of Jupiter is too dense to allow us to see the surface. The atmosphere has a distinct, banded structure, which consists of light zones that range from white to yellow and dark reddish-brown belts of various shades. The bands near the equator are fairly permanent, varying only in color and width. The most pronounced feature in the atmosphere is the Great Red Spot in the south tropical zone, which was first detected by Cassini almost 300 years ago and first photographed in 1879. A small red spot was photographed in the north tropical zone by Pioneer IV. These two spots are probably huge self-sustaining vortices like hurricanes on Earth.

From all the data gathered by the various techniques and probes described above, the following model of Jupiter's internal structure has been deduced. There is a small rocky, iron–silicate core no more than a few thousand miles in diameter, where the temperature is about 30,000°K and the pressure is millions of times greater than Earth's. This core is surrounded by a liquid metallic hydrogen layer or shell that extends out from the center to a distance of about 46,000 km. This fluid is an electrical conductor. Above this liquid metallic layer is a liquid layer of molecular hydrogen (nonconducting) that extends out to a distance of about 70,000 km from the center. This molecular layer in turn is surrounded by a thick atmosphere that extends 1000 km to the heavy clouds that we see. Ice crystals and water droplets constitute the lowest layer of the atmosphere above which is a layer of ammonium crystals and ammonium hydrosulfide crystals. Gaseous hydrogen, methane, helium, and hydrogen sulfides probably are mixed together to form the dense clouds and give them their coloration. The surface temperature of Jupiter right beneath its atmosphere is believed to be 120°K.

Jupiter has 14 satellites circling it with various periods; the 4 brightest and largest satellites were discovered by Galileo (the Galilean satellites). Three of them are larger than our moon with radii ranging from 1810 km to 2600 km. All of them seem to have masses that are too small to sustain an atmosphere. In 1675 the Danish astronomer O. Roemer first determined the speed of light by measuring the lag between successive eclipses (the increase in period) of the moons of Jupiter by Jupiter when Earth was moving away from Jupiter. When Earth is approaching Jupiter the interval between eclipses decreases. If the time between two successive eclipses of a Jovian moon is t, as seen by an observer fixed with respect to Jupiter, this interval becomes $t + (v/c)t$ if the observer is on Earth moving with a speed v away from Jupiter, where c is the speed of light. This is so because vt is the distance Earth moves between eclipses and vt/c is the additional time required by light from the eclipsed moon to reach Earth. Roemer noted that the total delay for all the eclipses that occurred between the time Earth was closest to Jupiter and farthest from Jupiter was 1000 sec. This represents the additional time it takes light from Jupiter to reach Earth when our distance from Jupiter has increased by 186,000,000 miles (from one side of the sun to the other). Dividing this distance by the 1000 sec, Roemer obtained 186,000 miles/sec for the speed of light, which is close to the accurately measured value of 29.979 billion cm/sec.

Saturn is the last of the naked eye planets, and except for its amazing family of rings, is quite similar in appearance to Jupiter. Its surface is marked by a system of rings and belts, which vary in color from creamy white to yellow; it is highly flattened (even more so than Jupiter) owing to its very rapid rotation—it spins around once every 10 hours 14 minutes; its density is 0.70 g/cm^3 (it would float if there were an ocean of water large enough to hold it); its mass is 95 times greater than Earth's mass; its diameter at the equator is 9.42 times greater than Earth's diameter; and it has at least ten satellites revolving around it. Saturn revolves around the sun once every 29.5 years (its sidereal period) and its synodic period is 378.09 days.

Saturn's internal structure is probably quite similar to that of Jupiter with one difference. The rocky core (consisting of iron and silicates) extends out 10,000 km from the center. This core is surrounded by a 5000-km layer of ice, which in turn is encased in an 8000-km layer of metallic hydrogen. Above this, and extending out to the surface is a thick layer of liquid molecular hydrogen. Saturn's atmosphere is quite similar to Jupiter's. Considering Jupiter and Saturn together, we may say that about 15% of their mass is contained in a rocky metallic core and hydrogen and helium make up the rest of their mass. Like Jupiter, Saturn has an internal source of heat that makes it radiate about twice as much heat as it receives from the sun every second. This heat, may be generated by a slow steady gravitational contraction of the two planets or by intense radioactivity.

The most remarkable feature of Saturn is its family of rings. The rings lie exactly in the plane of Saturn's equator and extend out to a distance of 85,000 miles from the planet's center. A large gap between the rings known as Cassini's division was found by Cassini in 1675. In 1850 the triple character of the rings was discovered when a hazy, crepe-like ring was found to lie close to the planet. Various aspects of the ring system are seen from Earth as Saturn revolves around the sun. The rings are not solid or liquid structures but consist of swarms of individual particles, revolving around the planet in accordance with Kepler's laws.

Uranus and Neptune, which were discovered during the last two centuries, have structures similar to Saturn's but they are much rockier. Each of these planets has an iron–silicate core that extends 8000 km out from the center and a layer of ice that extends out another 8000 km. There is no layer of metallic hydrogen but there is a layer of liquid molecular hydrogen that extends to the surface. The atmospheres of these two planets are similar to Saturn's. Uranus has five small satellites and Neptune has two satellites, one of which is quite large (radius 2000 km). Uranus circles the sun once every 84 years and Neptune does so in 165 years. Their synodic periods are 369.66 and 367 days, respectively.

In 1977 a system of five rings, like those around Saturn, were discovered circling Uranus. These rings were observed through a 36-inch telescope mounted in NASA's C-141 aircraft that was operating as a flying laboratory, above 75% of Earth's atmosphere, and also through Earth-bound telescopes in Perth, Australia, and Cape Town, South Africa. The rings were discovered when observers found that a star experienced five distinct eclipses during the 40-minute period before it passed behind Uranus (occulted) and five similar eclipses in the 40-minute period after it had emerged from behind Uranus. Dr. James Elliot used the 36-inch telescope to find the rings and Dr. Robert Millis observed them with the telescope at Perth. These rings, which consist of numerous boulders, are believed to lie in a band 11,000 to 16,000 miles above Uranus's atmosphere; they range in width from 6 miles (the inner ones) to 60 miles (the outer ones). Since the rings lie within the Roche limit of Uranus, they were probably formed when a former satellite of Uranus approached too close to the planet and was fragmented by the planet's tidal action.

Uranus was discovered quite accidentally in 1781 by the great English astronomer Sir William Herschel during a telescopic survey of the sky. The existence of Neptune was predicted and its position accurately determined from the gravitational effect it had on the motion of Uranus, which did not quite follow the orbit it would have if no planet beyond Uranus were present. Both John Adams in England in 1841 and Urbain Leverrier in France in 1845 carried out the necessary calculations and they obtained almost identical positions for the planet, which was found by the German astronomer Galle on September 23, 1946, less than 1° from its predicted position.

Pluto, the outermost planet in the solar system, at a mean distance of 39.44 astronomical units, circles the sun once every 248 years. Its mass is about 17% of Earth's, its radius is about half

Earth's, and it rotates once every 6 days 9 hours. Very little is known about its chemistry and internal structure, but they may be similar to those of the large satellites of Jupiter and Saturn. Titan, the large satellite of Saturn, may be similar to Pluto, for there is strong evidence that Pluto is an escaped satellite of Neptune.

Pluto was discovered in 1931 by C. Tombaugh at the Lowell Observatory. Its orbit had been calculated previously by Gaillot and Lowell who used the observed perturbations of the orbits of both Uranus and Neptune to obtain an orbit for Pluto.

In 1977 Voyagers I and II were launched toward Jupiter. Voyager I had its closest encounter with Jupiter, at a distance of 349,000 km, in March 1979; Voyager II had its closest encounter in July 1979. In the 98-day period from January 6, 1979, when Voyager I began its continuous measurements of the Jovian system, to April 13, 1979, when it completed its observations, it sent back thousands of photographs and vast quantities of data.

The Voyager probe revealed that the atmospheric dynamics of Jupiter are much more violent than had been indicated by earlier probes, with many spots, smaller than the famous Great Red Spot, interacting with each other and the Great Red Spot itself. The latter, with a diameter three times that of Earth, is cooler than its surrounding regions, and is topped by a substantially colder atmosphere than its neighboring regions. Its motion is anticyclonic and it rotates once every 6 days. Though the great spot is predominantly red, it contains blue and white counterrotating vortices. The upper cloud layers are marked by large lightning bolts similar to terrestrial superbolts. These bolts may be the source of the intense radio waves from Jupiter. The intense emission of ultraviolet radiation indicates atmospheric hot spots with temperatures on the order of 1000°K.

One of the most surprising features of the Jovian system of satellites is their intense volcanic activity. Seven active volcanoes were detected by the probes, with plumes extending 250 km above the surfaces of the satellites. The surface of Io, the second Galilean satellite, is marked by numerous spots of mottled yellows, reds, and dark browns. It is the most active (volcanic) solid body in the solar system; this activity may be caused by the imbalance produced in Io's interior by the very strong tidal action of Jupiter on it. One very noticeable region of volcanic activity on Io's surface is a hot spot that is 150°K warmer than the surrounding surface. The four Galilean moons of Jupiter are about the size of small planets. Io and

Europa are about as large and as massive as our moon, while Ganymede and Callisto are similar in size to Mercury but less massive.

One of the photographs obtained by Voyager I of the outer regions of the Jovian space against the distant stars reveals the faint trail of a ring of particles around Jupiter that cannot be viewed or photographed with Earth-based telescopes. The outer edge of this ring is about 128,000 km from the center of Jupiter and its thickness is less than 30 km.

An intriguing set of physical and chemical conditions in the space surrounding Jupiter were revealed by Voyager I. An electrical current of about 5 million amperes flows along the tube of magnetic flux connecting Jupiter and Io. Io itself is surrounded by a hot (100,000°K) plasma whose oxygen and sulfur ions are emitting ultraviolet radiation. The electron densities in this plasma exceed 4500 per cm^3.

The region out to a distance of six Jupiter radii (400,000 km) from Jupiter's surface contains a cold co-rotating plasma consisting of oxygen, sulfur, and sulfur dioxide ions. High-energy protons and sodium ions are also present. The vibrations and turbulences in this plasma generate long-wavelength (kilometric) radio waves, which have been detected with Earth-based radio telescopes.

Voyager I began recording the phenomena of Saturn in August 1980, when it was still 109,000,000 km from Saturn; its closest approach at a distance of 126,000 km occurred in November 1980, but before that encounter Voyager I had its closest approach to Saturn's satellite Titan, the largest in Saturn's system.

Voyager I found that Saturn's cloud layers are deeper in the atmosphere than those of Jupiter because Saturn is colder than Jupiter. The alternative dark belts and light zones extend to higher latitudes on Saturn than on Jupiter. At Saturn's equator the eastward cloud speeds, relative to the deep cloud layers, are 1100 miles per hour —four times the highest speeds of Jupiter's clouds. Saturn's atmosphere consists mostly of hydrogen with helium contributing only 11% to the total atmospheric mass; helium contributes 19% by mass to Jupiter's atmosphere. Methane, ammonia, ethane, acetylene, and phosphine are present in measurable quantities in Saturn's atmosphere, whose temperature decreases from 150°K in the very tenuous upper regions to 85°K midway down and then increases to 160°K near the surface.

Saturn's ring structure is far more complex than indicated by

Earth-based telescopes, which show only the three distinct A, B, and C rings with the famous Cassini division between the A and B rings. Voyager's data showed that the A and B rings consist of numerous dense ringlets situated at regular spacings. Three additional rings, the D, E, and F rings, were discovered beyond the A ring. Some of them are very narrow, having a multicomponent braided appearance.

Voyager I discovered three new Saturnian satellites, bringing to 15 the number of known satellites. These satellites are all icy structures and heavily cratered. Titan, the most massive and the largest of the satellites, and the only one known to have a substantial atmosphere, was thoroughly studied. The atmospheric pressure there is 60 percent greater than Earth's and its surface temperature ranges from about 80°K to 200°K. Its diameter is about 5760 km. The bulk of its atmosphere is nitrogen.

Voyager I studied Saturn's magnetosphere, which rotates with the planet at a speed of 150 km/sec at a distance of 17 Saturn radii. This field drags a plasma shell of ionized hydrogen and oxygen with it. The interaction of this plasma with Saturn's magnetic field generates bursts of radio waves.

Voyager II began its in-depth survey of Uranus in November 1985, when it was 10,300,000 km from the planet. Its closest approach to Uranus occurred in January 1986, when it was 107,000 km from the planet's center, and its observations of that planet ended in February 1986. Uranus's atmosphere resembles those of Saturn and Jupiter. The atmospheric temperature is 750°K in the extreme upper regions where the pressure is very low; it drops to 52°K at lower altitudes. The hydrogen and helium abundances in Uranus's atmosphere are about the same as in the sun and in the atmospheres of Saturn and Jupiter. Bands of methane clouds are present where the atmospheric pressure is about one Earth atmospheric pressure or higher.

A series of discrete concentric rings surrounds the equatorial region of Uranus, extending outward from a distance of 37,000 km from the planet's center to 51,000 km. The widths of the rings range from 2500 km to less than 1 km. Voyager II found ten new satellites orbiting Uranus below the orbit of Miranda, the closest of the five previously known satellites. The Voyager probe also found a dipole-type magnetosphere surrounding Uranus.

CHAPTER 7

Comets, Meteors, and Interplanetary Dust

When beggars die, there are no comets seen;
The heavens themselves blaze forth the death
of princes.

—WILLIAM SHAKESPEARE, *Julius Caesar*

The comets are the largest and most dramatic objects in the solar system. For a long time they were also the most mysterious, materializing suddenly as though from nothing, and engendering great concern and alarm. The ancients, who considered comets as objects of ill omen, had no way of determining their distances, and thus did not know whether they were atmospheric phenomena or objects beyond the planets. We know today that as a group they are the most distant objects in the solar system and spend most of their lives circling the sun in vast, almost circular orbits that lie halfway between the sun and the nearest star.

Occasionally, one of these distant objects, perturbed by the nearby stars, or for some other physical reason (a collision of some sort perhaps), loses some of its energy and drops in toward the sun in a very elongated orbit that may, after thousands or even hundreds of thousands of years, bring it to within a few million miles of the sun. It sweeps around the sun and, if one of the large planets (Jupiter or Saturn) does not disturb it gravitationally, returns to the point from which it descended. It will then not reappear again for thousands or hundreds of thousands of years. If the comet comes close to Jupiter or Saturn, its orbit may be so distorted and it may lose so much energy that it can never get back to its initial point, and it remains in our realm of the solar system, revolving around the sun in times that range from a few years to decades (3.3 years for

123

Encke's, and 76.1 years for Halley's Comet). These are the short-period comets that we generally think of when comets are mentioned. But all of these comets were originally at vast distances from us, revolving around the sun once every few million years.

A comet changes its appearance drastically as it approaches the sun owing to the action of the intense solar rays and solar wind as the distance between the sun and the comet decreases. Before the comet gets close enough to the sun to be altered to any extent, it consists of a central core or nucleus of icy material (mostly frozen water) whose diameter is about 10 km. This is surrounded by a head or coma whose size varies with distance from the sun; it attains a maximum size when the comet is about as far from the sun as Earth is (1 astronomical unit). The diameter of the coma varies from about 20,000 km when the comet is at Mercury's distance from the sun to 200,000 km when it is at Earth's distance. When first visible a comet rarely has a tail, but as it approaches the sun and comes to within 1.7 astronomical units, it develops a tail that in special cases may extend out to a distance of 150,000,000 km. The tail becomes visible when it is 10,000,000 km long.

When the comet is within Mars's distance from the sun, the temperature of the nucleus becomes high enough for the volatile material there to vaporize and form a gaseous envelope (the head or coma) around the core. This process is rapidly accelerated as the comet approaches the sun and the gases in the head are driven radially away from the sun by the sun's radiation and the stream of solar particles (solar wind). A tail is thus formed, which points away from the sun at all times (whether the comet is approaching the sun or receding from it) because of the outward propulsion by the solar wind and the solar radiation.

Owing to the formation of the tail, which consists of material that is driven away from the comet, the comet suffers continuous attrition as it moves in its orbit. As much as 1/200 of the mass of a comet evaporates during each perihelion passage of the comet, and this material remains in the comet's orbit circling the sun as it did when it was still part of the comet. In time the comet disappears completely and is replaced by a swarm of tiny particles, spread out over the entire orbit. These particles produce meteoric showers when Earth passes through or near them. If a comet comes close enough to the sun at perihelion, the tidal action of the sun may break it into two or more parts as happened to Biela's Comet in

1846 and to Taylor's Comet in 1916. In addition to water ice, a comet consists of the ices of methane and ammonia with particles of iron and nickel embedded. The silicates are also present as are sodium, calcium, and oxygen. About four new comets are discovered yearly, three of which return to their distant base and one of which remains close to Jupiter or Saturn. Of these four, only two are large enough to be visible. There may be as many as six million comets traveling in orbits that bring them within 1 astronomical unit of the sun. The mean perihelion distance of these comets is about 1.3 astronomical units. The masses of comets are very small, astronomically speaking; no comet has ever been observed with a mass greater than 1/1,000,000 of Earth's mass. If the mass of a comet were greater than that, it would have some gravitational effects on the orbits of the planets it passes, which is not the case. The collision of a comet with Earth would be very destructive. Such collisions should occur once every few million years.

In the past, comets were thought to be interlopers from regions beyond the solar system. Today we know this is not so, because as already noted, all the comets now in our part of the solar system came from a halo of comets (the Oort Cloud) surrounding the solar system. It has been estimated that there are about 300 billion potential comets in the Oort halo. We believe today that these potential comets are the primordial debris of the solar nebula that was left behind when this nebula contracted down to the sun some 5 billion years ago. Because of this it is important to study comets in as great detail as possible to obtain a picture of the initial state of the material from which our planetary system was formed.

During the 1986 perihelion passage of Halley's Comet, astronomers from observatories all over the world studied it intensively with Earth-based telescopes, with satellites, and with three specially designed space probes that passed right through the comet's nucleus or very close to it. Two of these probes were the Soviet spacecrafts Vega I and II and the third was the European Space Agency's Giotto probe. The two Vega probes encountered Halley's Comet on March 6 and 9, respectively, when the comet was at its closest approach to Earth—a distance of about 150,000 km. They observed Halley's nucleus from a distance of about 8500 km; their televised pictures show that the actual solid surface of the nucleus merged almost continuously into the bright overlying clouds of dust and gas. The nucleus itself was a source of many very active asymmetric jets of dust. The Vega data revealed that the nucleus of Halley's Comet is shaped like a peanut and rotates with a period of about 53 hours. The infrared detectors on the Vega probes determined that the nucleus was emitting high-intensity thermal infrared radiation at the time (at a distance of 0.8 astronomical unit from the sun) and that its temperature was higher than 300°K.

The Giotto probe encountered Halley's Comet a few days later and observed it in detail from a distance of 600 km; the televised images from Giotto showed jets of all kinds of particles from the nucleus. The particles in these jets that struck Giotto (at the rate of hundreds of thousands per second) ranged in size from microscopic to millimetric, but the typical particle had a diameter of 1 micrometer. The Giotto images showed an elongated nucleus with a major axis of 16 km and a minor axis of 8 km. The nuclear surface was irregular and cratered, with one hill clearly evident. One of the largest craters was shallow but with a diameter of 1.5 km. Two bright jets of dust, emerging from the sunlit side of the nucleus, were detected; other fainter jets were also recorded.

Both the Vega probes and the Giotto probe confirmed the deductions from Earth-based and satellite spectroscopic studies that comets consist primarily of water. Other molecules detected in the spectra of various parts of Halley's comet were carbon monoxide, carbon dioxide, methane, carbonic acid, and hydrogen cyanide. Finally, the Vega and Giotto data confirmed the hypothesis that cometary material is the same as the primordial material of which the solar system was formed.

"Falling stars" or "shooting stars," the sudden streaks of light across the sky that can be seen any night, are not stars at all but grains of interplanetary dust that strike our atmosphere at speeds of up to 20 miles/sec. At such speeds the friction of the air, even a few hundred miles above Earth's surface, is great enough to raise the temperature of the grains of dust so high that they are quickly vaporized and emit a stream of visible radiation. A typical meteor has a mass of about 1/1000 of a gram and a diameter of about 1 mm. It becomes luminous at a height of about 100 km above Earth's surface and travels about 20 km within the atmosphere before disappearing. At this height the atmosphere is so thin that any particular molecule of air travels, on the average, about 10 cm before it collides with another molecule of air. Thus, the air molecules strike the intruding dust grain or meteor like a stream of individual bullets. A single air molecule striking the surface of a meteor generates about 100 times as much energy as is required to tear a single molecule off the meteor. Hence, when the mass of all the molecules striking a meteor equals 1/100 the mass of the meteor, the meteor will have absorbed enough energy to be vaporized. Put differently, a typical meteor will be vaporized when it collides with 1/100,000 gram of atmosphere.

An observer at sea level on Earth can survey about 3000 km^2 of the sky, and over that area he will detect about 10 meteors per hour. Taking the entire Earth into account, we see that the number of meteors striking Earth's upper atmosphere every hour that become visible is about 2 million. These high-speed meteors contribute 10 tons of new material to the earth every day. To this we must add 400 tons of low-speed micrometeorite material. Thus, 410 tons of cosmic dust fall on Earth's surface each day. These meteors are half iron and half stone.

We have already noted that a comet disintegrates as it revolves around Earth leaving a trail of dust particles spread out over its entire orbit. About a dozen such meteor streams are known and when Earth passes through such a stream, a meteoric shower is visible. These streams are so arranged around the sun that a meteor shower occurs during 8 of the 12 calendar months. The name given to a meteor shower is the same as that of the constellation in which the point from which the shower seems to diverge lies. The number of meteors observed per hour in a shower ranges from 10 to 80, but in some cases thousands of meteors have been observed per hour.

When a large mass of interplanetary material enters Earth's atmosphere, it penetrates into the lower layers of the atmosphere and then explodes with a great burst of light and heat, scattering its fragments over a large area. Such a phenomenon is called a fireball. The most famous of them is the giant fireball that exploded on the steppes of Tunguska, Siberia on June 30, 1908. It destroyed all vegetation over a radius of 50 miles. A gigantic meteorite in prehistoric times created the large crater in Arizona that is 1220 m in diameter and 183 m deep. If the meteorite does not explode, it digs into Earth's surface and quickly cools off. A chemical analysis of the surface of such a meteorite shows the presence of many amino acids (19 of the 20 known amino acids have been found on such meteorites), which were formed in either interplanetary or interstellar space. There are four kinds of meteorites: metal, stone, sulfides, and tektites (glassy ones).

One important and visible manifestation of the interplanetary medium is the zodiacal light. This may be observed on a dark night as a faint band of light along the ecliptic whose intensity increases in the direction of the sun. Since the spectrum of the light in this luminous band is the same as that of sunlight, it follows that it consists of sunlight that has been scattered or diffracted by extremely tiny particles that trail Earth as it revolves around the sun. Another manifestation of interplanetary dust is the "Gegenschein" or "counterglow," which is a faint glow on the ecliptic opposite to the sun. It arises from sunlight that is scattered from the dust (solid particles) opposite to the sun in the sky the way the lights of a car are reflected by a fog.

Compared to interstellar space in our galaxy, the interplanetary space is fairly dense. It is estimated that the interplanetary medium in the vicinity of Earth contains, on the average, about ten particles (five protons and five electrons) per cubic centimeter. This is five times greater than the number of particles per cubic centimeter in interstellar space. In our atmosphere at sea level the number of particles or molecules per cubic centimeter is 27 trillion trillion. The protons and electrons in our interplanetary space come from the sun as a stream of charged particles (the solar wind). When these charged particles are captured by Earth's magnetic field, they produce in our upper atmosphere such phenomena as the aurora borealis (northern lights), the belt of charged particles (the Van Allen belt) girdling Earth, and disturbances of the ionosphere and of radio communications.

In addition to the charged particles in the solar wind in our interplanetary medium there is another class of charged particles, the cosmic rays, which travel at much higher speeds than do the solar wind particles. The cosmic rays that strike Earth come from the galaxy as a whole and from the sun. The galactic cosmic rays are much more energetic than are those from the sun that are emitted from solar flares.

CHAPTER 8

The Sun

The energies of our system will decay, the glory of the sun will be dimmed, and the earth, tideless and inert, will no longer tolerate the race which has for a moment disturbed its solitude. Man will go down into the pit, and all his thoughts will perish.

—A. J. BALFOUR, *The Foundations of Belief*

As viewed from Earth the sun appears to be almost the size of the moon (it subtends $\frac{1}{2}°$ in the sky), but since the sun is about 400 times as far from Earth as the moon is, its actual size is about 400 times greater than that of the moon. Knowing the distance of Earth from the sun we can measure the diameter of what we call its photosphere (the disk of the sun as faintly seen behind a thin cloud). The diameter of this disk is 1.4 million km or slightly less than 900 thousand miles. Its surface area is more than 10,000 times larger than Earth's, and its volume is more than 1 million Earth volumes.

Knowing the mean distance of the sun from Earth (the astronomical unit) and the length of the year, we can use Kepler's third law to calculate the mass of the sun. The formula for the sun's mass M is

$$M = \frac{4\pi^2 A^3}{GP^2}$$

where A is the astronomical unit, G is Newton's universal gravitational constant (equal to 6.67×10^{-8} cm^3/g sec^2), and P is the period of Earth. If A is expressed in centimeters, P in seconds, and G in the units given above, the mass of the sun is 2 billion trillion trillion g (2×10^{33} g) or 333,000 times that of Earth. From its mass and radius, we find that the mean density of the sun is 1.41 g/cm^3 (slightly more than Jupiter's) and the acceleration of gravity at its surface is 27,400 cm/sec^2, or about 28 times that on Earth's surface. A 100-pound person would weigh about 2800 pounds if he could stand on the sun. From the sun's mass and radius, the calculated speed of escape from its surface is 618 km/sec, which is to be compared to 11 km/sec from Earth's surface and 61 km/sec from Jupiter's surface.

As seen from Earth the sun rotates once every 27 days at the equator and once every 31 days near the poles. The sun is not fixed at the center of the solar system as one might be led to believe on first learning about the heliocentric (sun at the center) theory of the solar system. The center of the sun moves in a very small, complicated orbit around the center of mass of the solar system owing to the pull of all the planets on it. The largest contribution to this gravitational pull comes from Jupiter, but since Jupiter's mass is only 1/1000 of the sun's mass, the motion of the sun and its orbit is 1/1000 of Jupiter's. In other words, the center of the sun moves around the center of mass of the solar system (the center of the solar system) in an orbit whose mean distance from the center of mass is about 500,000 miles. The sun's equator is inclined 7°10.5′ with respect to the plane of Earth's orbit around the sun.

The sun, a star, is the only self-luminous body in the solar system. A gaseous sphere whose photospheric temperature (surface temperature) is about 6000°K, it emits energy into space in all directions equally at a rate of 4 billion trillion trillion (4 × 10^{33}) ergs/sec, of which Earth receives less than a two-billionth. The amount of solar energy emitted per second is enough to melt a column of ice about 4 miles wide and about 93 million miles long. Each square centimeter of the sun's photosphere emits energy at a rate of 62.5 billion (62.5 × 10^9) ergs/sec, which means that the power equivalent of each square yard of the sun's photosphere is 70,000 horsepower. Each square centimeter of the top of Earth's atmosphere receives 2 calories of solar radiation every minute. From this number, which is called the solar constant, and from the distance of the sun, we can calculate the total energy radiated per second from the sun (the solar luminosity), which we have already given. The solar constant does not equal the amount of solar energy that a square centimeter on the surface of Earth receives every minute; some of the solar radiation striking the top of the atmosphere is reflected and some is absorbed by the atmosphere so that less strikes Earth's surface every second than strikes the upper atmosphere.

When we direct sunlight through a prism (as Newton first did), it is spread out into an array of colors (the solar spectrum) that, at first sight, ranges continuously from deep red to violet (the well-known rainbow). But a closer examination of this array of colors shows that it is not continuous but broken up or crossed by many faint, sharp dark lines. We call these lines (absorption lines) Fraunhofer lines in honor of the German physicist J. Fraunhofer who first observed them and studied 700 of them in the 19th century. These Fraunhofer lines in the sun's spectrum give us very important information about the sun's structure and chemistry. If the sun ended at its photosphere (if there were nothing above the sun's photosphere) there would be no dark lines crossing its spectrum, and the latter would indeed be a perfectly continuous array of colors. A hot opaque surface like the sun's photosphere (or like a hot electric light filament) can emit only a continuous array of colors (continuous spectrum) whose intensities depend on the temperature of the surface. Hot surfaces favor more intense blue colors and cool surfaces favor more intense red colors. With regard to the sun, the most intense colors are in the yellow part of the spectrum, which is what one expects from a surface whose temperature is on the order of 6000°K.

The sharp dark lines in the sun's spectrum tell us that a cool gas of some sort (the solar atmosphere) surrounds the hot photosphere. As the thermal radiation, with its continuous array of colors, passes through the cool (by comparison with the photosphere) solar atmosphere, the various atoms in the atmosphere interact with this radiation in such a way as to superimpose the observed absorption lines on the continuous spectrum. Each species of atom (e.g., hydrogen, helium, iron) responds to definite colors only, so that, depending on the kinds of atoms that are present in the atmosphere, only a discrete set of absorption lines, superimposed at definite regions of the spectrum (at definite colors), is present. The absorption lines in the sun's spectrum arise mostly from the atoms of hydrogen, sodium, magnesium, silicon, calcium, and iron. By studying the intensities of the absorption lines and the kind of absorption lines that are present in the sun's spectrum, we deduce that the temperature of the sun's atmosphere varies from 4200°K near the photosphere to higher values as we move out. The lowest layer of the atmosphere is called the chromosphere; most of the absorption lines arise in the chromosphere.

The chromosphere extends outward about 2000 km, where the temperature is about 9000°K. At this height the chromosphere gives way, rather sharply, to the solar corona, which can only be seen during a total eclipse of the sun. Although the transition from the chromosphere to the corona is fairly sharp, chromospheric material often projects into the corona for tens of thousands of kilometers in the form of great prominences and spicules. The average prominence, which consists mostly of hot gaseous hydrogen, ejected violently from the photosphere, and which forms a kind of bridge between two points on the photosphere, has a height of about 30,000 km, a length of 200,000 km, a thickness of 5000 km, and a volume of 10,000 trillion trillion cm^3. The unusually large eruptive prominence of August 21, 1973 extended more than 400,000 km above the sun's surface, and the June 4, 1946 prominence extended more than 1 million miles above the sun's photosphere. As much as 10^{17} g (200 trillion pounds) of material, which leaves the surface at speeds of up to 450 km/sec, may be contained in a large prominence. The whole phenomenon may last a month or more.

The solar corona is a very tenuous, luminous halo that surrounds the sun and extends out to about 1 million miles. It varies in shape (depending on sunspot activity) from a spherical shell (at maximum sunspot activity) to an elongated elliptical structure (at sunspot minimum). The elongation is always in the equatorial plane of the sun. At maximum sunspot activity the total luminosity of the corona is 1.3-millionths of the sun's luminosity (about 0.6 of a full moon) and at sunspot minimum its luminosity is 0.8-millionth of the sun's luminosity (about 0.35 of a full moon). The spectral lines of the corona show the presence of the highly ionized atoms (atoms with electrons missing) of iron, magnesium, silicon, oxygen, calcium, and nickel. There are many free electrons in the corona moving at average speeds on the order of a few hundred million centimeters per second. Such average speeds correspond to temperatures on the order of 1 million degrees, so we say that the coronal temperature is about 1,000,000°K.

Taking into account all the data about the chromosphere and the corona, we find that the total mass of the sun's atmosphere is about 100 thousand trillion tons. This is only one twenty-billionth of the total solar mass and about one ten-thousandth of Earth's mass. The sun's atmosphere is a much narrower layer of gas, relative to the total sun, than our atmosphere is relative to our planet. If we scaled the sun down to Earth's size (i.e., reduced its diameter by a factor of 100), the chromosphere would be just about 12 miles thick. Since the depth of the photosphere is how far down into the sun's atmosphere we can look, we see that the sun's atmosphere is very opaque—much more so than Earth's atmosphere.

The sun's surface is not at all quiescent but extremely active, as indicated by the occurrence of prominences, solar flares, sunspots, and other disturbances. A flare is a very intense eruption of hot gases in the neighborhood of a sunspot that is often so bright that it can be seen against the brilliant solar background. A single flare may last as long as 30 minutes, and, from an area one ten thousandth that of the total surface area of the sun, emit, in one second, one-thousandth of the total energy the sun emits per second. Flares are generally associated with increases in solar X rays, cosmic rays striking Earth, spectacular auroras, and vast storms in Earth's magnetic field. The temperature within a flare is on the order of 15,000°K. The total energy released by a flare is sufficient to supply all of our energy needs for 100,000 years.

Many of the violent features on the photosphere of the sun and in its chromosphere and corona can be understood and explained in terms of magnetic fields on the sun's surface and the vast vertical convective currents under the surface that dredge solar gases up to the surface (the photosphere) from depths of about 100,000 km. These convective currents are like the circulation of water in a hot water heating system or in a large caldron of boiling water. These subphotosphere currents account for the flares, the sunspots, and the high temperature (1 million degrees) in the solar corona. In the high-temperature regions of the sun, where the convective currents begin, the hydrogen and helium atoms, which account for about 98% of all the atoms in the sun (75% hydrogen and 23% helium), are ionized (atoms with their electrons torn away). Owing to this ionization of the gases in the convective regions (the ionized gas is called a plasma), the free charged particles (electrons and ions) trap the internal magnetic fields of the sun, distort them in various ways, and transport them to the surface. As these separate magnetic fields rise to the surface, they may reinforce themselves at certain points of the photosphere, gain enormously in strength, and trap the free electrons and ions, forcing these charged particles to circulate around the magnetic fields the way air does in a tornado. The currents created by these circulating charged particles generate magnetic fields of their own, which increase the intensity of the fields already present that were brought up by the convective currents. As the strengths of these local magnetic fields increase, the fields and the swirling ions block the hot gases from below so that they cannot emerge into these magnetic vortices. A sunspot is thus formed, which is cooler and therefore darker than the surface regions surrounding the vortex. Since no hot underlying gases can emerge through the vortex, the sunspot looks just like a hole in the surface of the sun. In a sense it is a hole because it is a shallow depression whose surface lies a few hundred kilometers below the sun's photosphere. The average temperature of a sunspot is about 3900°K whereas the temperature of the surrounding region is about 6000°K.

The intensity of the magnetic field in a sunspot is about 6000 times larger than the intensity of Earth's magnetic field. Since sunspots have diameters that range from about 10,000 miles to 80,000 miles, the magnetic energy associated with a sunspot is enormous. Sunspots come in pairs, with one member of the pair corresponding to the north pole of a magnet and the other member to a south pole. Such pairs of spots occur in groups of various sizes in both hemispheres of the sun, and as one group disappears after about 6 days (average life of a group) another group arises. Some large spots in a group may exist for a month and a half. Sunspot groups occur in cycles of 11 years in the sense that the minimum and maximum sunspot activities repeat themselves every 11 years.

Just as sunspots are caused by the vast turbulence (convective currents) under the photosphere, so, too, are the solar wind, the granules on the photosphere, the prominences, and all the other complex phenomena observed on the sun's surface.

Below the convective regions, starting about 100,000 km below the photosphere, is a region dominated by intense radiation that is generated by thermonuclear reactions in the core of the sun. This stream of radiation, together with the gravitational field in the sun and the chemical composition of the matter at each point in the sun, determines the size and surface temperature of the sun and the physical conditions (temperature, pressure, concentration of mass, and the rate at which energy flows outwardly) at each point inside the sun. The sun must so adjust itself that there is a steady, orderly outflow of energy from the sun's surface. The rate at which energy leaves the sun's surface (the luminosity of the sun) must equal exactly the rate at which energy is generated throughout the sun. This means that the size of the sun (its radius), its surface temperature, and the temperature at each point inside the sun must be just right to achieve this orderly outflow of energy. The stream of radiation from the sun's interior also determines the point below the sun's surface where convection sets in and takes over the role (from radiation) of transporting the solar energy to the surface.

As we have noted, the convective zone of the sun extends downward about 100,000 km below the photosphere. At that point the temperature is 650,000°K, the density is about 0.006 g/cm³, and the pressure is about 100,000 times greater than Earth's atmospheric pressure. Here all the atoms are ionized so that the magnetic fields are intertwined as the charged particles rush about.

As we go down toward the center of the sun, the pressure, density, and temperature rise quite rapidly until we reach the thermonuclear core, which begins when we are 70% of the way down from the photosphere (the core extends 30% of the way out from the center). At this point the temperature is about 8,000,000°K and the thermonuclear generation of energy begins. The density here is about 40 g/cm³ and the pressure is 10 billion Earth atmospheres. The temperature and density are just about right for protons to produce energy slowly by coalescing and forming helium. In this thermonuclear process, four protons (the nuclei of four hydrogen atoms) coalesce to form a single helium nucleus with the transformation of mass into energy according to Einstein's equation $E = mc^2$ (see Chapter 4). The mass of a single proton is 1.008 atomic mass units so that the mass of four protons is just 4.032 atomic mass units. The mass of a single helium nucleus, however, is very nearly 4 atomic units. Thus, when four protons coalesce to form a single helium nucleus, 0.032 unit of mass is changed into energy. This is called the binding energy of helium. We obtain the amount of energy released in this thermonuclear process by multiplying 0.032 by c^2. When 4.032 g of hydrogen fuse to form 4 g of helium, 0.032 g of mass is transformed into energy. Using Einstein's equation we find that this equals 720 billion calories—enough heat to melt 9 billion g (20 million pounds) of ice. We describe just how this thermonuclear transformation of hydrogen to helium occurs in a later chapter.

The sun is radiating energy from its surface at the observed rate because inside the sun's core 564.5 million tons of hydrogen is transformed into 560 million tons of helium every second. As we move from the edge of the sun's core downward to the center of the sun, the temperature, pressure, and density increase rapidly, and the rate at which hydrogen is transformed into helium rises rapidly also. At the very center of the sun the temperature is 15.5 million degrees, the density is 160 g/cm^3, and the pressure is about a trillion times greater than Earth's atmospheric pressure. The chemistry of the sun's core is considerably different from that of the rest of the sun because the thermonuclear fusion of hydrogen to helium there has been going on for billions of years. When the sun began its life, it was chemically homogeneous with the same combination and abundances of chemical elements throughout as are found in its atmosphere now—about 75% hydrogen, about 23% helium, and about 2% of the heavier elements. Because of the billions of years of thermonuclear fusion of hydrogen into helium in the core, 50% of the core's original hydrogen has been transformed into helium. The core now consists of 37% hydrogen, 61% helium, and about 2% heavy elements.

As the thermonuclear fusion of hydrogen in the core continues, and more and more helium is produced, the sun's entire structure will undergo important changes. It will become redder, cooler, and larger, and when almost all of the hydrogen has been changed to helium, the sun will become a red giant star, swallowing Mercury, Venus, and Earth in the process. We discuss the evolution of stars later and see why stars like the sun must become red giants.

When the energy generated by the thermonuclear fusion of hydrogen is released inside the sun's core, it is quite different from the benign radiation we receive from the sun's surface. Inside the core the energy is produced in the form of intense gamma rays, which would immediately destroy all life on Earth if they reached us 8 minutes (the time it would take them to travel from the sun to Earth) after they were produced. Fortunately for us the radiation produced in the core at any moment does not reach the sun's surface until some 20 million years after it is produced. During those years, the radiation leaving the core is absorbed and reemitted very many times by the electrons in the sun. This endless process of absorption and reemission changes the character of the core radiation drastically until it is the kind of radiation we now receive from the sun. We have spoken only of the visible radiation coming from the sun, but in addition to that we receive radio waves, infrared rays, ultraviolet rays, neutrinos, and X rays. Taken together, these constitute a small fraction of the visible radiation from the sun.

CHAPTER 9

The Stars

Their Distances, Luminosities, and Colors

He that strives to touch the stars
Oft stumbles at a straw.

—EDMUND SPENSER, *The Shepheardes Calender*

We have discussed the sun in some detail not only because it is the source of the gravitational force that governs the motions of the planets, but also because it is the only star whose surface we can see and whose behavior we can study in great detail from moment to moment. If we did not have the sun to show us what a star actually looks like, we could never be sure that the theoretical models we construct of stars give us true pictures of the stars. In the previous chapter we described the sun's interior as though we actually had instruments making measurements there—an obvious absurdity. Our knowledge about the conditions inside the sun comes from our application of the known physical laws to the structure of the sun and from our insistence that these laws lead to a solar model that reproduces the observed solar characteristics. These characteristics are such observable quantities as the luminosity of the sun, its size, its mass, and its chemical composition. We now propose to do the same thing for stars in general, which, like the sun, are hot, self-luminous spheres of gas.

Just as we are guided in constructing a model of the sun's interior by its luminosity, by the conditions on its surface, and by its mass and size, so are we guided in constructing models of stars in general by their overall, observable characteristics. But before we can measure or determine such things as a star's luminosity, its surface temperature, its size or mass, we must know its distance. Newton was aware that the distances from the sun to even the closest stars are many times greater than the distance from the sun to the farthest planet. He reasoned that their complete lack of response to the sun's gravitational attraction means that the stars are millions of times farther from the sun than is Earth. From this he concluded that the stars are at least as luminous as the sun, for otherwise they would not be visible to the naked eye of an observer here on Earth. Galileo had already looked at the Milky Way with his homemade telescope and had observed that the Milky Way consists of thousands of individual faint stars, so Newton's conclusion applies to the Milky Way stars as well as to the visible naked eye stars.

In 1838, about 200 years after the birth of Newton, the German astronomer Bessel, after a few years of careful observations of the star 61 Cygni (a binary star), determined its distance by measuring the amount by which its apparent direction from Earth changes every 6 months (its annual parallax). If you hold an object at a given distance from your eyes, you see that it appears to shift back and forth as you view it first with one eye and then with the other. The closer to your eyes the object is held, the greater is this apparent shift. As the distance of the object from your eyes is increased, this apparent shift decreases. If you know the distance between your eyes and you measure the apparent shift of the object (the angle formed by the two lines from your eyes to the object), you can calculate the distance of the object by simple geometry. This same procedure, on a much larger scale, when applied to the stars (it is called the method of trigonometric parallaxes) yields the distances of the stars. As Earth moves around the sun, it changes its position relative to the stars by 186,000,000 miles every 6 months so that the apparent direction of a given star changes every 6 months. This change is most pronounced for the nearby stars, becoming smaller and smaller as we go to the more distant stars.

Half the angle, expressed in seconds of arc, through which a star appears to shift in 6 months is called the star's parallax. It is the angle subtended at the star by the semidiameter (radius) of Earth's orbit (the astronomical unit). Stellar parallaxes are most easily measured with very long focal-length telescopes because the longer the focal length of a telescope is, the greater are the displacements of stellar images on photographs taken 6 months apart with such a telescope. As an example we note that a displacement of 0.053 mm of a stellar image on a photograph taken with the 36-

inch refractor at the Sproul Observatory means that the apparent direction of the star has changed by $1''$ of arc. Examples of the parallaxes of a few bright stars are Polaris $0''004$, Mira $0''025$, Aldebaran $0''048$, Capella $0''071$, Rigel $0''004$, Betelgeuse $0''005$, Sirius $0''377$, Procyon $0''286$, Canopus $0''017$, Arcturus $0''091$, Antares $0''008$, Vega $0''123$, and Rigel Kent (Alpha Centauri) $0''745$.

We can calculate the distance of a star by dividing its parallax into the astronomical unit and then multiplying the result by an appropriate scale factor to give us the distance in the units that we want. Thus, if we divide any star's parallax into the product 206,265 × 93,000,000, we obtain the distance of the star in miles. As an example, the distance of Alpha Centauri is about 26 trillion miles. Since the distances of the stars expressed in miles are very large numbers, it is more convenient to use a much larger yardstick than a mile to express stellar distances. The light-year is one such yardstick. It is the distance light travels in 1 year. Since light travels 186,000 miles in 1 second and there are about 31.6 million seconds in a year, a light-year is about 6 trillion miles. We obtain the distance of any star in light-years by dividing its parallax into the number 3.26. The distances in light-years of the stars listed above are (in the order given) 815, 130, 68, 45, 815, 650, 9, 12, 195, 36, 407, 26, 4.5.

An interesting consequence of the enormous distances that the astronomer has to work with is that he can never get a true picture of the universe, or even of a small fraction of it, as it is at any one moment. As he looks out into space to greater and greater distances, he is also looking into the past. All we can say about the very distant objects is that what we learn about them at this moment is the way they were very many years ago. But as to what they are like now, or where they are at this moment, or even if some of them still exist we have no way of finding out by direct observation.

It is, of course, possible to make predictions by assuming that stars evolve gradually from one state to another, and by applying the laws of nature, as we know them, to the entire universe; however, we cannot be sure that we are right, because there is definite evidence of stars that do not evolve gradually, but rather undergo sudden and drastic changes.

As we go out to the more distant stars, the method of trigonometric parallaxes described above for measuring stellar distances cannot be used because the apparent semiannual displacements of the stars become so small that the measurements are quite unreliable. The errors are as large as or larger than the values of the parallaxes for these distant stars so that the measured values have no meaning. The method of trigonometric parallaxes ceases to be reliable for stars whose distances exceed 500 light-years. For such stars other methods are available, which start with knowledge of the star's intrinsic luminosity and then comparing this luminosity with the star's apparent brightness.

The apparent brightness of a star is defined as the radiation from that star intercepted in 1 second on Earth by 1 cm^2 of a surface held at right angles to the line to the star. Instead of expressing the apparent brightness of the stars in the usual units of brightness, astronomers use a quantity that is not the brightness itself, but is related to it in a definite way. This quantity is the *apparent magnitude* of the star, not to be confused with the size of the star. The apparent magnitude is a number that is assigned to a star to represent its position or rank on a scale of brightness first introduced by Hipparchus in 134 B.C., and then extended and made more precise by Ptolemy almost 2000 years ago. On this scale small numbers are assigned to stars that appear bright and larger numbers are given to the fainter stars. These numbers may be negative as well as positive. The brighter a star appears to us, the smaller or more negative is its number on this scale. The magnitude scale was set up in such a way that a first-magnitude star is exactly 100 times brighter than a sixth-magnitude star, which is just visible to the naked eye. Two stars differing by one step on the magnitude scale differ in brightness by a factor of very nearly 2.512. The exact factor is the fifth root of 100. The apparent magnitudes of the stars listed earlier in the order given are 2.3, 2, 0.85, 0.08, 0.11, 0.8, −1.45, 0.35, −0.73, −0.06, 1.0, 0.04, −0.1. The sun's apparent magnitude is −26.84, which means that every second it sends us about 14 billion times as much light as does Sirius (it is 25.4 magnitudes brighter, in appearance, than Sirius; every 5 magnitude steps equal a factor of 100).

The apparent brightness or apparent magnitude (m) of a star does not tell us very much about the intrinsic brightness (luminosity) of a star. A star may appear bright because it is very luminous or because it is very close. For example, Sirius, which is the brightest naked eye star in the sky, is far less luminous intrinsically than Betelgeuse, which appears much fainter than Sirius. The important factor here is the star's distance or its parallax. We note that whereas Sirius is 9 light-years away, Betelgeuse is 650 light-years away— more than 70 times farther than Sirius.

If we know the apparent magnitude of a star and its parallax, we can find its intrinsic luminosity, that is, the total radiation it emits per second. We do this by first finding the star's absolute magnitude, which is the magnitude a star would have if it were 32.6 light-years away. This is a standard distance (it is called 10 parsecs, 1 parsec being 3.26 light-years) that is used to define the absolute magnitude (M) of a star. We imagine the star at this distance (10 parsecs) and then calculate how bright it would appear when placed there. We can do this if we know the star's apparent magnitude and its true distance. We can thus assign an absolute magnitude to every star whose apparent magnitude and parallax are known. The absolute magnitudes of the stars listed earlier are 4.6, 1.0, -0.7, -0.6, -7, -6, 1.41, 2.65, -417, -0.2, -4.7, 0.5, 4.3. The absolute magnitude of the sun, which we give for comparison, is 4.8. This means that if the sun were removed to a distance of 32.6 light-years, it would be about 2.5 times as bright as a star that is barely visible. Its brightness would be decreased by 31.64 magnitudes because at 10 parsecs of 32.6 light-years it would be about 2 million times farther from us than it is now.

Since we know the actual luminosity of the sun, 4×10^{33} ergs/sec, and also its absolute magnitude, 4.8, we can find the luminosity of any star by comparing its absolute magnitude with that of the sun and then either multiplying by 2.512 or dividing by this number for each magnitude step difference. If we do this with the stars listed above, we see that most of them are many times more luminous than the sun. In fact, most of the naked eye stars in the sky are more luminous than the sun; they are visible because they are so luminous. Comparing the absolute magnitudes of the stars listed earlier with the absolute magnitude of the sun, we see that their luminosities are (we take the sun's luminosity as our unit of luminosity): 5700, 210, 166, 145, 62,000, 20,900, 23, 7.3, 6310, 100, 6310, 53.5, 1.6.

Of the 13 stars listed, only one, Alpha Centauri, is as luminous as the sun; all the others are more luminous and some are incredibly luminous. For example, the blue-white star Rigel in Orion emits more than 62,000 times more energy into space every second than does the sun. We note also that Polaris, Betelgeuse, Canopus, and

Antares are thousands of times more luminous than the sun. Such stars are referred to as supergiants, and stars that are a hundred or a few hundred times more luminous than the sun are called giants. Stars less luminous than the sun are called dwarfs.

The rate at which Rigel is pouring out energy tells us something very important about this star (and other stars like it): it must be very young, probably no more than a few million years old. The reason for this is that stars like Rigel exist on the same nuclear fuel (hydrogen) as does the sun, and they are using up this fuel at a tremendous rate. Since the amount of hydrogen with which a star begins its life is determined by the mass of the star, and stars rarely have masses that exceed 50 times the sun's mass, it is clear that Rigel does not have more than 60 times as much hydrogen as does the sun. But if Rigel is using up its hydrogen 60,000 times as fast as the sun is, its age can be no more than about one-thousandth of the sun's age or about 5 million years. Young stars like Rigel, which are constantly being born in our part of the galaxy, are called population I stars. The sun, too, is a population I star.

The red supergiants Betelgeuse and Antares are not young stars even though they are thousands of times more luminous than the sun and are using up nuclear fuel at a rapid rate. They are, in fact, very old stars that began their lives as blue-white stars like Rigel. These red supergiants have used up all their hydrogen fuel and are living on heavier nuclear fuel (e.g., helium, carbon). In time, Rigel (within about 100 million years) will become a red supergiant like Betelgeuse. We discuss the evolution of stars more fully in a later chapter.

We have spoken about blue-white stars and red stars as though we could assign a definite color to each star, but this is not at all so. The light from a star, like the light from the sun, consists of a continuous array of colors ranging from deep red to violet, as can easily be demonstrated by passing the light from any star through a prism and thus obtaining the star's spectrum. But in spite of this similarity of the spectra of stars, we can still speak of the different colors of stars because the intensities of the various colors in the total light from different stars are not the same. In general, we designate a star's color by the most intense color present in its radiation. For example, we say that the sun is a yellow star because its yellow light is more intense than any of its other light. In the same way we call Betelgeuse and Antares red stars because they radiate more red light than they do blue or violet light. A careful inspection of the naked eye stars on a clear moonless night in a region of Earth where there is little dust in the atmosphere reveals the variation in color as we look from star to star. The astronomer is not content with merely noting that stars vary in color; he introduces a precise way of measuring color, which we describe below.

Since we want to be as accurate as possible when we assign a color to a star, we must do this numerically just as we assign a parallax or a magnitude to a star. But how does one measure colors of objects numerically? It may seem that only poets and painters can give the true measure of a color, so how can the astronomer reduce this apparently aesthetic concept to a number? The procedure is very simple and depends upon the different effects that different colors produce on the eye and on the photographic plate. Every amateur photographer knows that he can use either a red or yellow light in his darkroom without spoiling his exposed photographic film. This is so because the active rays of light that deposit silver grains on a film are the blue and violet rays. On the other hand, we also know that the red and yellow rays of light have a greater effect on the eye than do the blue and violet rays.

This means that if a source of blue light and one of red light of equal intrinsic intensity are measured with a photometer, the response of the instrument is not the same for both; the difference in the response depends on whether the photometer uses the eye or a photographic plate. This also applies to the apparent brightnesses of the stars. If one looks at a star like Betelgeuse (a red star), it appears brighter to the eye than it does on a photographic plate, whereas Rigel (a blue-white star) appears brighter on a photographic plate than it does to the eye. The astronomer uses this difference to measure the color of a star.

The procedure for determining the color of a star is quite simple and is based on measuring its magnitude. First, you get the image on an ordinary photographic plate and determine its apparent magnitude (this is called the *photographic magnitude*). Next, you measure the apparent magnitude visually (the *visual magnitude* of the star; this may also be done photographically by using special panchromatic plates). The color is now given by the number obtained when you subtract the visual magnitude from the photographic magnitude. This number is called the *color index* of the star.

If the color index of a star is zero, its blues and reds are equally intense and the star is called a white star; if it is less than zero, that is, negative, the star is on the blue side; and if the color index is greater than zero, the star is on the reddish side. For example, Sirius, which has a color index of zero, is a white star, and the sun with a color index of about 0.81 is a yellow star, while Betelgeuse and Antares are red stars with a color index larger than 1.5. The smallest color index ever found for a star (very blue) is about -0.62, and this appears to be about as small as the color index can get for any star.

Determining the color index is not just a matter of finding one more characteristic that can be used to describe a star; the color is useful in helping us discover other important properties that aid us in our investigations into the nature of stars. In particular, we use it to determine the surface temperatures of stars.

CHAPTER 10

Heat, Radiation, and the Temperatures of Stars

> *Bright star, would I were steadfast as thou*
> *art—*
> *Not in lone splendor hung aloft the night*
> *And watching, with eternal lids apart,*
> *Like Nature's patient, sleepless Eremite,*
> *The moving waters at their priest-like task*
> *Of pure ablution round earth's human shores.*
>
> —JOHN KEATS, *Bright Star*

We have reached a stage in our study of stars at which we can begin to deduce the numerical values of important surface and internal stellar characteristics that we cannot measure directly. Among these characteristics, whose numerical values we can calculate by applying the known laws of physics to stellar data, is the surface temperature of a star. Like the sun, all the stars around us are hot, self-luminous bodies. But just how hot are the stars and how do we measure their surface temperatures, which are measures of their hotness? In our discussion of the sun we gave its surface temperature as about 6000°K. How did we obtain this value and can we apply the same procedure to determine the temperatures of stars in general? The answer is yes to the last part of the question and we can answer the first part once we understand the meaning of temperature and the nature of stellar radiation.

Most of the information we have about stars has come to us from the radiation they send us; from this radiation we determine the surface temperatures of stars. We know from our general experience that the nature of the radiation emitted by a hot body depends in some way on how hot the body is (its temperature). For example, we know that red-hot coals are not as hot as white-hot coals, and, in general, that the color of the radiation emitted by a hot body changes from dull red, to yellow, and finally to white as the body's temperature increases. We therefore surmise that the surface temperatures of the stars are related in some way to their colors and that the sun is a yellowish star because its surface temperature is about 6000°K. But to see just how we obtain the temperature of a star from its radiation, we must have a precise definition of and a clear understanding of the temperature concept and we must understand the basic laws of radiation. We begin with the temperature concept.

So accustomed are we to looking at thermometers and hearing about the temperature of our atmosphere from day to day that we believe we know exactly what temperature means. And yet most people would be hard put to give an adequate definition of this important idea; they tend to confuse temperature with heat, and yet a little thought shows immediately that these are two quite different things. The temperature of water boiling in a kettle is considerably higher than that of the ocean, but the amount of heat in the ocean is immeasurably greater than that in the boiling water in the kettle, for the waters of the oceans can melt millions of tons of ice whereas the boiling water in the kettle can melt, at most, a pound or so of ice.

We see from this example that before we can get a clear understanding of the concept of temperature, we must first understand the nature of heat. Physicists first clearly recognized in the mid-19th century that heat is just another form of energy. Up to that time the only kind of energy that was known was mechanical energy, either in the form of motion (kinetic energy) or position (potential energy). We know from the laws of planetary motion that in an isolated system, the total mechanical energy is conserved; that is, the total amount never changes. This principle keeps the planets in their orbits.

What happens in a system that is not isolated, such as a pendulum? We know from experience that the total mechanical energy does not remain constant because the pendulum ultimately comes to rest. But does this mean that the principle of the conservation of energy does not apply to such a system? The answer is no. We still have conservation of energy, but we must now take another form of energy into account to keep the energy bookkeeping straight. This form of energy is heat.

A little thought shows that, fundamentally, heat is really not different in form from the kinetic energy of the pendulum. The only, but very important, difference between heat and the kinetic energy of the bob of the pendulum is that heat is kinetic energy distributed over many molecular particles moving in random directions. The difference between the kinetic energy of a large moving body and the heat obtained when the body is brought to rest by friction is the same as the difference between the wealth concentrated in the hands of a multimillionaire, and the same wealth after his heirs have paid the inheritance taxes and the money has been distributed according to his will; it is distributed among many people.

When a pendulum is brought to rest by the frictional resistance of the air, the energy that was present in the bob of the pendulum has been distributed among millions and millions of particles all moving in different directions. The energy is still kinetic energy, but it has become disorganized and is no longer as useful as when it was concentrated in the bob of the pendulum.

One way of characterizing heat is to say that it is disorganized kinetic energy. This point of view introduces the molecular concept of the structure of matter, but one can get at the concept of heat without bringing molecules into the picture.

We consider a cylinder filled with a gas and having a movable piston at the top of the cylinder. This piston can be used either to compress the gas, thus increasing the pressure inside the cylinder, or to reduce the pressure in the gas. We can keep the pressure in the cylinder constant by placing a set of fixed weights on the piston. If we suddenly compress the gas by pushing down the piston, we do work on the gas, and the kinetic energy of the molecules of the gas increases, or, as the physicist expresses it, the internal energy of the gas increases. This change is indicated by the rise in the temperature of the gas. If the cylinder is made of metal so that we can conduct heat away from the gas, we discover that the total increase in the internal energy of the gas (kinetic energy of the molecules) resulting from the work done on it is smaller than the amount of work done. This difference in internal energy of the gas and the work done on it is what the physicist defines as heat.

We have just stated what is known as the *first law of thermodynamics*; it is nothing more than an extension of the principle of the conservation of energy to systems in which heat, internal energy, and mechanical energy are present. This principle can also be stated by saying that there can be no perpetual motion machine; a system can be made to do work only if energy (heat, for example) is supplied to it. (This is just the reverse of the procedure outlined above. One applies a flame to the cylinder, heating the gas and causing the piston to rise; one thus gets work done by the gas.) Heat is thus transformed into work but not completely; some of the heat remains trapped in the gas as internal energy.

In a sense, we might characterize the first law of thermodynamics as the law of optimism because it holds out to mankind a great promise of ultimate release from the drudgery of manual labor: it says that we can get as much work done for us as we please just by supplying heat to mechanical systems. And there are vast amounts of heat all around us just for the asking. If this were the only law that governed the relationship between heat and work, we would, indeed, be enjoying a paradise, for all we would have to do would be to use the heat that is present in the air, the waters of the oceans, and the soil to generate electrical power of limitless quantities. But nature has played us a cruel trick and imposed a law of pessimism that places a very strict limit on the kind of heat that can be used to drive machines. This restriction is known as the *second law of thermodynamics*, or the *law of entropy*. The concept of temperature is developed from this second law.

There are various ways of introducing the second law, and so important is it in the scheme of nature that it is worth spending some time considering these different aspects. We have already noted that the difference between the ordinary kinetic energy of a large body and the heat into which it is dissipated by friction is that the heat is disorganized kinetic energy and therefore not as available for doing work as was the original kinetic energy. We already see in this the germ of the idea that is fully expressed by the law of entropy, which states that not all heat is useful for doing work. Before we say more about the second law itself, we must be a bit more precise about the notion of organized and disorganized energy.

Nature not only keeps a careful account of the total amount of energy in the universe, but also keeps track of the state of organization of this energy so that if one part of the universe gains organization, some other part must lose organization. At no time can the total amount of organization increase. At best it may remain constant, but actually the total amount of organization is getting smaller and smaller. But how do we know when energy is highly organized or disorganized? We say that the more concentrated heat energy is, the more highly organized it is, so that if a gram of water and a half gram of water have the same amount of heat energy, it is more highly organized in the half gram than in the gram. The second law can now be stated: heat can do work for us only by passing from a given state of organization to one of lower organization. The quantity that measures the state of organization of the heat energy in a system is called the *entropy* of the system (the smaller the entropy, the higher the organization). The second law of thermodynamics says that in any isolated system the entropy can never decrease.

There is still another and very instructive way of discussing the second law of thermodynamics, and this is in terms of the lack of homogeneity or the amount of differentiation in the universe. The more homogeneous the universe or an isolated system becomes, the more disorganized we shall say it is and the higher is its entropy. A system in which bodies are at different temperatures (some hotter and some colder) is more highly differentiated than one in which the bodies are at the same temperature. As heat flows from the hot bodies in this system to the cooler ones, the system becomes more and more homogeneous (less highly organized or more undifferentiated) and its entropy increases. The second law of thermodynamics can then be interpreted by saying that as time passes, the universe must become more and more homogeneous, just the way black and white grains of sand in a spinning container become more and more homogeneously distributed among themselves.

The second law is of great importance in that it separates the events in nature into two groups. In one group are all the events that occur in such a way that they are reversible, i.e., events that can proceed in a given direction or in the opposite direction if all the conditions are reversed. As an example of a reversible process we may consider the exceedingly slow compression of a gas. If the force acting on the piston is reversed, the gas slowly expands to its original volume.

In the second group of events are all those that are irreversible, and these encompass most of the events in the universe. In fact all natural events are irreversible. The recognition that irreversible events occur in nature led to the introduction of the second law, which is sometimes expressed by saying that irreversible processes exist in nature. It is easy to show that whenever irreversible processes occur, the entropy of the universe increases i.e., irreversible processes lead to a state of greater disorganization.

If you were a physicist and wanted to state the second law of thermodynamics in terms of an irreversible process that is easily understood and recognized by everyone, which, of all the many such events in nature, would you choose? It would have to be a simple event that can be explained in simple terms and described by easily measurable quantities. This is the problem that faced the physicist Clausius, and he decided to express the law of entropy by considering the flow of heat from one body to another. His statement of the law in its simplest form is that if two bodies are brought together, heat always flows spontaneously from the hotter to the colder body. This seems so obvious to us that we hardly consider it worth noting and wonder why it should be stated as a special law. Actually, we take this law for granted because it is part of our daily experience, but it is one of the most profound statements in all of physics.

Lord Kelvin expressed the second law somewhat differently by stating that heat can never be transformed entirely into work if everything else is left unchanged in the universe; some of the heat always remains unavailable. This unavailability is very important for the evolution of our universe, because in almost every process that occurs in nature some energy is transformed into heat. This means that ultimately the entire energy of the universe will have been transformed into completely disorganized, and therefore unavailable, heat energy. In a sense, this would represent the death of the universe since nothing further could take place. This would be the state of maximum entropy or maximum disorganization (also maximum probability); any departure from this state would be contrary to the second law of thermodynamics and to the laws of probability.

Another feature of the second law that is worth noting is its relationship to the direction of flow of time. From our daily experience we are accustomed to the fact that time flows only in one direction—from the past to the future. We have learned to differentiate between the past, the present, and the future by observing events that we know can unfold in only one way. If we are shown two pictures of the same person, one being an infant and the other an adult, we immediately say that the picture of the adult was taken later in time than the other picture. But how can we be sure that time does flow in only one direction in our universe, and what law of nature can serve as the finger of time?

As far as the first part of the question is concerned, we cannot be sure that all events in the universe need be interpreted in terms of a single direction for the flow of time. As a matter of fact, in the light of modern theories of the structure and properties of the electron, certain events can be looked upon as though they were progressing from the future into the past. But this applies only to the behavior of certain isolated particles (e.g., the motion of positrons) and not to matter in bulk. When we deal with ordinary events, time is unidirectional, and the second law of thermodynamics differentiates the past from the future. We can say that in any isolated system—e.g., the entire universe—the "later" event is associated with increased disorganization or increased entropy.

We see that time has meaning in an isolated system only if there are differences of organization within it, for if the system were completely homogeneous, there would be no way of differentiating one instant of time from any other. Time will thus have meaning in our universe only as long as our universe is in the process of decay. Once it has run down to the state of maximum entropy, there will be no way of separating the past, the present, and the future.

But what has all this to do with measuring the temperature of a star? The answer is that the second law enables us to find the temperature of a glowing object just by looking at it, without having to place a thermometer in contact with it. We consider a vast, homogeneous reservoir of heat, e.g., a very large furnace. Suppose that we know its entropy, or the state of organization of its energy. If we now remove a small amount of heat from the furnace, the entropy of the furnace will change (it will, in fact, decrease), so that we can compare the change in the amount of entropy with the amount of heat that is removed. The temperature of the furnace is defined as the amount of heat removed, divided by the change in the entropy.

We can apply this to a star and by measuring its luminosity determine how much heat it loses per second. But how do we know the state of organization of the stellar energy? Because we must know this before we can apply the ideas developed in the preceding paragraphs, we study the nature of the radiation emitted by the star.

Before doing that, however, we introduce the temperature scale used in astronomy.

We have seen that heat and temperature are quite different things even though they are related. Temperature is really a measure of how concentrated the heat is in a given substance, and the instrument that measures this is a thermometer. There are various kinds of thermometers, but they all depend on the fact that a material body changes its dimensions when heat is either added or subtracted from it.

A simple way to introduce a temperature scale is the following: Take the thermometric substance—e.g., a tube of mercury—and immerse it in ice and water. Mark the tube at the height to which the mercury rises. Call this mark zero. Do the same thing again with the tube immersed in boiling water and call this mark 100. Now put 99 equally spaced marks between the zero and the 100 mark and call each of these a degree. This is called the centigrade scale. Any other scale can be introduced in the same manner.

The spacings of the marks on the thermometric substance that define the temperature scale depend on the particular substance used, but there is one situation in which the scale is the same for all substances. This is of such great importance in physics and astronomy that it is worth devoting some time to it. The situation arises whenever the thermometric substance is a gas in which the molecules are not too closely crowded together. A more exact way of specifying a gas for which this is true is to say that the gas is a perfect gas. This simply means that the gas molecules are so widely separated that they have no effect on one another. Of course, there is some interaction between molecules because they collide every now and then, but this occurs only very rarely compared to the time during which the molecules do not collide.

Another way of defining a perfect gas is to give the relationship between the volume, the pressure, and the temperature of the gas. To do this we first define the concept of pressure. Whenever a gas is enclosed in a container, it pushes against the walls of the container owing to the bombardment of the walls every second by vast numbers of gas molecules. The pressure is not the total push, but rather is the push or force divided by the area. We can illustrate the difference between force and pressure by considering what happens when one walks on deep snow without and with snowshoes. In the first case, one's weight is distributed over the area of one's shoes and the pressure (force per unit area of the snow) is quite large so that one sinks into the snow. In the second case, because of the large area of the snowshoes, the pressure on the snow is much smaller and one can walk without sinking. Pressure may be expressed as pounds per square inch or as dynes per square centimeter.

A perfect gas is defined as one whose volume increases by 1/273 of its value at 0°C for each degree centigrade rise in temperature if the pressure of the gas is kept constant. If the volume of a perfect gas is held constant, then the pressure increases by this same amount with increasing temperature. Because of this, all perfect gases give the same temperature scale, called the absolute scale of temperature. To see how this scale is related to the centigrade scale, we consider a perfect gas that is cooled down as far as possible, assuming that it always remains perfect no matter how cold it gets. If we start with the gas at 0°C, then from what was said above, we see that when we get to −273°C the gas should have no volume at all. Theoretically, this is the lowest attainable temperature.

Although no substance remains gaseous down to very low temperatures, we find, from other considerations, that −273°C is the absolute zero temperature. Using this fact, we can define the absolute temperature scale as the centigrade scale, but with the zero point at 273° below the temperature of freezing water.

We now characterize a perfect gas as follows: If the pressure of a perfect gas is multiplied by its volume, and this product is then divided by the absolute temperature of the gas, the result is always a constant. This constant is the same for all perfect gases regardless of their chemical nature, provided that the number of molecules in the given quantity of gas is always the same. This means that, for a given temperature and a given volume of gas, only the number of molecules in the gas determines the pressure regardless of the mass of the particular molecule. This fact is of extreme importance in studying the internal structures of stars because we know that the material in stellar interiors is in a perfect-gas state right down to the very center of the star.

A simple way to find the pressure in a perfect gas is to multiply the density of the gas by its absolute temperature, divide this product by the molecular weight of the gas, and then multiply this result by the universal constant 82,700,000. The pressure is then given in dynes per square centimeter if the density is measured in grams per cubic centimeter. This formula is the one generally used to determine the gas pressure at any point in the interior of a star. We define the molecular weight later.

Now that we have introduced the absolute temperature scale, we shall see how knowledge of the nature of the radiation emitted by a star enables us to determine the absolute temperature of its surface. Conceptions of the nature of radiation in the past have alternated between the corpuscle and the wave theory, each trying to dominate the field exclusively, but today a kind of compromise reigns in which both theories play an important role in explaining the behavior of light. Newton was the first to introduce a particle theory of radiation, in which the particles, or corpuscles, are of a mechanical nature with the same kind of mass and inertial properties as any other particles. This theory suffers from a fatal defect: it can account for the way the path of light bends when it passes from one medium into a denser medium only by assuming that light travels faster in the dense medium than it does in the less dense medium (faster in water than in air, for example). Since the experimental evidence contradicts this assumption, the Newtonian corpuscular theory was discarded.

The theory that was universally accepted for more than 200 years after the failure of the Newtonian theory is the wave theory of light as developed by Christian Huygens, Thomas Young, Augustin Fresnel, and finally, James Clerk Maxwell. This theory accounts for most of the observations such as refraction, interference, diffraction (the bending of light around corners), and propagation along straight lines. It reached the peak of its success and greatest development at the hands of Maxwell who proved mathematically that light is a particular form of electricity and magnetism, propagated through space at a constant speed.

Although Maxwell was a physicist, his work on the nature of electricity and magnetism was invaluable to all branches of science. He was born on November 13, 1831, in Edinburgh, Scotland.[1] His father, John Clerk Maxwell, was trained as a lawyer but made his living managing lands and country estates. The Maxwell family lived on a small estate that James's father had inherited from his own grandfather. Although James was an only child who was doted on by his parents, he was an affectionate, inquisitive boy who always asked questions about how things, particularly mechanical devices, worked.

James's mother died of cancer when he was 9 years old. The following year his father decided that it was time for James to begin his formal education so he enrolled his son in the Edinburgh Academy. James created quite a stir when he first arrived at school because he showed up wearing an unusual outfit, designed by his father, that was trimmed with lace. This grand entrance made an unfortunate first impression on his classmates and he was the butt of many of their merciless jokes, but his buoyant personality and active mind gradually earned their respect. Although he did not particularly enjoy his subjects in the beginning, James soon developed a fondness for mathematics and later won a medal at the Academy for his mastery of it. He also composed a thoughtful paper on the construction of ovals, which was read to the Royal Society at Edinburgh with both Maxwell and his father in attendance.

At the age of 16, Maxwell entered the University of Edinburgh where he continued his studies in mathematics. He also wrote two papers on curves and the equilibrium of elastic solids, which were published in the *Transactions* of the Edinburgh Royal Society. He spent much of his spare time conducting experiments with electricity and magnetism and constructing various mechanical devices. Three years later Maxwell left the University of Edinburgh to enroll in Cambridge University where he became a student of the distinguished tutor William Hopkins, whose job was to prepare his students for the mathematical examinations that would be given to them at the end of their undergraduate careers. After 3 years of intensive study of mathematics and the physical sciences, Maxwell was finally ready to take the tripos but he suffered an acute illness in the summer of 1853 that incapacitated him for weeks. Although he slowly recovered his health that autumn, the illness did encourage in him a greater sense of Calvinist religiosity. The following January Maxwell sat for the mathematical tripos in the Cambridge Senate House with his legs bundled in blankets to protect against the chill in the unheated building. Despite his still somewhat fragile physical condition, he impressed his examiners greatly and finished as second wrangler. He also tied for first place in another mathematics competition for the Smith's Prize.

After receiving his degree from Cambridge, Maxwell supported himself by tutoring university students. He also conducted numerous experiments in optics—particularly color sensation—which eventually led to his being awarded the Rumford Medal of the Royal Society in 1860. Maxwell also became more interested in the electrical experiments of Michael Faraday and began to wonder how Faraday's results might be explained mathematically.

In 1855 Maxwell was elected a fellow of Trinity College at Cambridge, but the deteriorating health of his father caused him to give up his position and return to Scotland. His father died the following year, shortly before Maxwell was appointed to the chair of natural philosophy at Marischal College in Aberdeen. Although Maxwell may have regretted his decision to leave Cambridge, he seems to have enjoyed his life in Aberdeen, dividing his time between giving poorly-attended lectures and experimental demonstrations, carrying out his own experimental and theoretical work in optics, electricity, and magnetism, and, finally, managing the family estate. However, the high point of his Aberdeen career was his participation in the competition for the Adams Prize. He was required to write an essay explaining the structure of Saturn's rings. In his paper Maxwell assumed that the structure of the rings was determined by Newton's law of gravitation and thus demonstrated that the rings could not be solid or fluid. He thus concluded that the rings consist of numerous minute particles. His essay won the competition and established his reputation as one of the foremost mathematicians in Europe.

Maxwell then turned his study to the dynamic properties of gases and completed a now-famous paper on the subject in 1859. Unlike many of his predecessors, Maxwell did not make the patently false assumption that all molecules of a gas move at the same speed. The difficulty of characterizing the behavior of numerous molecules colliding against each other led Maxwell to develop a statistical method that presupposed that the gas molecules can be divided into groups (depending on their velocity) whose velocities approximate Gauss's famous bell-shaped distribution curve. While this mathematical innovation was not completely precise, it adequately described the aggregate behavior of molecules in a gas. This work led Maxwell to formulate an equation for gas pressure that is still applicable to many branches of physics. His model of the behavior of gases, the analysis of which was based on Newtonian mechanics, explained many properties of gases. Although some of his assumptions about the nature of gas molecules have subsequently been criticized, even his severest detractors have acknowledged the utility of his model. Unfortunately, neither Maxwell nor his contemporary and friendly rival, the Austrian physicist Ludwig Boltzmann, developed a completely satisfactory theory of gases: their equations failed to give the correct specific heats of gases. This deficiency remained until Max Planck offered his quantum theory of radiation in 1900.

In 1858 Maxwell married Katherine Dewar, the daughter of the principal of Marischal College. Two years later his position at Marischal was eliminated, but Maxwell was not unemployed for long; he soon joined the faculty of King's College, London, as professor of natural philosophy and remained for 5 years. Although Maxwell had an exhausting teaching schedule, he still found time to complete important papers on color theory, the theory of gases, and his theory of the electromagnetic field. In the latter paper, he presented his celebrated equations mathematically describing the electromagnetic phenomena observed by Michael Faraday. By the time Maxwell wrote his *Treatise on Electricity and Magnetism* in 1873, he had purged his conception of the electromagnetic field of most of its mechanical characteristics.

By 1870 Maxwell had cut back greatly on his academic duties and had retreated to the family estate where he continued at his leisure his work on the topics to which he had already devoted most of his scientific career. He also wrote a textbook on heat and maintained an extensive scientific correspondence. He occasionally traveled to Cambridge to serve as an examiner for the mathematics examinations. However, his quaint life as a country gentleman was not destined to last, because in 1871 Cambridge University created its first chair in experimental physics. This act followed the submission to the Chancellor of the University, the Duke of Devonshire, a report criticizing the university's shortcomings in this area. The duke responded quickly by offering a substantial sum from his own fortune to pay for both the construction of what became the Cavendish Laboratory and the endowment of the aforementioned chair. Although Maxwell was not the first choice of the hiring committee, he did show his interest in the position after William Thomson, the preferred candidate, rejected the offer. Maxwell was then offered the position and he readily accepted it. As a condition of employment, he also agreed to edit a number of manuscripts written by the experimental physicist Henry Cavendish, an ancestor of the

Duke of Devonshire, which, after their publication in 1879, established the fine scientific reputation that Cavendish now enjoys. However, the preparation of Cavendish's papers took up much of Maxwell's few remaining years and greatly restricted the amount of time he could devote to his own scientific work.

While at Cambridge Maxwell taught courses on electricity, magnetism, and heat. He also gave humorous public lectures, which, unlike his upper-level course lectures, were very popular. His university classes attracted the brightest students but he could not resist playing an occasional prank on his students to make a point. Maxwell was also deeply involved in the administration of the Cavendish Laboratory and worked tirelessly to solicit contributions from well-heeled alumni; he also donated a great deal of his own money to it.

Maxwell began to suffer from a painful illness in 1877 but did nothing about it for nearly 2 years even though his health continued to decline. During that time he spent all of his spare time caring for his wife who was now bedridden owing to poor health. Maxwell tried to continue his scientific work, however, he was increasingly weakened by what was later diagnosed as stomach cancer. Maxwell seemed to realize that his death was inevitable but he faced it calmly, believing that his primary responsibility lay in taking care of his wife. He died on November 5, 1879, and was buried near his home.

With Heinrich Hertz's experimental verification of Maxwell's theory in the last decade of the 19th century, the Newtonian corpuscular theory appeared to be as dead as Ptolemy, and yet in the very process of proving Maxwell's theory Hertz was laying the basis for its ultimate downfall. But before we go further into this phase of the story, let us stop for a moment and see how much useful information we can get from the wave theory.

According to the wave theory of light, we must consider radiation as a periodic electromagnetic vibration that is propagated through space at a speed of 186,000 miles per second in a vacuum. We can represent these vibrations by means of a wavy line, a sinusoidal curve, the properties of which have been extensively studied by mathematicians for many centuries. Of course, light is not actually propagated in this elementary form but is rather a mixture of many such simple forms called harmonics. The spectroscope is an instrument that analyzes a beam of light into its simple harmonic components.

Associated with any simple harmonic wave are three important numbers λ, v, and c, which are related by an elementary formula. The Greek letter λ (lambda) stands for the wavelength of the light and is the distance between any two successive crests of the wave. The wavelength of the light determines the color of light that we see. For visible radiation the wavelength is so small that it is convenient to replace the usual unit of length, the centimeter, by the *angstrom*, which is 100,000,000 times smaller than the centimeter.

The shorter the wavelength of the light, the bluer the light appears; the long wavelengths correspond to the red colors. The reddest colors have a wavelength of approximately 7000 angstroms and the deep violet colors have a wavelength of about 3500 angstroms. The eye is not sensitive to wavelengths longer than 7000 angstroms or shorter than 3500 angstroms, but this does not mean that no radiation exists beyond these wavelength limits. With the development of Maxwell's electromagnetic theory of light, it became clear that visible light itself is a minute fraction of a vast radiation spectrum that ranges all the way from the mile-long radio waves to the very penetrating gamma radiation having wavelengths equal to a fraction of an angstrom.

The wavelengths of the radiation from the sun and stars range from a few meters (radar-like waves) down to a fraction of an angstrom (gamma radiation associated with the cosmic rays). The study of the long-wave radiation during the last few years has developed rapidly and now constitutes an entirely new branch of astronomy called *radio astronomy*.

The Greek letter v (nu) represents the frequency of the radiation and is just the number of crests of the wave that pass any given point in 1 second. We can get a visual picture of this by imagining a water wave moving across the surface of a lake upon which a duck is sitting. The duck oscillates up and down as the wave advances; the frequency of the wave is the number of times the duck moves up and down in a given period of time. For light, the number of vibrations per second is extremely large. The bluer the light is, the larger the number of vibrations per second. As an example of this we note that when we see what we call violet, the retina of our eye receives about 1,000,000,000,000,000 alternating electric and magnetic vibrations per second. We can see different colors because many millions of years ago nature built into our bodies (the retina of the eye) special molecules that have just the right kind of structure

to enable them to vibrate in unison with the electromagnetic pulsations of light. The more slowly these retinal molecules vibrate,
the redder the sensation they convey to the brain. We shall see later
that the frequency of light plays a fundamental role in the modern
corpuscular (quantum or photon) theory of radiation.

The letter c stands for the speed of light; as already noted, it
is one of the fundamental constants of nature. A basic feature of
Maxwell's theory of light is that the speed of light in a vacuum is
the same for all wavelengths. Experiments have also shown that
the speed is the same for all observers and independent of the motion
of the source. As we have seen, this fact, inherent in Maxwell's
theory, plays a dominant role in Einstein's special theory of relativity.

We come now to the relationship between λ, v and c, which
we visualize by comparing a monochromatic wave to an infinitely
long passenger train composed of identical cars. The wavelength
corresponds to the length of a single car and the frequency is the
number of cars that pass any point in a given time interval. From
this it is easy to see how the speed of the train is related to the
wavelength and the frequency. Since the speed is just the distance
traveled by any car in a unit of time, it is the total number of cars
that have passed in the unit of time multiplied by the length of a
single car. This means that the speed of the train (light) is found by
multiplying the wavelength by its frequency. The mathematician
writes this as follows: $\lambda v = c$.

Even while the wave theory of light was enjoying all its successes, seeds of discontent with its rule were being quietly sown in
unexpected quarters. It was in connection with the radiation emitted
by a hot body (continuous spectrum) that the first doubts concerning
the universal validity of the wave theory were raised. We know
from experience that if we slowly heat a piece of metal in a flame,
it begins to glow with a dull cherry red color and then as it gets
hotter the color becomes yellow and finally almost white. This
shows us that as an object becomes hotter it begins to radiate more
and more of the blue colors as compared to the reds. Any theory
of radiation must predict this if it is to be accepted, and of course,
the Maxwell wave theory does predict something of the sort. The
trouble, however, is that the wave theory does not correspond to
the observed facts in all their details.

To see the difficulties we consider the emission of radiation from hot bodies in somewhat more detail. It is convenient to limit ourselves to the investigation of special kinds of bodies because the radiation emitted by such bodies has certain universal properties that can be easily expressed by means of the laws of thermodynamics. In addition, it appears that stars radiate energy the way these special bodies do. The kind of body we have in mind is what the physicist calls a *blackbody*, and the radiation emitted by these bodies is called *blackbody* or *thermal radiation*.

We can get at the concept of a blackbody by considering what happens when radiation falls on any opaque surface. Two things happen: part of the radiation is absorbed by the surface, and part of it is reflected. If all the radiation were reflected, the surface would be called a perfect reflector and it would be perfectly white. On the other hand, if all the radiation were absorbed, the surface would be called a blackbody or perfect absorber, which is also the most efficient kind of emitter of radiation. This must be so because the faster the blackbody absorbs radiation, the hotter it gets, and therefore the more rapidly it must radiate. A blackbody takes radiation of any kind and spreads it out over the entire continuous spectrum. We get blackbody radiation simply by heating a perfect blackbody.

To study blackbody radiation, we must have a perfect blackbody; but it is not very easy to determine the blackness of a body. Instead of trying to work with an object that looks black, we get the result we want by punching a tiny hole in a furnace and studying the radiation coming from this hole. The smaller the hole is, the closer the radiation approximates blackbody radiation because the smaller a hole is, the blacker it is, as we know from experience.

What is it that we want to find out about blackbody radiation? We want to determine how much of each particular color is present in the stream of radiation that comes per second out of the little hole in the furnace. Since this depends on how hot the furnace is, we are really interested in determining how the distribution of the emitted radiation among the various wavelengths, or frequencies, depends on the temperature of the furnace. We may illustrate this by means of a simple analogy.

Suppose that instead of a furnace emitting radiation we had a marble (marbles that children play with) factory that manufactured marbles of all colors ranging from red to violet. Suppose there were a sorting machine that placed each marble as it came out into a bin marked with the appropriate color. At the end of each day we could then count the number of marbles in each bin and in that way determine the distribution of the marbles among the various colors. We could then establish some kind of law of marble distribution, which, of course, would depend on the particular tastes for colored marbles among the children of the community and the total number of children present.

It is easy enough to do this experimentally with blackbody radiation by allowing the radiation to pass through a prism. This spreads the colors out into a continuous spectrum, and we can then measure the intensity of each color. We can make a graph of this, plotting the intensity of each color against the wavelength; such a graph is called a distribution curve.

Certain properties of the radiation are at once evident from an examination of this curve. To begin with, not all colors are present in equal intensity, since the very long and very short wavelengths fall off to zero intensity. The curve reaches a maximum value for one particular color and then falls off on both sides of this color. If the temperature of the furnace is increased, two things happen: the curve is shifted upward so that it encloses a greater area above the horizontal axis, and the point of maximum intensity is shifted toward the shorter wavelengths.

These two facts are fairly easy to interpret. The first means that as the temperature of the furnace is increased, the amount of energy radiated per second from the small opening increases; the second means that the intensity of the blue colors becomes relatively greater as the temperature of the furnace increases.

Did the wave theory of light predict these characteristics of blackbody radiation? Yes and no. When it came to predicting the total amount of energy emitted per second (the total number of marbles manufactured per day, regardless of color) or the wavelength of maximum intensity (the color of the largest number of marbles) the theory proved admirable and agreed exactly with the radiation curve. The wave theory predicted, in agreement with the observations, that the amount of energy emitted per second (the area under the radiation curve) should increase with the fourth power of the absolute temperature of the furnace. If this temperature is doubled, the total energy emitted per second is 16 times greater, and if the temperature is tripled, the rate of emission increases by a factor of 81, and so on. This is known as the Stefan–Boltzmann law in honor of the two physicists who discovered it.

The wave theory also predicts that the wavelength in which the greatest amount of energy is emitted by the furnace per second is cut in half if the temperature is doubled, or, more generally, the theory correctly states that the wavelength of maximum intensity varies inversely with the absolute temperature. This is known as Wien's displacement law in honor of its discoverer. But this is as far as the classical wave theory could go, and these successes could do no more than delay its ultimate demise.

Before we consider just wherein the wave theory of Maxwell failed, we note that we can use the two features of the theory described above to determine the surface temperatures of stars either by measuring the total energy emitted per second from a star or by measuring the wavelength of greatest intensity in the radiation. This is permissible only if we assume that stars radiate like blackbodies, which is a valid assumption.

The classical wave theory, however, does not describe correctly the behavior of the radiation curve over its entire range. No matter how one tries to force the classical theory into the mold of the observations, it refuses to fit, and at some point deviates strikingly from the observed data. In one attempt to match theory and observation, it was found that the two agreed nicely for the long wavelengths but disagreed violently for the short waves. In fact, according to the theory, a blackbody should emit all of its energy in the very short wavelength region in one violent outburst. This is known as the ultraviolet catastrophe. Another attempt at matching theory and observation led to agreement at the short-wave end of the spectrum but to disagreement at the long-wave end.

By the end of the 19th century, it became clear that nothing short of a completely new and revolutionary approach to the whole problem of radiation could lead to a theory that agrees with experiment. This revolution was not long in coming. In 1900, the famous theoretical physicist, Max Planck, showed mathematically that as long as one assumes that a blackbody emits radiation continuously in the form of waves, the ultraviolet catastrophe is inevitable.

How was one to correct the situation without injuring the wave theory too much? Planck finally decided, much against his will, that the only solution was to assume that radiation is emitted in the form of little pellets (quanta, as he called them, hence the birth of the quantum theory), each with its own wavelength and frequency. This gives the analogy with the marble factory much more meaning.

Planck's attempt at reconciling the quantum concept with the wave theory by assuming that a quantum has only a temporary existence and changes into a wave was short-lived when Einstein showed that not only is radiation emitted in the form of pellets but that it continues and exists as a collection of corpuscles at all times. Einstein called these corpuscles photons. This development was the final and decisive blow against the wave theory of electromagnetic radiation. Not only was this concept a great departure from the venerable and powerful wave theory, but it also did violence to the classical picture of a particle. Einstein's photon is not a Newtonian corpuscle because it can have no mass when at rest. The mass of a photon would vanish, or become part of some ordinary particle if it were brought to rest. A photon has meaning only when it is moving with the speed of light. Moreover, the photon has a wavelength and a frequency—which is like having, in biology, a plant that has arms and legs and a central nervous system. Here was a pretty state of affairs, indeed, and yet there was no getting around it even though Planck first assumed that the photons (his quanta) have only a temporary existence at the moment when they are emitted, and then turn quickly into waves.

Max Planck was born in Kiel, Germany, on April 23, 1858, the son of a law professor at Kiel University. He was the descendant of a distinguished line of lawyers and scholars. Given his family background, it was thought that he would become a lawyer. This apparent destiny did not come to pass because soon after Max entered the Maximilian Gymnasium at the age of 9, he began to attend the classes of a sharp-witted mathematician named Herman Müller. What separated Müller from so many of his fellow teachers was his ability to demonstrate abstract physical principles using simple examples that his young students could understand and remember. Planck was apparently deeply influenced by Müller because it was through the latter's lectures that he began to see science as an unending search for the fundamental laws of nature. Not surprisingly, Planck's view of science had deep philosophical roots; he believed that the universe was guided by subtle yet encompassing physical principles. In his *Scientific Biography*, Planck later recalled how he was first struck by a lecture on one of the most fundamental of nature's laws—the principle of the conservation of energy:

My mind absorbed avidly, like a revelation, the first law I knew to possess absolute, universal validity, independently from all human agency: The principle of conservation of energy. I shall never forget the graphic story Müller told us, at his *raconteur's* best, of the bricklayer lifting with great effort a heavy block of stone to the roof of a house. The work he thus performs does not get lost; it remains stored up, perhaps for many years, undiminished and latent in the block of stone, until one day the block is perhaps loosened and drops on the head of some passerby.[2]

After Planck graduated from the Gymnasium, he attended the University of Munich for 3 years and then went to Berlin for an additional year. When he first entered the university, he had been torn between his growing interest in science and his love for music. For a time he gave serious consideration to the pursuit of a musical career but the lure of contributing to humanity's knowledge about the natural world proved to be too strong for Planck to resist. He studied mathematics and experimental physics but Planck later admitted that his real education did not begin until his frustration with the unsatisfactory lecturing styles of his Berlin professors Kirchhoff and Helmholtz drove him to spend much of his spare time digesting the papers of many important physicists, most notably Rudolf Clausius, the discoverer of the second law of thermodynamics.

Planck became so immersed in Clausius's work on heat that he devoted his doctoral research to the second law of thermodynamics. He later remarked that the effect of his work on his professors was "nil" as he suspected that neither Kirchhoff nor Helmholtz had bothered to read it carefully if at all.[3] Undeterred, Planck continued his work in thermodynamics and published a series of papers from 1880 to 1892 that yielded fruitful results, even though he later found out that much of his earlier work had been anticipated by the American physicist Josiah Willard Gibbs.

In 1887 Planck was offered a chair in theoretical physics at the University of Kiel; he considered this appointment to be one of the happiest events of his life. Soon after setting up a household in Kiel, Planck was awarded second prize by the Philosophical Society of Göttingen for a paper he wrote about "the nature of energy." Although Planck's appointment may have been partly due to his father's friendship with one of the physics professors at Kiel, Planck soon showed the merit of his selection and published four important papers on thermodynamics.

After the death of his mentor Kirchhoff in 1889, Planck was

invited to take his place at Berlin and teach theoretical physics. He became especially good friends with Helmholtz and also made the acquaintance of other important scientists such as Wilhelm von Bezold and August Kundt. During the next few years Planck became more deeply involved in an ongoing debate between Wilhelm Ostwald, who argued that the whole of physics and chemistry could be derived from the law of energy, and Ludwig Boltzmann, who denied that. Planck supported Boltzmann in a series of impressively argued essays but the decisive defeat for Ostwald's "energetics" school came only after Boltzmann had proposed his molecular theory.

Planck became interested in thermal radiation owing to the experimental work being done at the National Physical Laboratory in Berlin by two teams of experimental physicists on the spectral distribution of blackbody radiation. Planck conducted his own series of theoretical investigations into the nature of blackbody radiation and eventually developed a new radiation formula, which he submitted to the Berlin Physical Society on October 19, 1900. This formula became the basis for "the most revolutionary idea which ever has shaken physics."[4] Less than 2 months later, he gave a lecture to the German Physical Society and proposed his "elementary quantum of action," which directly challenged classical physics by asserting that all thermal radiation is atomistic in the sense that it is emitted from a blackbody only in multiples of some basic entity known as Planck's constant of action and designated by the symbol h. Previously, physicists had assumed that all radiation is emitted continuously and not in lumps.

Despite the tremendous philosophical and physical implications of Planck's theory, very little attention was paid to it by physicists in the first years of the 20th century. That situation began to change in 1905 when Einstein published his paper on radiation in which he explained the photoelectric effect using a corpuscular interpretation of light and showed that all radiation—not just thermal radiation—is emitted in the form of quanta and retains this form and its wave properties at all times. Planck, however, was more concerned with trying to fit the elementary quantum of action into the classical theory and he devoted a number of years to this effort. He did not regret the time he spent in this largely futile attempt because of the additional insights he gained along the way into the nature of radiation. Planck later wrote that "I now knew for a fact that the elementary quantum of action played a far more significant part in physics than I had originally been inclined to suspect, and this recognition made me see clearly the need for the introduction of totally new methods of analysis and reasoning in the treatment of atomic problems."[5] Planck credited Bohr and Schrödinger with introducing the conceptual tools needed to the quantum theory from the classical theory.

Planck was also one of the earliest scientists to recognize the brilliance of Einstein's theory of relativity, which appeared in the *Annalen der Physik* in 1905. Planck used Einstein's relativistic principles in a series of remarkable papers on the dynamics of moving systems and hinted in one of them at the tremendous latent energies bound up in all matter that would make possible atomic energy: "Though the actual production of such a 'radical' process might have appeared extremely small only a decade ago, it is now in the range of the possible, through the discovery of radioactive elements and their transmutation, and in fact the observation of continuous production of heat of radioactive substances is direct evidence for the assumption that the source of this heat is just nothing else than the latent energy of the atoms."

Planck was also instrumental in convincing Einstein to come to Berlin from Switzerland in 1913. Although the two men were completely opposite in personality and outlook, they shared a common fascination with the physical world. Both were convinced that there are only a few simple laws in nature and that any theory that could not be constructed in a straightforward and accessible manner was probably not correct. Planck and Einstein also shared a fondness for music and often played impromptu concerts together. Their shared interests in physics and music led them to become great friends. Planck persuaded Einstein on several occasions in the 1920s not to leave Germany when the latter became the scene of vicious anti-Semitic attacks. In any event, the presence of both Einstein and Planck in Berlin made it the greatest center of theoretical physics in the world. Despite the anemic state of the German economy after the Treaty of Versailles, the German university system boasted one of the most impressive groups of scientists ever assembled.

For Planck the Great War was a personal as well as a national tragedy. His first wife had died in 1909 and three of his four children from that marriage died while Germany was at war. His son Karl perished in battle in France in 1916 and his twin daughters, Emma and Margarete, died while giving birth in 1917 and 1918, respectively. These losses dwarfed the personal and professional satisfaction that came with Planck's receipt of the Nobel Prize in physics in 1919.

Planck sought refuge from the pain in his family and his nation by returning to his scientific work. He contributed greatly to the efforts to apply Bohr's method of quantization to systems having more than one electron. He also published many papers on thermodynamics and statistical mechanics in the 1920s.

Planck retired in 1928 and was succeeded by Erwin Schrödinger. Not one to spend his retirement in a rocking chair, Planck continued his theoretical investigations and his activities at the Berlin Academy. During this period he was especially concerned about the attacks then being mounted by many of the younger physicists on the law of causality. According to Max Born, "Planck's own child, quantum theory, had grown beyond all expectation and now dominated the whole of physics; but it had taken a direction which led straight away from Planck's fundamental convictions [because] causality and strict determinism, even the assumption of an external objective world independent of ourselves, became problematic." Planck had welcomed Schrödinger's wave mechanics as a solution to these difficulties though the objections of his dissenting colleagues became more persistent with time. Doubtless much of Planck's view on the subject was conditioned by his innate conservatism in scientific matters and his strong religious views. Because Planck saw the universe as having been made by a rational creator, science and religion were not fundamentally opposed to each other.

When Hitler came to power in 1933, he began to persecute Jews in general and Jewish scientists in particular. Planck paid a personal visit to Hitler to try to persuade him not to sack Fritz Haber, whose work on nitrogen during World War I had directly contributed to the German war effort, but it was to no avail. His pleas on behalf of Haber prompted such a violent outburst from Hitler that Planck gave up any hope of being able to dissuade the Nazis from destroying much of the German scientific establishment in the name of "racial purity." Like Heisenberg, Planck was too devoted to Germany to emigrate and he believed that he had to stay behind if only to save the German scientific establishment from total destruction.

Despite Planck's unparalleled rank in the German scientific hierarchy, his only remaining child from his first marriage, Erwin, who held an important post in the German government, was shot in 1944 for his alleged role in the assassination attempt on Hitler. Planck's house was also destroyed along with all of its contents during an Allied air raid on Berlin. However, Planck bore these tragedies with the same stoic demeanor that had carried him through all of his earlier trials and tribulations. After the war ended, Planck moved to Göttingen, but his health gradually declined and he died on October 4, 1947, a few months shy of his ninetieth birthday.

Einstein's important contribution to the quantum theory of radiation stemmed from an interesting discovery made by Heinrich Hertz, who had devised a series of simple experiments to prove Maxwell's electromagnetic theory of light that involved using an induction coil to produce sparks between two metal knobs. These sparks produced electromagnetic waves that Hertz showed had all the properties of light waves. While engaged in this work, Hertz observed that he could produce sparks much more readily if the knobs were bathed in ultraviolet radiation. This meant that the ultraviolet radiation shining on the knobs somehow caused the air between the knobs to become electrically charged and therefore able to produce an electrical discharge from the knobs. It was at once obvious to Hertz that the air between the knobs became charged because the metallic knobs emitted electrons when they were illuminated by the ultraviolet radiation. This marked the discovery of what we now call the *photoelectric effect*. This effect can be produced by visible light as well as by ultraviolet rays, but if the frequency of the light that is used is too low, no electrons are emitted.

One of the interesting features about the photoelectric effect that cannot be understood in terms of Maxwell's wave theory is the relationship between the speed of the emitted electrons and the color of light that is used to produce the effect. Since the energy of an emitted electron is measured by its speed when it leaves the metal surface, one would, on the basis of the wave theory, expect the speed of the electrons to fall off as the intensity of the light irradiating the surface is reduced since this reduction in intensity reduces the energy in the beam. But this is not what happens at all. The speed of the emitted electrons depends only on the frequency of the light used, not on its intensity. The higher the frequency, or the bluer the light, the faster the electrons move as they come out of the metal surface. If the intensity of the light is reduced, the speed of the electrons is not affected. They still come out with the same speed, but fewer of them are emitted per second. In fact, no matter how weak the beam of light may be, if the frequency is high enough, electrons start coming out immediately as though the beam of light consisted of little bundles of energy, each one of which kicks out its own electron with just the right speed.

After Planck had introduced the quantum concept, Einstein quickly saw that the photon possesses just the correct properties to explain the photoelectric effect. Using the quantum theory and extending it to cover the behavior of light at all times, Einstein obtained a simple formula that correctly accounts for the energy of the emitted electrons. Einstein received the Nobel Prize for physics in 1921 for explaining the photoelectric effect.

What is the nature of the photon? Aside from the fact that it appears to be an electromagnetic particle with a frequency and a wavelength, all we know about it is how much energy it has and that it is spinning. According to Planck's theory, we must assign to each photon an amount of energy that depends in a simple way on the color of the photon—the bluer the photon, the greater its energy. The relationship between energy and color introduced by Planck can be expressed as follows: To find the energy of a photon, take the frequency of the photon and multiply it by the number 6.624 divided by one thousand trillion trillion (the number 1 followed by 27 zeros). This number, which is written as 6.624×10^{-27}, is called the *Planck constant of action* and is one of the famous constants in nature. It is always denoted by the letter h. The mathematical form of the Planck relation is $e = h\nu$, where e is the energy of a photon and ν is its frequency.

Planck's theory is amazingly accurate in predicting the behavior of blackbody radiation over the entire range of the spectrum. With its aid we can measure the surface temperatures of the stars.

From all that we have said up to now we see that we can use three different methods to get the surface temperature of a star, but the answers are not quite the same for each because stars are not perfect radiators (black bodies), but only approximately so. One of these methods depends on comparing the total amount (i.e., the amount in all colors) of energy emitted every second from 1 cm^2 of a star's surface with the total amount of energy emitted every second from a 1-cm^2 opening in a furnace. If the total luminosity of a star and its area are known, this can be done, since the luminosity divided by the area gives the energy emitted per square centimeter per second. This leads to what is called the *effective temperature* of the surface of the star. The other two methods depend on analyzing the star's color.

The total energy emitted per second by 1 cm^2 of the star's surface in all frequencies gives the surface temperature via the Stefan–Boltzmann law. This number equals the absolute temperature to which a furnace has to be raised so that the total energy emitted per second through a 1-cm^2 opening equals the energy emitted per second from 1 cm^2 of the surface of the star. To find the effective temperature of a star we must know its luminosity and also its surface area. If we have this information, we get the effective temperature as follows: Express the luminosity in ergs per second, and divide this quantity by the area of the star expressed in centimeters squared. Divide this number by the universal constant (the Stefan–Boltzmann constant) 0.00005672 and then take the fourth root. The result is the effective temperature of the star.

Instead of using Stefan's law, we can use Wien's displacement law and obtain what is known as the color temperature of the star. Here, too, the procedure is usually fairly easy, provided that we can pick out the wavelength of maximum intensity in the radiation. For the very cool stars the radiation curve is quite flat, and it is difficult to pick out the maximum intensity; and for the hotter stars the maximum falls deep in the ultraviolet and again is hard to find. If, however, we have this wavelength of maximum intensity expressed in centimeters, we can find the temperature of the star by dividing this number into another universal constant, 0.28971. The temperature we get in this way is the star's color temperature, a value somewhat different from the effective temperature.

The third method, and the one most extensively used, is to compare the curve of the star's radiation with the curve that is theoretically deduced from Planck's radiation law. This corresponds to matching the color of a star over its entire spectrum with that of a blackbody of a given temperature. Since we have seen that the color of a star is given by its color index, we can determine the star's temperature from its color index. If to the color index we add the number 0.64 and divide this sum into 7200, the number thus obtained is the absolute surface temperature of the star. This temperature is also referred to as the star's color temperature.

By applying these various methods to the stars, we have found that the temperature increases steadily as we go from red to blue stars. The coolest red stars have temperatures of about 2500°K, whereas blue stars have temperatures well in excess of 50,000°K.

A star like Arcturus, which is an orange star, has a temperature of approximately 3800°K, whereas yellow stars (e.g., the sun) have temperatures in the neighborhood of 5500–6000°K. Procyon has a temperature of 7100°K and typical white stars have temperatures of 10,500°K. As we go from white to blue-white stars the temperature goes up more rapidly; the hottest blue-white stars have temperatures that range from 20,000 to 30,000°K.

From these figures we see at once that the temperature plays a very important role in the mechanism that produces a star's spectrum. We shall presently see just what that mechanism is and what effect the temperature has on it. But before we do so, we list the numbers that we have assigned to the stars thus far.

In all, five measurable quantities are associated with stars: the parallax, the apparent magnitude, the absolute magnitude, the color index, and the surface temperature. We have already seen that some of these are related to each other in a simple way, like the parallax, the apparent magnitude, and the absolute magnitude; or the color index and the temperature. We now introduce a sixth characteristic of a star, which is intimately related to the star's surface temperature but which we cannot specify by a single number. This characteristic, called the spectrum of the star, permits us to arrange the stars into spectral groups.

CHAPTER 11

Spectral Classes of Stars

The spacious firmament on high,
And all the blue ethereal sky,
And spangled heavens, a shining frame,
Their great Original proclaim.

—JOSEPH ADDISON, *The Spectator*

In the previous chapter we saw that a star's color (its color index) reveals its surface temperature, but the color gives us only a rough picture of the nature of the light emitted by stars. Two stars may have the same color index and yet the detailed features of the radiation they emit may differ considerably. The reason for this is that the color of the star's radiation is determined by the temperature of its photosphere, but before the light from the photosphere reaches us it passes through the star's atmosphere, and each atom in this atmosphere leaves its imprint on this radiation. Two stars with the same surface temperatures may have atmospheres whose physical and chemical properties differ considerably. Thus, the pressures and the abundances of the various chemical elements in the two atmospheres may be quite different so that the effects of the two atmospheres on the radiation streaming through them will be different. For many years astronomers have investigated the interaction of a star's atmosphere with the star's radiation by passing the radiation through a spectroscope and studying the spectrum so obtained. We noted in the chapter on the sun that the spectrum of the light coming from a star consists of a continuous array of colors on which a series of dark lines (the absorption or Fraunhofer lines) is superimposed. These dark lines reveal the physical and chemical properties of the star's atmosphere. We shall see that, in general, the temperature of the radiation coming from the star's photosphere determines the overall characteristics of the star's spectrum and the intensities of the various lines, but there are important details in which the lines may differ from star to star though the surface temperatures are the same (same color).

As we have seen, the spectrum of the light from a hot dense body (e.g., a solid) consists of a continuous array of colors whose relative intensities depend only on the temperature of the body and not on its chemical nature. This is called a continuous spectrum; such a spectrum tells us nothing about the chemical nature of the hot body. If, however, we excite a gas (stimulate it to emit radiation) consisting of atoms of a particular kind (e.g., hydrogen) to a point where it glows, we find that the spectrum is drastically different from that of the hot solid body. Instead of a continuous array of colors, the spectrum of the radiation emitted by the gas consists of a series of sharp discrete colored lines, which may be few or many in number, depending on the nature of the atoms in the gas. Such a spectrum is called an emission or bright line spectrum. Each type of atom has its own characteristic emission spectrum, which distinguishes that particular type of atom from any other type. No two types (e.g., hydrogen and helium) have the same kinds of emission spectra (the same set of bright lines). The bright line emission spectrum thus identifies a species of atoms unambiguously and uniquely. If a gas, such as the atmosphere of a planet or a star, consists of a mixture of different atom types, the emission spectrum produced by such a gas contains the emission lines of all the atom types in the gas.

Each bright line in the emission spectrum of a gas occupies a definite and characteristic position that is determined precisely by the wavelength of the light that produced that line. We may then say that the emission spectrum of a gas consists of a series of bright lines to each of which a precise wavelength is assigned. We shall see later, after having discussed the structure of atoms, how the wavelengths (or frequencies) of the spectral lines of an atom are calculated.

To see how an absorption spectrum arises (the type of spectrum emitted by a star) we consider a hot surface surrounded by a cool gas that may consist of one atom type or many. If the cool gas were not present, the spectrum of the radiation would be continuous; but with the gas present the continuous spectrum is replaced by an absorption spectrum. The dark lines in this absorption spectrum fall precisely at those positions in the spectrum that would be occupied by the bright lines if the emission spectrum of the gas were produced. In other words, the wavelengths (and frequencies) of the emission lines and absorption lines of any given atom type are identical.

To understand how absorption lines arise, we note that an atom consists of electrically charged particles (electrons) that respond to the vibrations of an electromagnetic wave (e.g., light, X rays, radio waves) the way a cork in water responds to the vibrations of a surface wave. An atom subjected to electromagnetic waves may, indeed, be compared to a group of radio sets each of which is tuned to a different station. If a continuum of radio waves (a group of electromagnetic waves consisting of all possible wavelengths) impinged on these radio sets, the latter would respond only to a discrete group of waves (those whose wavelengths equal the wavelengths to which the sets are tuned or, put differently, the wavelengths of the signals sent out by the radio stations to which the sets are tuned). A person receiving the continuum of radio waves after these waves had passed the radio sets would detect a diminution in the intensity of the radio waves at precisely those wavelengths to which the radio sets are tuned. These decreased intensities represent absorption lines (energy has been absorbed by the radio sets from the incident radio waves) in the radio wave continuum.

Since each different type of atom consists of a different number of electrons (e.g., hydrogen has one electron, helium has two, lithium, three) and the electrons in these various atoms have different modes of vibration, these various atoms respond to, or resonate to different sets of electromagnetic vibrations incident on them. Thus, we may say that hydrogen (like a given radio set) is tuned to one set of electromagnetic waves, helium to another set, and so on. If, then, a hot incandescent source of light (like the photosphere of the sun or an electric light filament) that emits electromagnetic waves of all wavelengths is surrounded by a cool gas consisting of various kinds of atoms, each of these atoms vibrates in response to those and only those electromagnetic waves whose frequencies correspond to the characteristic vibrational frequencies of the atom itself. Thus, all the atoms of a given kind (e.g., hydrogen) remove (or reduce the intensity of) a discrete set of wavelengths in the hot radiation passing through the cool surrounding gas. These diminutions in intensity show up as dark lines (absorption lines) in the spectrum of the radiation, and each kind of atom contributes its own set of absorption lines to this absorption spectrum. These dark lines thus identify the kinds of atoms that are present in the cool gas.

If the surface or photospheric temperatures and the luminosities of all stars were the same, and if their atmospheres contained the same abundances of the various chemical elements, the absorption spectra of all stars would be the same. But this is not so; the spectra of stars differ just as their colors do, and the variations in stellar spectra are related to the variations in stellar colors. Since, as we have seen, the color (color index) of a star depends on the star's surface temperature (the hotter the star, the bluer it is), we correctly infer that the surface temperature is most important in the formation of the absorption spectrum of a star. Since hydrogen and helium are by far the most abundant elements in stellar atmospheres, and these abundances do not differ greatly from star to star, the abundances of the various chemical elements play only a minor role in the variations of stellar spectra; nor are stellar spectra very sensitive to luminosity either. In spite of their vast range in luminosities and surface temperatures, stars can be grouped into a very few spectral classes. *These classes are not sharply differentiated and the spectral characteristics vary continuously as we go from the very hottest to the very coolest stars.* But since there are clear distinctions from one class to the next, this type of classification has great practical use. We find, as we go from one spectral class to the next, that there is a continuous variation in the intensities of some of the outstanding spectral lines. In this classification the spectra of stars are divided into seven major groups, designated as O, B, A, F, G, K, and M, and each of these is divided into subgroups so that there is a continuous sequence of intermediate spectral types ranging from O to M.

The guiding principle used in setting up this classification was the falling off in the intensity of the lines of helium and hydrogen. The spectrum of hydrogen had been studied in laboratories, but helium was first discovered when its bright lines were found in the solar spectrum during a total eclipse. The hydrogen and the helium spectral lines are much more intense in the white and blue-white stars than in the yellow, orange, and red stars, whereas the lines of the metals are quite prominent in the latter but quite insignificant in the blue-white stars. We need not list all the lines in the spectrum of a star belonging to a particular class but only the strongest. In the spectra of O-type stars, the strongest lines are those of helium atoms that have lost one unit of negative electricity. Such atoms are called ionized atoms. The B-type stars are characterized by intense lines of neutral helium atoms with the lines of hydrogen also quite prominent. As one goes from B- to A-type stars, the helium lines become weaker and the hydrogen lines begin to dominate, becoming most pronounced for stars like Sirius and Vega. From A- down to G-type stars, the hydrogen lines become fainter and fainter while the lines of the metals, first those of the ionized metal atoms and then those of the neutral metal atoms, grow more and more intense. Thus, in the spectrum of the sun, which is a typical G-type star, the strongest metal lines are those of iron, and calcium; the hydrogen lines are only moderately strong. Helium lines are not present in the sun's absorption spectrum.

By the time we get to the M-type stars, e.g., Betelgeuse and Antares, even the lines of the metals are no longer prominent and the spectrum is dominated by wide dark bands instead of sharp narrow lines. These bands are produced by the molecules in the star's atmosphere, primarily those of titanium oxide and the hydroxyl radical OH.

Before we explain these peculiarities of stellar spectra, we point out a few things that are apparent at once. As we go from O- to M-type stars, we pass continuously from the blue-white to the red stars, so that helium lines are associated with the former and molecular bands with the latter. Another point of interest is that hydrogen lines are present with considerable intensity in the spectra of all stars. These two features are of great importance in understanding the physical and chemical structure of stars, for the first feature confirms the importance of the star's surface temperature in the formation of the star's absorption spectrum, and the second confirms the overwhelming abundance of hydrogen in all stellar atmospheres. Indeed, hydrogen is so abundant that its lines show up under even the most unfavorable circumstances.

To see why the surface temperature of a star plays the dominant role in the formation of its absorption spectrum, we first note that the charged particles (i.e., electrons) are not equally tightly bound in different atoms, such as hydrogen and helium. They are thus not equally free to respond to radiation of a given frequency and to go from one mode of vibration to another. The more tightly bound an electron is inside an atom (to the nucleus by electrostatic forces), the more energy must be transferred to the atom from the radiation to propel the electron to a higher mode of vibration. The more energetic the radiation from the star—that is, the higher the star's surface temperature—the more effective this radiation is in changing the charged particles' modes of vibration, thereby producing different absorption lines. For this reason stars having different surface temperatures exhibit the absorption lines of different atoms. *In general, the more tightly bound the charged particles of an atom are, the higher the star's surface temperature—and hence the hotter the radiation from the photosphere—must be to change the atom's modes of vibration.* This means that the absorption lines of helium are present with appreciable intensities only in the spectra of the hottest stars since the helium electrons are more tightly bound than the electrons in metals or in hydrogen.

One other fact should be mentioned before we consider the spectral classification itself: *If the radiation passing through the atmosphere is energetic enough, it may tear a charged particle out of an atom entirely. We then say that the atom is ionized. If such an ionized atom interacts with the radiation passing it, we obtain absorption lines of the ionized atom, which are quite different from those of the un-ionized atom.* Since hydrogen has only one charged particle in its atom, ionized hydrogen can have no absorption lines, for when it loses its single charged particle, the ionized atom (the proton) has no modes of vibration. All other atoms do form absorption lines when they are ionized, and the absorption lines of various ionized atoms are found in the spectra of stars. In general, the lines of ionized atoms are present in the spectra of hot stars, and the lines of un-ionized (i.e., neutral) atoms appear in the spectra of cooler stars.

We now consider the correlation between stellar surface temperatures and the presence of the absorption lines of various atoms in stellar spectra as we go from one spectral class to another. When a star like the sun is first formed by gravitational contraction from a cloud of dust and gas, it is rather cool and its atmosphere consists of atoms and molecules. As the cloud contracts, its gravitational potential energy is slowly transformed into the random kinetic energy of its atoms and molecules. Half of this released potential energy is radiated away into space, but the other half remains trapped in the cloud, raising the temperature of the cloud. This increase in temperature causes the cloud to radiate away still more energy with the result that the cloud continues to contract and to get hotter. Depending on its mass, the cloud stops contracting sooner or later; stars less massive than the sun are cool and reddish when they stop contracting, whereas stars more massive than the sun are white or blue-white when they stop contracting.

Owing to the low temperature in the atmospheres of the red or M-type stars (about 2500°K), the collisions among molecules such as OH and other oxides are not violent enough to dissociate them into their atoms so that the spectra of these molecules appear, but the temperatures are too low for the hydrogen spectrum to show up. The temperatures in the atmospheres of the G-type yellow stars (like the sun) are too high for even the most tightly bound molecules to survive, and molecular absorption bands are thus not present in the spectra of such stars, but the spectral lines of some of the ionized metals are present. The hydrogen absorption lines are also present even though the surface temperatures of yellow stars are below the temperature most favorable for the hydrogen spectrum. In the atmospheres of A-type stars the temperature is high enough to excite the hydrogen atoms to a point where the Balmer hydrogen spectral lines are dominant and most intense, but the temperature is still too low for the helium lines to be very strong. The helium lines are most intense for the hot blue-white B-type stars. The surface temperatures of these stars are so high, however, that many of the hydrogen atoms are ionized so that the hydrogen lines are much weaker in the spectra of B-type stars than in the spectra of A-type stars.

This brief analysis shows us how the surface temperature of a star determines the overall features of its spectrum, but certain important details of stellar spectra depend on stellar characteristics other than the temperature. To see why this is so, we note that the surface temperature of a star is related to two characteristics, the star's luminosity and radius: Two stars having the same surface temperature (and hence belonging to the same spectral class) may have different luminosities because one is bigger than the other and this difference can influence the fine details of the spectrum. The temperature determines the rate at which each square centimeter of the star's surface emits energy, and the total luminosity is measured by this rate times the total number of square centimeters in the surface. Because of this relationship involving the luminosity, the radius, and the surface temperature, stars of the same spectral class can fall into two main groups. The overall spectral features of the stars in the two groups are the same, but the luminosities and radii of the stars in the two groups are different, and there are slight spectral differences between the groups, which enable astronomers to recognize the group to which a star belongs just by inspecting its spectrum.

The discovery that stars of the same spectral class fall into two luminosity groups goes back to the Danish astronomer E. Hertzsprung, who found that not all red stars (spectral class M) are the same. This in itself was not surprising because one might have expected to find many different sizes and therefore different luminosities among these stars. But what was completely unexpected was the discovery that there are just two different groups of red stars. In one group are a few stars of large luminosity, which Hertzsprung called *giant stars*. In the other group are large numbers of low luminosity stars, which he called *dwarfs*. In 1913 H. N. Russell, the American astrophysicist, extended Hertzsprung's ideas and showed that the spectral classes K, G, and F can also be divided into large- and small-luminosity stars.

These results can best be represented by a graph that is known today as the Hertzsprung–Russell (H–R) diagram. In this graph the spectral classes O, B, A, F, G, K, and M are equally spaced along the *abscissa* (horizontal axis) and the absolute magnitudes (luminosities) are plotted along the *ordinate* (the vertical axis). Instead of the spectral classes, one can also use the temperatures or the color indices (owing to this, the H–R diagram is also called a luminosity–color diagram or a luminosity–temperature diagram). Suppose now that we pick a thousand stars at random and plot them on this diagram. Since each star belongs to a definite spectral class and has a definite absolute magnitude, it is represented by a point on the graph. Do these points form a definite pattern or are they distributed in a random, haphazard fashion?

If one had no previous knowledge of stellar structure, one might assume that the stars would be randomly distributed over the diagram. But this is not true at all. Most of the stars fall into two well-defined groups that form two branches that meet in the neighborhood of the A-type stars. The diagonally running branch extending all the way from O- and B-type stars (in the left upper part of the diagram downward and to the right) to M-type stars and forming a kind of S shape is called the main sequence; it contains more than 90% of all the stars. This branch extends from the very hot luminous stars that are called blue giants, to the faint cool stars that are called red dwarfs (the original dwarfs of Hertzsprung). The second branch is almost horizontal and runs to the right from spectral class A to class M. This branch tilts up somewhat at the M end of the graph. The stars in this branch, which are on the average about five magnitudes (about 100-fold) more luminous than the main-sequence stars of the same spectral class, are the giant stars.

Ejnar Hertzsprung was born in Frederiksberg, Denmark, on October 8, 1873.[1] His father, Severin, was trained as an astronomer but had declined to pursue a scientific career, believing there to be greater job security with the Danish government's Department of Finances. Severin's business skills enabled him to earn a comfortable living, but he tried to instill in his son an appreciation for the beauty of the heavens. Ejnar grew up in a household in which the topics of conversation ranged from the structure of the Milky Way to the economic policies of the Danish government. Because Severin did not try to push his son into a particular occupation, believing it better that his son select the vocation for which he was best suited, Ejnar chose the middle road between pure science and business and began to study chemistry and engineering while still a young boy. He received his degree from the Polytechnical Institute in Copenhagen in 1898 and then journeyed to Russia to hone his experimental skills in St. Petersburg. Three years later, he left Russia for Germany, then the acknowledged leader in chemical research, and worked as an assistant in Wilhelm Ostwald's laboratory at Leipzig. Hertzsprung apparently became bored with chemistry, however, because when he returned to Denmark the next year he began to study astronomy.

Hertzsprung became interested in using photography to improve observational techniques even though astronomers had been using cameras for several decades to map star positions. Because of Hertzsprung's background in chemistry, he was more familiar with the theoretical aspects of the photographic process than were many of his fellow astronomers and more adept at taking photographs. Although his knowledge of observational astronomy was somewhat limited, he gradually overcame this deficiency by reading all of the relevant scientific journals and corresponding with other astronomers. He developed his observational skills by working at two observatories in Denmark for several years. His work provided him with the background needed to write several papers on the use of photography in astronomy; in some of these papers, he suggested that the luminosities of stars can be measured by photographing their spectra. His detailed analyses of astronomical photographs not only led him to discover giant and dwarf stars but also to develop (independently of Russell) what is now known as the Hertzsprung–Russell diagram, plotting the absolute magnitudes of a sample of stars against their temperatures. This diagram formed the basis for much of the subsequent work that has been done by astrophysicists on the evolution of stars.

During the next few years Hertzsprung carried on an extensive scientific correspondence with Karl Schwarzschild, the German astronomer at Göttingen, who was the first to solve a special case of Einstein's gravitational field equations. Schwarzschild encouraged Hertzsprung's researches and eventually asked him to come to Göttingen in 1909. Hertzsprung arrived at the University of Göttingen soon thereafter and accepted an appointment as associate professor of astronomy. He later moved to the Potsdam Observatory as an astronomer after Schwarzschild was appointed its director.

In 1912 Hertzsprung went to the Mount Wilson Observatory and oversaw extensive photographic studies of several star clusters. This work was a natural outgrowth of his earlier work at Göttingen during which he had determined that the star Polaris has a variable magnitude, thus confirming a hypothesis that had been vigorously debated for nearly 20 years by astronomers but which had evaded direct efforts to prove it. In 1913 Hertzsprung successfully calculated the distance to the Small Magellanic Cloud and was awarded the gold medal by the Royal Astronomical Society. Hertzsprung's method, which depended in large part on the research of the Harvard astronomer Henrietta S. Leavitt, remains the basic method for determining the distances to other nearby galaxies. Hertzsprung's later researches dealt with the distribution of Cepheid variable stars in the Milky Way and the motions of double star systems. He was also very interested in the relationship between color and luminosity of stars.

Hertzsprung became an associate professor at the University of Leiden in 1919. He was also appointed the associate director of the observatory at Leiden and eventually became its director in 1935. Despite his administrative duties, he continued to take thousands of photographs of stars and to make many of the tedious measurements of their relative positions by himself. Before Hertzsprung retired in 1944 to return to Denmark, he was awarded a number of honorary doctorates and elected to nearly a dozen scientific societies in Europe and North America.

Henry Norris Russell, whose name will forever be linked with that of Hertzsprung owing to the famous diagram that bears their names, was born in Oyster Bay, New York, on October 25, 1877.[2] His father, Alexander, was a Presbyterian minister. Although Henry's parents recognized the importance of education, they kept their son at home until he reached the age of 12, when he was sent to a boarding school near Princeton University to prepare for college. Despite his lack of formal education during his primary school years, Henry was a fine student who excelled in mathematics and the sciences. His distinguished performance there made his admission to Princeton a foregone conclusion; he graduated with honors from Princeton in 1897 and immediately enrolled in the graduate astronomy program there. He was so adept in his coursework and research activities that he completed his doctoral requirement in 3 years. He received his Ph.D. in 1900.

This grueling pace exacted a toll on Russell's health, however, because soon after his graduation he became seriously ill and was forced to return to Oyster Bay to recuperate. His recovery was slow and only after nearly a year had passed was he able to go back to Princeton for a brief period of time. His stay there was followed by a 3-year stint at Cambridge University, where he worked in the Cavendish Laboratory as a research assistant. In 1905 he returned to Princeton to work as an instructor in astronomy. Six years later he became a professor of astronomy there and enjoyed a long and productive career that continued until his retirement 36 years later.

From the beginning of his professional career, Russell was concerned with celestial mechanics and eclipsing binary stars. He was aided by a graduate student named Harlow Shapley, who became famous in his own right for his studies of extragalactic nebulae. Like Hertzsprung, Russell recognized the potential applications of photography to astronomy and used it to develop a new technique for determining stellar parallaxes. This work gradually led Russell to his theory of stellar evolution, which, while now dated since nothing was known then of thermonuclear fusion, seemed to provide the best explanation for the development of stars. Russell also took numerous photographs of the moon to determine more precisely its position relative to Earth, and calculated the age of our planet to be about 4 billion years using the rate of decay of uranium as a sort of geological clock.

In the early 1920s Russell studied the spectra of the sun's sur- face and its sunspots and demonstrated that this procedure can be used for analyzing stellar atmospheres. He also used spectral anal- ysis to study the atomic structures of various elements such as cal- cium and barium. Improved spectral techniques enabled Russell to calculate the relative abundances of various elements in the sun's atmosphere and to argue convincingly that hydrogen is the primary element in all stars. Russell's studies of the sun led him to conclude that the mass and chemical composition of a star are the two most important factors in determining its subsequent development.

During the latter decades of his career, Russell gradually re- duced his research efforts and devoted more time to scientific cor- respondences and writing. His textbook on general astronomy was a standard volume for many decades and helped to train two gen- erations of American astronomers. Russell also wrote a monthly astronomy column for *Scientific American* from 1900 to 1943. After his retirement in 1947, he continued to keep up with new devel- opments in astrophysics and became an informal senior advisor to many astronomers both in the United States and abroad. Before his death on February 18, 1957, he was elected to the National Academy of Sciences, the Royal Astronomical Society, and the American Association for the Advancement of Science. He was also a member of several foreign science academies and the recipient of a number of medals and prizes. Although Russell made significant contribu- tions to nearly every branch of astronomy and astrophysics during his long career, he will always be remembered for developing what we now call the Hertzsprung–Russell diagram.

A few things about this diagram need emphasizing. To begin with, it was found, shortly after Russell's work, that in addition to the ordinary giants and dwarfs, a group of unusually luminous stars (only very few in number) falls on another horizontal branch about four magnitudes above the giants. These stars are called *super- giants*. In addition, a group of fairly numerous stars was discovered that lie in the lower left-hand corner of the diagram. These are called white dwarfs because they are mostly of spectral class A, which means that they are quite hot; but their absolute magnitudes are fairly large so they have small luminosities. The red-dwarf end of the main-sequence branch of the diagram becomes the most thickly populated part of the diagram as ever larger numbers of nearby stars are included.

The various branches in the H–R diagram are not thin lines but rather broad bands, which may be due to two things: To begin with, it is clear that whenever we place a star on this graph, its position there is not precisely known owing to the errors involved in determining its absolute magnitude. This imprecision gives rise to a certain amount of scattering of the points, and therefore contributes to the broadening of the branches. However, part of the broadening is undoubtedly due to certain differences in the internal structures of stars that lie on the same branch and belong to the same spectral class. The evidence for this is obtained by constructing an H–R diagram for stars that have had a common birth and a common evolutionary development, for example, the stars belonging to a cluster in our Milky Way. For such clusters the stars arrange themselves in the H–R diagram along thin lines like beads on a string. The study of the H–R diagrams of stellar clusters has greatly elucidated the way stars evolve as they age.

Astronomers had long suspected that the branches in the H–R diagram are related in some way to stellar evolution but, all attempts to unravel the clues contained in this diagram as to the evolution of stars were unsuccessful until the theory of the internal structure of stars showed two important things: (1) the structure of a star at any moment in its history is determined by its mass and chemical composition (essentially, the amount of hydrogen and helium it contains); (2) the chemical composition of a star changes as it ages because it generates energy by fusing hydrogen into helium and thus altering its chemistry. This means that the position of any star in the H–R diagram is determined by three numbers: its mass, its chemical composition, and its age. Now the H–R diagram of a few thousand naked eye stars (the nearby stars), chosen at random, certainly includes a large range of masses, a variety of initial chemical compositions (the chemical compositions of the stars when they are born), and a great variety of ages. The H–R diagram of a group of stars representing such a mixture of masses, chemical compositions, and ages presents no simple clues to the evolutionary tracks along which points in the H–R diagram move as the stars they represent age.

Astronomers discovered a way out of this difficulty when they began to study the physical properties of stars in clusters, i.e., stars that live together and move through space together. There are two kinds of star clusters: the famous globular clusters, which consist of hundreds of thousands of the very oldest stars (population II stars) and form a halo around our galaxy, and the open clusters (also called galactic clusters), which lie in the plane of our galaxy and consist, in general, of a few hundred (in some instances as many as a few thousand and in others as few as 30 or 40) population I (second generation) stars (intermixed with gas and dust). The stars in an open cluster form a loosely knit gravitational structure and move together through our galaxy (around its core), thus showing us that they form a single generic family and were born together from the same chemical mixture. Whether we study the stars in open clusters or in globular clusters, we detect a wide range in their luminosities and colors, and we see at once that this range is the clue that we are seeking to the evolutionary track along which a star travels in the H–R diagram as it ages. Since all the stars in a cluster were formed from the same chemical mixture (the original chemically homogeneous cloud of gas and dust that fragmented into the stars that now form the cluster), each of them had the same initial chemical composition and was chemically homogeneous throughout. Moreover, all the stars in a cluster are the same age because they were formed at the same time from the original cloud of dust and gas. These two statements eliminate from our consideration two of the parameters—the chemical composition and the age—that determine the position of a star in the H–R diagram, and leave only the initial mass of the star to be dealt with. We may state this somewhat differently and perhaps more clearly, as follows: If the positions in the H–R diagram of two stars belonging to the same star cluster are different, this difference can only be due to the difference in their initial (zero-age) masses since their initial chemical compositions were and their ages are the same. In other words, if we study the H–R diagram of the stars in a single cluster, we have a clear picture of how the initial mass of a star affects its evolution in a given period of time, and this determines where it lies in the H–R diagram now since the initial mass is the only thing that differs from star to star in the cluster.

To see how the study of the H–R diagrams of both open (galactic) clusters and globular clusters has given us an empirical insight into the evolutionary tracks of stars in the H–R diagram, we consider the various H–R diagrams of such clusters. These diagrams show that every cluster (open and globular) has many stars that lie along the main sequence and some stars that lie along a branch that turns upwardly and to the right of the main sequence. The very old clusters M67 (open) and M3 (globular) have well-developed branches that extend from the main sequence up into the red giant and supergiant regions of the H–R diagram and then back down again. Since the stars in any one cluster are the same age and each started with the same initial chemical composition, these different branches must represent different stages in a star's evolution as determined by its mass. If we accept this as a fact, and we shall soon see why we must, we are faced with the following question: on which branch of the H–R diagram does a star begin its life—on the main sequence or on the red giant branch?

The answer to this question is found in the theory of the internal structure of stars, which astrophysicists have fashioned into a re-markably powerful analytical tool for taking a star apart mathe-matically and seeing exactly how it is built and operates. This type of analysis has shown that a star that is chemically homogeneous throughout must lie on the main sequence; such stars cannot be red giant stars or lie far from the main sequence. The mathematical analysis (based on physical laws) shows that red giant stars cannot be chemically homogeneous structures; they are chemically inhom-ogeneous, with dense helium cores and hydrogen-rich shells sur-rounding the cores. From these two statements we see that when stars are formed from the gravitational condensations in a homo-geneous mixture of gas and dust, they begin their active lives as stars on the main sequence because they are chemically homoge-neous structures. For this reason the main sequence is also called the zero-age line of stars. After spending some time on the main sequence, the stars become chemically inhomogeneous as they transform hydrogen into helium in the core, and evolve into red giant stars and beyond. The more massive a star is, the faster this happens.

This picture of the way stars evolve and change as they grow older is one of the most beautiful examples of the way observational astronomy and theoretical astronomy or astrophysics (the part that deals with the theory of stellar structure) have collaborated to reveal a sequence of events in the lives of stars that spans billions of years. The empirical analysis of the evolutionary data from the H–R diagrams of open clusters permits us to deduce the ages of clusters from the diagram in a very ingenious way. The older a cluster is, the larger is the number of stars in it that have evolved away from the main sequence and the more extensive is its non-main-sequence branch. This means that in old clusters the point along the main sequence where the stars begin to turn away from the main sequence is lower than for young clusters. One more fact is easily deduced from the H–R diagrams of star clusters. Since all the stars in the cluster began their lives on the main sequence, those that are now red giants must have started out as massive, luminous blue-white giants, high up to the left on the main sequence and moved from left to right in the H–R diagram as it evolved.

We have seen that we must know the distance of a star before we can learn very much more about it, and the only direct way of measuring the distance is by the method of trigonometric parallaxes. Since this method is limited to the nearby stars, it is essential that we develop other indirect methods that can be applied to the distant stars if we are to obtain a correct picture of the evolution of stars and of the structure of our Milky Way. All indirect methods involve measuring or observing some stellar property that enables one to determine the absolute magnitude of a star. From this determination and a measurement of the star's apparent magnitude, its parallax can be found by reversing the procedure that was used to obtain the absolute magnitude from the apparent magnitude and the parallax. A hint as to a possible approach to the development of such a method is contained in the H–R diagram.

Suppose that we want to determine the parallax of a star that is so far away that the direct trigonometric method gives grossly inaccurate results. How are we to proceed? As a first step we determine its spectral class, and then, without further ado, place it on the main sequence of the H–R diagram in the average position for stars of that spectral class. Since most of the stars in the sky belong to the main sequence, the chances are good that this star will be in its correct position. If that is so, we can read off the absolute magnitude of the star directly from the ordinate scale of the H–R diagram, which then gives us the parallax.

But in about 10% of the cases we may be wrong because the star may really belong to the giant, supergiant, or white-dwarf part of the diagram. This method is therefore reliable only if we work with large numbers of stars and obtain mean parallaxes. If we wish to apply this approach to a single star, we must have some method for determining to which branch of the H–R diagram the star belongs. Fortunately, this can be done by carefully comparing the fine details in the spectrum of one star in a given spectral class with those of another in the same spectral class.

Two stars in the same spectral class have in general the same overall spectral characteristics, but the two spectra may differ in certain details. To see how this happens, we need only keep in mind that two stars that belong to the same spectral class may differ greatly in size. This introduces slight variations in the spectral lines. We can understand how this variation occurs if we use an important characteristic of stars (which we discuss in more detail later): the masses of stars are not very different as we go from one branch of the H–R diagram to the other branches, whereas the radii differ widely. Thus in going from the G-type main-sequence stars to the G-type giant stars, we find that the mass increases by about a factor of 4, but the diameter increases by a factor of 20. This means that the pressure in the atmospheres of the giant stars is considerably smaller than that in the atmospheres of the main-sequence stars like the sun, because of the smaller gravitational fields on the surface of the giant stars.

But how does this affect the spectrum of the star? Since the material in the star's atmosphere is composed of atoms that can move about and collide with one another, the spectrum will be determined by what these atoms are doing and under what conditions they are absorbing and radiating energy. To begin with, an atom sends out or absorbs its characteristic spectrum only if it is not too violently interfered with by other atoms. If, while an atom is in the process of radiating or absorbing energy, another atom collides with it or even comes very close to it, the first atom is disturbed and this disturbance shows up as a broadening of the emission or absorption lines. This occurs for atoms in atmospheres where the pressures are large because then the atoms are crowded together and collide frequently. Thus, we expect to find that the absorption lines in the spectra of giant and supergiant stars are sharper than those in the spectra of main-sequence stars. Very sharp lines mean supergiant stars.

Another consequence of the difference in the atmospheric pressures of stars is in the intensities of lines that arise from ionized atoms. If an atom has one of its electrons torn away, the atom is said to be ionized, and its spectrum is different from that of the unionized atom. The ionization occurs because of collisions among the atoms of the atmosphere and also because of the radiation coming from the interior of the star. If an ionized atom can move around without meeting very many electrons, it has a good chance of remaining ionized and therefore of exhibiting the ionized spectrum. But if the pressure in the atmosphere is high, the atom collides with many electrons and does not remain ionized very long, so that the ionized lines in the spectrum are weak. The presence of strong ionized lines as compared to the neutral lines of an atom indicates a low-pressure atmosphere, and therefore a giant or supergiant star.

In 1914 the procedure for determining the absolute magnitude of a star from its spectral lines was developed by the American astronomer Walter S. Adams of Mount Wilson and the German astronomer Arnold Kohlschütter. They set up a criterion for differentiating between giants and dwarfs by comparing the ionized lines of certain metals with the neutral lines of other metals. By drawing graphs of the intensity ratios of different groups of lines, they established a scale of absolute magnitude based upon these ratios. If, for a given star, the intensity ratio for these groups of lines is measured, the absolute magnitude of the star can be found from the graph.

This method of finding absolute magnitudes, and therefore parallaxes, is called the method of *spectroscopic parallaxes*. Since this method can be applied as long as one can obtain a reasonably good spectrum of a star, it is applicable to very distant stars for which the trigonometric method breaks down. With this method the parallaxes of a few hundred thousand stars have been determined. The accuracy of this method is about 15% so that this gives much better results for stars of very small parallaxes than the trigonometric method does. The two methods give results of comparable accuracy for stars that are about 65 light-years away.

CHAPTER 12

Atomic Structure and Stellar Spectra

When the morning stars sang together,
And all the sons of God shouted for joy.

—Job 38:17

It may seem, owing to the stars' great distances that it is practically impossible to find out what is happening deep in their interiors or even in their outer atmospheres. However, the very high temperatures in stellar interiors make it easier to analyze their structures than to study the internal structure of Earth. We use what Eddington called an "intellectual boring machine" that metaphorically dredges up material from the deep interior of a star and subjects it to the analysis afforded by our knowledge of the physical laws that apply to such material. If we deduce the behavior of such material in this way, and are then led by these deductions to the correct properties of the observed radiation emitted by the star, we know that we are on the right track.

One may object that we have no right to apply the laws that we have discovered here on Earth to objects that are at such great distances from us. It is true that we have no direct evidence that the laws of physics are the same for the interior of stars as they are under laboratory conditions on Earth, but we make this assumption because it is the simplest thing to do; if we did not, we could make no progress at all. In any case, we have found that this procedure leads to consistent results and is in complete agreement with observations. To some extent we do have direct evidence that the laws as we know them apply to stellar interiors because we have reproduced stellar interior conditions during nuclear bomb explosions and have tested the laws under these conditions.

We divide our investigation of stellar structures into two parts: in the first we deal with the atmosphere of the star, in the second with the deep interior. Actually, there is no sharp demarcation between these two parts, but the spectrum of a star, as described in our discussion of the sun indicates that there is a main body of a star from the surface of which the radiation is emitted in the form of a continuous spectrum, with no individual spectral lines present, and an extended shell of gas above this surface in which the absorption spectrum is produced.

The radiation that comes to us from a star originates in the deep interior, not too far from the center, in the form of highly penetrating gamma radiation as the result of thermonuclear reactions. The flow of this energy from the deep interior ultimately determines the physical conditions in the star's atmosphere although it may spend but a few seconds in passing through this atmosphere. During these last few moments in the life of the radiation before it leaves the star forever, each atom in the atmosphere leaves its imprint on it in the form of the absorption lines that we observe in stellar spectra.

To understand what happens in the atmosphere of a star, we must first have some knowledge of the structure of the atom, since the behavior of the atoms in the stars determines the properties of the radiation that we receive. The picture we have today of the interior of an atom is vastly different from that which the 19th century physicists had, and is based on entirely new concepts of the nature of fundamental particles.

It was already known in the middle of the 19th century, as the result of Faraday's experiments on electrolytic solutions, that atoms consist of both positive and negative electrical charges in equal quantities. Faraday showed that when a salt is placed in solution, each molecule of the salt breaks up into negative and positive ions, the electrical charge on each ion always being a whole multiple of a unit charge.

Near the end of the 19th century, Sir Joseph Thomson extracted the unit negative charges from various atoms in the form of a beam. He did this by constructing a cathode-ray tube (the forerunner of the modern television tube) and sweeping the negative charges away from the hot cathode by means of a high electrical voltage. By applying electric and magnetic fields to this beam, Thomson demonstrated that it was composed of negatively charged particles. He showed that these particles are about 2000 times lighter than the lightest known atoms (hydrogen) and therefore cannot be considered as atoms themselves. He called these particles *electrons*. In

1905, Millikan measured the electric charge on an electron and showed that it corresponds to the electrolytic unit of charge found by Faraday.

Soon after the discovery of the electron, the fundamental positive charge, the *proton*, was discovered. It was shown that this particle is a hydrogen atom with its electron missing. With the discovery of the proton and the electron, it was possible to construct a model of the atom that could account for its chemical and optical properties. All one had to do, it seemed, was to arrange an equal number of protons and electrons in a stable configuration, with enough protons introduced to account for the mass of the atom. Just how the protons and electrons were to be arranged was suggested by an important experiment performed by Lord Rutherford.

Using the rapidly moving alpha particles—doubly ionized helium atoms, or the helium nucleus—that are emitted spontaneously by uranium, Rutherford bombarded very thin gold foil and studied the tracks of the alpha particles that bounced off the foil. These tracks show that all of the protons in an atom are concentrated in a single central nucleus and that the electrons swarm like a cloud around this nucleus. With these experimental data as a basis, Rutherford suggested that the atom is constructed on a planetary model with the nucleus playing the part of the sun and the electrons moving in elliptical orbits around the nucleus.

One can apply the laws of motion to such an electrical system if one replaces Newton's law of gravity by Coulomb's law of electrical attraction (which is similar in form to Newton's law); one then gets the same kinds of orbits for the electrons in the atom as one has for the planets in our solar system. The electron in an orbit must move with just the right speed at each point of its orbit so that the centrifugal force on it is just balanced by the force of electrical attraction exerted on it by the positively charged nucleus.

At first sight this model seemed to give just what was wanted; in addition, it had the satisfying property of reproducing on a minute scale the features of nature on a large scale (the solar system). However, in terms of electromagnetic theory, as it was known at that time, this model of the atom had a fatal defect: it could not exist. The reason for this stems from one of the consequences of Maxwell's great electromagnetic theory. One can show that according to this theory, a charged particle that is accelerated must lose kinetic energy by radiating. This means that a charged particle moving in a circular or elliptical orbit must slow down because it is accelerated and sends out electromagnetic waves. According to Maxwell's theory, an electron in an atom cannot stay in one orbit but must spiral closer and closer to the nucleus continuously, sending out energy all the time as it does so. This classical behavior of the electron has two consequences that are in direct contradiction to observation. First, all of the electrons in an atom would finally end up in the nucleus and atoms as we know them would cease to exist. Second, there could be no bright-line emission spectra because as the electrons spiraled into the nucleus they would emit energy continuously with a range of wavelengths and we would always get a continuous spectrum.

This was so serious a strike against the Rutherford model of the atom that nobody at that time gave it any chance of survival. But great things were happening in physics and a new champion of the planetary model was even then laying the foundation for an atomic theory that was to be the forerunner of our present theory of the atom.

If one were to rank the importance of the physicists who have lived in the 20th century, Albert Einstein would certainly be ranked first, but only slightly above Niels Bohr, who was in large part responsible for both our modern conception of the atom and the movement from the causal orientation of classical physics to the statistical basis of quantum mechanics. Unlike Einstein, who preferred to work by himself, Bohr gathered around him a devoted group of leading physicists and engaged in many discussions that enhanced and focused their theoretical researches. Bohr enjoyed the company of his fellow scientists and inspired them to their best efforts by his imaginative suggestions and his penetrating but thoughtful criticisms.

Niels Bohr was born on October 7, 1885, in Copenhagen. Like many of his fellow scientists, he was the son of a university professor. Although his father taught physiology at the University of Copenhagen, he counted among his friends scholars in all areas of the sciences and the humanities. The Bohr home received a steady stream of distinguished visitors, many of them his father's colleagues from the university. One of Niels's fondest memories was listening to the weekly conversations that took place in the parlor between his father and whoever happened to be visiting that evening. The topics ranged from politics and science to the arts and religion. In a way, these conversations provided Niels with a detailed yet scattered knowledge about the world that increased his thirst for knowledge and indirectly helped to prepare him for the rigors of his formal education.

Niels was an outstanding high school student who excelled both in his classes and on the playing field. He was fond of all sports but he particularly enjoyed soccer and sailing. Although he was most often at home with his schoolbooks, he did not retreat into a self-imposed isolation of academic endeavor. He often took long walks or bicycled at breakneck speed down the narrow streets of Copenhagen.

Bohr began his undergraduate education in 1903 at the University of Copenhagen and was an outstanding student. He was particularly adept in the physical sciences and mathematics but also enjoyed literature, music, and poetry. Despite his humanistic leanings, however, he never doubted that he would become a scientist. His graduate research dealt with the electron theory of metals; after he received his doctorate in 1911 he left Denmark to work for a short time with J. J. Thomson at Cambridge followed by a 2-year stint with Ernest Rutherford at Manchester. Bohr found Rutherford to be the ideal teacher; Rutherford provided guidance but encouraged his students to develop their own ideas. By the time Bohr left Manchester in 1913 to return to Denmark, he had finished the first of his papers that revolutionized atomic physics and led to what is now known as the Bohr theory of the hydrogen atom, which pictured the electron as revolving around a proton nucleus (this work preceded the discovery of the neutron by James Chadwick in 1932). Unlike classical physicists, however, Bohr argued that the electron could only radiate energy in discrete quantities called quanta (an amount equal to Planck's constant of action multiplied by the frequency of the emitted radiation). According to Bohr's theory, an electron emits a quantum of energy in dropping to a lower energy level (orbital shell) or it absorbs a quantum to move to a higher energy level.

Although Bohr was convinced of the value of his work, he found that few of his fellow physicists were interested in hearing about his ideas. At this time Bohr, who had married the year before, was lecturing at the University of Copenhagen and trying to make ends meet on his meager salary. Feeling that the time was not right to pursue his work in Denmark, Bohr returned to Manchester for two more years where he worked as a reader in mathematical physics while continuing his theoretical research. During this period abroad news of his work began to circulate more widely in scientific circles.

Bohr returned to Denmark in 1916 at the height of the Great War to accept a professorship in theoretical physics that had been created for him at Copenhagen. Although his office was cramped and he did not have an assistant to help him with his work or correspondence, Bohr pursued his work without distraction and pressed his suggestion that an institute for theoretical physics be created to coordinate Danish research programs and attract foreign scientists. Bohr realized that Denmark's small size precluded it from competing with the major industrialized states in all areas of scientific research but he believed that if the Danish government concentrated its efforts on selected areas of science, it could develop internationally renowned scientific research facilities.

Bohr's dream of a world-class institution was realized in 1921 when the Institute for Theoretical Physics was dedicated. Under Bohr's leadership, the institute quickly became one of the centers of theoretical research in Europe. It attracted both outstanding professional physicists and the most brilliant graduate students. Much of the prestige of the institute was due to the increasingly high regard with which the scientific community viewed Bohr's theory of atomic structure. By the time his institute was founded, Bohr was recognized as one of the most important physicists in the world. His status was confirmed when he received the Nobel Prize in physics in 1922 for his work in atomic theory, the year after Einstein was awarded the same for his work on the photoelectric effect.

Bohr's receipt of the Nobel Prize helped make his name a household word in Denmark and much of Europe. As Denmark's most celebrated and honored scientist, he was revered by both scholars and the general public. Much of the country's affection for Bohr undoubtedly stemmed from Bohr's evident concern for his country and countrymen as well as his tireless efforts to boost the Danish scientific establishment. Bohr seemed to embody the best qualities of the Danish people and was well aware that he—not the Danish prime minister, for example—was the symbol for many persons of the modern Dane.

During the 1920s and 1930s, Bohr received numerous honors and awards. Because of his great skills as a lecturer, he was often asked to give speeches. As with his scientific papers, he would polish the drafts of his popular and technical lectures repeatedly until he felt that they could no longer be improved even though this perfectionism often left him reworking phrases until he walked up to the podium. Bohr's lectures took him to many foreign countries. He made several trips to the United States and lectured at the University of Chicago and the California Institute of Technology. During this time Bohr began to probe the philosophical features of modern physics and developed his principle of complementarity (also known as the Copenhagen interpretation of quantum theory), which he propounded with unflagging energy until it was accepted by most of the leading physicists. Much to Bohr's regret, however, he was never able to persuade Einstein of the merits of his argument because Einstein was convinced that the universal law of causality must be maintained even if it meant denying the worth of quantum mechanics, a subject to which Einstein himself had made substantial contributions. Their differing opinions did spark a series of vigorous and lively debates in which Einstein offered several brilliant quantum-mechanical paradoxes and Bohr resolved them, sometimes using Einstein's theory of relativity or his principle of invariance. Their debates helped to iron out some of the details of quantum mechanics and entrench it more firmly than ever in the firmament of modern physics.

Bohr's researches during the 1930s were geared toward applying quantum mechanics to the electromagnetic field and developing his concepts of the ''compound'' nucleus and the liquid-drop model of the nucleus to understand better the dynamics of collisions between atomic nuclei and nuclear reactions. His interest in nuclear physics was prompted in part by his desire to determine how new types of nuclei are formed during these collisions.

Although Bohr continued his scientific work in Copenhagen, he became increasingly worried about the political situation in Europe after Hitler became the ruler of Germany in 1933. As the true extent of Hitler's ambitions and the depravity of his tactics became clear, Bohr realized that it was probably just a matter of time before the Continent would once again be drawn into a catastrophic war. Bohr opened up his institute to scientists fleeing Germany and continued his admirable efforts on behalf of these displaced men and women until Germany invaded Denmark in 1940. Bohr was in Copenhagen when the Nazis occupied the country, but he refused to collaborate with the new regime in any way. Soon thereafter, he escaped to Sweden and then was flown to England and later the United States, where he spent the duration of the war working on the Manhattan Project at Los Alamos.

After the war, Bohr returned to Denmark and began to advocate the establishment of an international agency to regulate the uses of nuclear energy. Unlike many politicians and military leaders, Bohr saw that the atomic bomb monopoly held by the United States in 1945 would not last very long because the knowledge needed to build the bomb could not be forever kept secret by a single power. His predictions were confirmed 4 years later when the Soviets exploded their own atomic bomb.

Bohr's pleas for the internationalization of atomic energy fell on deaf ears, in part because neither the United States nor the Soviet Union was particularly interested in ceding control of atomic energy and weapons production to a third party. Undaunted, Bohr continued to speak out on such issues as nuclear proliferation and nuclear war as well as the need for greater scientific cooperation among nations until his death on November 18, 1962.

In 1913 Niels Bohr salvaged the Rutherford planetary model of the atom by introducing a few simple but quite revolutionary assumptions about the way in which the electrons behave while they are inside an atom. These assumptions are the following:

1. An electron does not radiate as long as it is moving in an elliptical orbit around the nucleus. This assumption is diametrically opposed to the Maxwellian theory.
2. Only a discrete set of elliptical orbits is permissible.
3. An electron emits or absorbs energy only while it is passing from one permitted orbit to another. If an electron jumps from a more distant orbit to one closer to the nucleus, it emits just one photon whose energy exactly equals the difference in the electron's energy in the two orbits; when it jumps from an orbit closer to the nucleus to a more distant orbit, it does so by absorbing one photon of just the right energy.

Using these assumptions, Bohr derived most of the properties of the spectrum of hydrogen and computed the wavelengths of the emission lines of this spectrum. Before we see just what his result was for hydrogen, we must clarify some of the aspects of the three assumptions above. It is important, first, to keep in mind that these assumptions were introduced quite arbitrarily with no reason other than that they led to correct results, although in arriving at these assumptions Bohr was greatly influenced by the work of Planck and Einstein that we have already discussed.

In his first assumption, Bohr introduced the concept of stationary states for electrons inside atoms. As long as an electron moves in one of these states (orbits), it need not obey Maxwell's equations; it can be accelerated without radiating. In his second assumption, Bohr imposed the idea of discreteness in the geometrical arrangement of the electron's orbits. This is necessary to obtain discrete lines in the spectrum of an atom. The third assumption, dealing with the way an atom absorbs or emits radiation, introduces the concept of the photon directly into atomic theory. Only while an electron goes from one discrete orbit to another does it emit or absorb radiation, and only one photon at a time, of just the right energy, can be emitted or absorbed.

To use these assumptions, Bohr had to devise a set of rules for determining the permissible orbits. These rules are rather complicated for elliptical orbits; but for circular orbits only one rule that gives the circumference, and therefore the radius, of each permissible orbit is needed. The rule is the following for the smallest orbit in the hydrogen atom: take Planck's constant h (the number 6.624 $\times$ 10^{-27}), square it, and divide by 4; now divide by the square of π (pi, the number that gives the number of times the diameter of a circle is contained in its circumference); and finally divide this result by the square of the charge on the electron (the charge on the electron is 4.8 $\times$ 10^{-10} units). The number we get is about one one-hundred-millionth of a centimeter and is referred to as the Bohr radius or the radius of the first Bohr orbit.

This orbit, which is closest to the nucleus of the atom, is called the ground state. The radius of the next higher orbit is obtained from this ground-state radius by mulitplying by 4, and those following are found by multiplying by the numbers 9, 16, 25, 36, etc. We see then that the radii of all of the orbits can be found by multiplying a fundamental constant of nature by the squares of the numbers 1, 2, 3, etc. These numbers define the circular orbits completely and are referred to as the *principal quantum numbers* of the electron in the atom. It is customary to assign the letters K, L, M, etc. to these orbits.

For elliptical orbits, the situation is more complex because an ellipse is not completely determined by its size alone, the way a circle is. For a given position of an ellipse in space, three numbers are needed to describe it: the size of its major axis (the largest diameter in the ellipse), which is given by the same principal quantum number as above and is a measure of the energy of the electron in its motion around the ellipse; the shape of the ellipse (this is equivalent to the eccentricity of the orbit of a planet), which determines the angular momentum of the electron in its orbit and is given by a positive number called the *azimuthal* or *orbital* quantum number; and the orientation of the plane of the ellipse in space, which is given by a set of negative and positive integers called the magnetic quantum numbers. These numbers are a measure of how the energy of an electron changes when the atom is placed in a magnetic field. When this happens, each line in the spectrum of an atom splits into a number of lines, known as the Zeeman lines.

These sets of quantum numbers cannot be arbitrarily chosen but must stand in certain simple relationships to each other. Thus, if n, l, and m represent the three sets of quantum numbers described above, then n, the principal quantum number, can take on all integer values starting with 1; but for a given value of n (an ellipse of a given size) the orbital quantum numbers can only have integer values less than n. The magnetic quantum number m may then take on any positive or negative integer value between and including $-l$ and $+l$. These rules are sufficient to determine the essential features of the orbits of the electrons even in complex atoms, and to calculate the wavelength of the atom's spectral lines.

As an example of the way these quantum numbers are to be assigned to electrons inside an atom, we consider the first and second permissible Bohr orbits in the hydrogen atom. The first, or K orbit, in which the electron has the smallest amount of energy, has for its principal quantum number the value *one*. There can therefore be only one value for the orbital quantum number, namely, *zero*; and from this it follows that there can be only the value *zero* for the magnetic quantum number.

In the second, or L orbit, the electron's principal quantum number is 2 so that its orbital quantum number may either be *0* or *1*. If the orbital quantum number has the value *zero*, the magnetic quantum number of the electron must also be *zero*; but if the value of the orbital quantum number is *1*, the magnetic quantum number may have any one of the three values -1, 0, or $+1$. We see, then, that although there is only one possible set of quantum numbers for the electron in the lowest orbit, there are four possible sets of quantum numbers for the electron in the second orbit, and larger and larger sets as we go to still higher orbits. Thus, each Bohr orbit really corresponds to substates that increase in number as we go to higher states of energy inside the atom.

But how does one go from these rules for determining the orbits of the electrons in the atom to the wavelengths of the lines that these atoms may emit? To answer this we go back to Bohr's third assumption about the way an electron emits or absorbs energy. Under ordinary conditions the electrons in an atom try to occupy the lowest permissible orbits because these are the orbits in which their energy is the smallest. In this respect, electrons behave like water running downhill or rocks rolling down a hillside. We shall see a bit later, however, that there is a rule that allows only two electrons to occupy the ground state (the smallest orbit) of an atom even though all the electrons would like to do so.

To see how Bohr used his third assumption, we consider the hydrogen atom. In this atom the nucleus is a single proton, and moving around it in the lowest orbit is a single electron. As long as nothing happens to disturb this atom, the electron remains in the ground state, and the atom emits no radiation. However, the electron can be forced out of this ground state in two ways: it may be shaken out as the result of a collision between the atom and some other particle (e.g., an electron, atom), or it may absorb some passing photon (quantum of energy) and skip off. In other words, it can get out of this lowest orbit only by acquiring some additional energy.

When this happens, the electron goes to some other higher orbit, provided it has just the right amount of energy. This discreteness is one of the interesting features of this picture of the atom. The electron becomes selective and absorbs only that amount of energy that brings it to a definite permitted orbit. If a photon passes by and has either too much or too little energy to take the electron into one of the higher stationary states (another name for a permissible orbit), the electron does not absorb that photon. Suppose now that the right kind of photon or collision comes along to force the electron into a higher orbit. How long will it stay there? In general, it will spend about a hundred millionth of a second there, during which time it will revolve around the proton about 10,000,000 times, and then jump down to the ground state again by emitting a quantum of energy (a photon). If initially it jumps into the third or fourth or fifth or some higher orbit, it may cascade down to the ground state, emitting one photon each time it changes from a higher to a lower orbit.

If the atom receives a very violent jolt or if the photon absorbed by the electron is energetic enough, the electron is torn completely away from the proton and the atom is then ionized. The energy necessary for this is called the ionization energy of the atom for the electron in that orbit. The ionization energy differs from atom to atom, and the higher this ionization energy is, the more violent the collision must be (i.e., the higher the temperature of the gas must be) for ionization to occur.

Now how do we determine the wavelengths or frequencies of the lines emitted by an excited hydrogen atom (an atom with its electron in one of the higher energy levels, i.e., in a high orbit)? Before we can begin, we must know the energy of the electron when it is in any one of the permitted orbits. This energy is taken as negative because work has to be done to tear the electron out of any such orbit. If W is the amount of energy needed to remove an electron from the atom when the electron is in the lowest orbit, the energy of the electron when it is in any allowed orbit is obtained by dividing W by the square of the principal quantum number of the given orbit and taking the negative of this quantity. Thus, the energies of the electron in the permitted orbits, starting from the lowest, are $-W$, $-W/4$, $-W/9$, etc.

If an electron is in any orbit and jumps down to a lower orbit, the frequency of the emitted photon is obtained as follows: subtract the energy of the electron when it is in the lower orbit from that when it is in the higher orbit and divide this result by the Planck constant. If we divide this quantity into the speed of light, we obtain the wavelength of the radiation.

This simple procedure enabled Bohr to account for the various lines that are observed in the spectrum of the hydrogen atom. These lines are arranged in different series depending on the lower orbit to which the electron jumps down. If the electron always jumps from the higher orbits to the very lowest orbit (the K orbit), the series of lines is called the Lyman series; most of the lines in this series lie in the ultraviolet region of the spectrum. If the lowest orbit is the L orbit (the second orbit), the lines form the well-known Balmer series in the visible part of the spectrum.

Before we can apply the ideas that we have outlined above to the analysis of stellar atmospheres, we must see how well Bohr's theory accounts for the properties of the heavier atoms. Although it became clear soon after Bohr announced his theory that it suffered from many defects in accounting for the fine details of atomic spectra, it was surprisingly successful in explaining the gross behavior of even complex atoms. Even though most physicists were convinced by 1920 that the Bohr theory would have to be replaced by a new theory that would depart completely from the classical mechanics and the classical ideas of space and time, the Bohr theory was an excellent tool for studying and understanding the structure of atoms. Before we can discuss this point, we must introduce certain additional simple concepts about atoms.

With the discovery of electrons and protons and the introduction of the Rutherford–Bohr model of the atom, it was possible to account for the chemical properties of the different elements that are found in nature as well as for their gross spectral features. In particular, one could explain the periodicity in the chemical properties of the elements. During the last quarter of the 19th century, the Russian chemist Dmitri Mendeleev discovered that families of elements exist which have the same or very similar chemical properties, and that elements in a given family are related to each other in a simple way. To understand this we note that, according to the Bohr theory, the chemical behavior of an atom is determined by the electrons in the outermost orbit of an atom; if the electron pattern in the outer orbits is the same for different elements, the chemical properties are similar. This can be made clear by introducing the concept of the *atomic number* of an atom, which is the number of electrons in the atom revolving around the nucleus. This number plays an important role in the periodicity of the chemical elements.

We can now describe Mendeleev's periodic chart of the chemical elements as follows: Arrange all the elements, starting with hydrogen, according to their atomic numbers, i.e., according to the number of electrons circulating around the nucleus. We then find that the chemical properties of the elements repeat themselves in definite cycles. Thus, the elements helium, neon, and argon with atomic numbers 2, 10, and 18, respectively, are all inert gases. We see that the atomic number cycle for these elements is 8. Krypton, xenon, and radon, with atomic numbers 36, 54, and 86, also belong to this family of inert gases. In this entire group of gases the atomic numbers of the first three differ by 8 as we go from one to the next one; the atomic numbers of the third, fourth, and fifth differ by 18; and the atomic number of the last one differs by 32 from the one preceding it. If we note that helium, with atomic number 2, is the first member of this group, we are struck by the fact that the numbers 2, 8, 18, 32, etc. play an important role in determining the cyclical properties of the chemical elements. This is found to be true not only for the inert gases but also for the alkali metals such as sodium and potassium, and the other families of elements as well.

From what we have just said, it is clear that the electrons in the different atoms must arrange themselves in groups that are determined by the sequence of numbers 2, 8, 18, etc., which can be obtained by taking the squares of the integers 1, 2, 3, etc., and doubling them. In other words, the arrangement of the electrons in the various atoms is determined by the numbers $2n^2$ where n is the principal quantum number of the electrons. Before we discuss this further, we define *atomic weight*.

If we assign the number 16 to the atomic weight of the oxygen atom, then the atomic weight of any atom is the number on this scale that tells how much the atom weighs relative to oxygen. On this scale the atomic weight of hydrogen is 1.008 and that of helium is 4. The atomic weight is very nearly twice the atomic number except for ordinary hydrogen, but increases faster than twice the atomic number as we go to the heavier atoms. Since the atomic weight is larger than the atomic number, and a single proton has very nearly unit atomic weight, it follows that there must be some heavy particles in addition to the protons in the nucleus. As an example of this, we note that since the helium nucleus has an atomic weight of 4 and an atomic number of 2, its nucleus must contain two protons and two other electrically neutral particles that have about the same mass as the proton.

For a long time the nature of these other nuclear particles was not known, but it was known that they had to be about as massive as protons and capable of changing into protons by emitting electrons. A particle of this sort was discovered in 1933 by Frédéric Joliot and Irène Curie in France and by James Chadwick in England. This particle is the *neutron*, and we now know that all nuclei consist of neutrons and protons only. There are no electrons in the nucleus even though some nuclei emit electrons spontaneously (radioactive elements). This emission occurs when a neutron within a nucleus changes into a proton; at that moment the electron is born and escapes from the nucleus. These electrons are called *beta rays*. Taking the neutron into account, we now define the atomic weight as the sum of the number of protons and neutrons in the nucleus.

As we go to the heavier atoms, we find that the external electrons arrange themselves so that only 2 electrons move in the lowest orbit (the K orbit of principal quantum number *1*), no more than 8 electrons in the second orbit, 18 electrons in the third orbit, and so on until all the electrons are accounted for. Just why the electrons fall into this pattern will be understood after we have discussed one more feature of the electron.

We have already noted that although the Bohr theory of the atom was quite successful in explaining the broad features of atomic spectra, it failed to agree with, or explain the refined details of the spectra. Each of the lines in the spectra of atoms like sodium and potassium is not single but consists of two lines close together. The Bohr theory cannot explain such doublets, so it was necessary to introduce a new assumption about the electron. This was done in 1925 by Goudsmit and Uhlenbeck, two Dutch physicists, who assigned a spin to the electron. They assumed that each electron is a tiny top spinning about an axis, the way Earth does, with a total amount of rotational motion (angular momentum) equal numerically to one-half of the Planck constant divided by 2π. In addition, they imposed the restriction on the spin that the axis of spin can either line up parallel to the axis of revolution of the electron around its orbit, or antiparallel to this direction: The electron can spin in the same sense in which it is revolving or in the opposite sense. Since a spinning electron is equivalent to a small magnet, this means that the energy of the electron when its spin is lined up parallel to its revolution is different from its energy when its spin is antiparallel, so that we get two different energy levels very close together.

To specify that an electron may spin in two different directions within an atom, an additional quantum number S, called the *spin quantum number*, was introduced with the restriction imposed on it that it can have only two values, $+1$ or -1 ($+1$ when it is spinning in the same sense as it is revolving and -1 other wise). This number enables physicists to explain many other features of complex atomic spectra. We can also explain now why there cannot be more than a certain number of electrons in each orbit. The basic principle upon which this explanation rests was first stated by the physicist W. Pauli.

We have noted that no more than two electrons can move in the first orbit, eight in the second, and so on. The reason for this was expressed by Pauli in 1927 in the form of a law of nature that has since come to be known as the Pauli exclusion principle. This principle states that no two electrons within the same atom can have the same set of four quantum numbers. In other words, if two electrons are to find orbits in the same atom, the numerical values of at least one of the quantum numbers describing the orbits must be different for both of them. Since there are four such quantum numbers for each electron, governed by certain rules, we see that only certain combinations of these numbers satisfy the exclusion principle. This limits the number of electrons that may move in an orbit having a definite principal quantum number. A simple calculation shows that the maximum number of electrons in such an orbit is twice the square of the principal quantum number. This agrees with the observed arrangement of the electrons in any atom and accounts for the periodic table and the periodicity of the chemical properties of the elements.

In spite of all the success of the Bohr model, it became more and more obvious that it would soon have to give way to an entirely new concept of the nature of matter. As early as 1920, most physicists were aware of a new wind that was blowing across the plains of knowledge carrying the germs of new, revolutionary ideas. We can give only a brief discussion here of these discoveries concerning the nature of matter.

With Planck's and Einstein's discovery of the corpuscular nature of radiation, it appeared that waves were to be forever banished from the realm of physics. But even at that moment a theory was being born that was to bring the wave concept back to its former prominence. In the theory of relativity the adherents of the wave theory found a new source of inspiration that was finally to impose the wave concept so firmly on our view of nature that the very idea of a particle was to lose its precise meaning.

In the early 1920s the French physicist Louis de Broglie was greatly struck by the Einstein mass–energy relationship, and he saw in it a basis for a wave theory of matter. He reasoned that if mass is equivalent to energy, it must possess some of the properties of a photon (which is pure energy). By equating the Einstein expression for the energy of mass to the Planck expression for the energy of a photon, he deduced a frequency for a mass particle. From this he showed that a particle in motion must have associated with it a wavelength equal to Planck's constant divided by the product of the mass of the particle and its velocity, i.e., its momentum.

Nobody paid too much attention to this theory until two things happened, one an experimental discovery and the other a theoretical development. Two American physicists, Davisson and Germer, in an investigation of the way electrons are reflected by metallic surfaces, discovered that an electron interferes with itself in just the way a wave does, and when they measured the wavelength that each electron seemed to have, they obtained the de Broglie result.

At about the same time, the Austrian physicist Erwin Schrödinger decided, after reading de Broglie's paper, that if electrons behave like waves, there should be a wave equation describing this behavior. He set for himself the task of doing for electrons what Maxwell had done for light (photons). In a sense this was reversing Planck's procedure. Planck started from a wave picture of energy (the Maxwell wave equations of radiation) and constructed a particle theory (the quantum theory of radiation), whereas Schrödinger, with the help of de Broglie, started from a particle picture of matter (Newtonian mechanics) and derived a wave theory of matter (the Schrödinger wave equation). This wave equation for an electron inside an atom is one of the most remarkable discoveries in physics.

This one equation eliminated almost all the troubles that had beset the Bohr model of the atom. Not only did it describe the wave properties of an electron according to de Broglie's concept, but it also gave a correct description of the energy levels of an electron inside an atom (the orbits introduced by Bohr). It was no longer necessary to introduce any assumptions about orbits and the radiation of an electron. All of this followed automatically from the Schrödinger equation. Moreover, three integers appear in the equation automatically, which are precisely the three quantum numbers that had been introduced arbitrarily in the Bohr theory.

Erwin Schrödinger was born on August 12, 1887, in Vienna. His father had studied chemistry but passed up the opportunity to become a professional chemist, finding Italian painting and botany more to his liking. Erwin's mother was the daughter of a professor of chemistry at the Vienna Institute of Technology. Both of Erwin's parents were imbued with the literature and music of 19th-century Austria; they tried to instill in their son an appreciation for the grandeur and tradition of classical European culture. Erwin's father's interest in botany led him to write several papers on plant phylogeny that contained valuable insights on the subject. This interest in the organic world was passed down to Erwin who was fascinated with the philosophical and scientific questions about the nature of life. In fact, one of his consuming ambitions during his professional career was to use the basic principles of physics to explain how life begins and evolves.

Erwin's early years were both happy and exciting because his parents spent many afternoons with him visiting the museums and historical sights in Vienna. He was fortunate enough to attend the Gymnasium in Vienna as a young boy where he was schooled in the classics and the sciences. He also developed a fondness for literature and enjoyed memorizing long passages of poetry. Although his restless intellect and scholastic aptitude were apparent at an early age, Erwin was not pressured by his parents to succeed in his subjects; his excellent academic performance stemmed from his own perfectionism. Despite being one of the most talented students at the Gymnasium, however, Erwin despised the structured curriculum and the requirement that selected assignments be memorized for no particular reason.

When Schrödinger entered the University of Vienna in 1906, he had decided to pursue a scientific career. Although Max Planck had announced his quantum theory in 1900 [which stated that radiation can only be emitted in bundles (quanta) described as multiples of some fundamental quantity now known as the Planck constant of action] and Albert Einstein had published his special theory of relativity in 1905 (which had significantly amended the Newtonian concepts of space and time by tying them together into a four-dimensional space-time continuum), much of the European scientific community had yet to feel the impact of these two intellectual revolutions. Most of the physics students in the first few years of the 20th century were trained almost exclusively in the classical physics that had begun with Galileo and Newton and culminated with the work of Faraday and Maxwell. Their instructors did not pay much attention at first to the new ideas of Planck and Einstein; only after the start of the second decade of the new century were their ideas gradually accepted by the majority of physicists.

Schrödinger's own education focused on the works of the classical physicists; his most influential teacher was Fritz Hasenöhrl, a disciple of Ludwig Boltzmann. Hasenöhrl encouraged Schrödinger's interest in the physics of continuous media, which provided him with the theoretical training he later needed to formulate his wave equation of the electron. In 1910 Schrödinger graduated from the university with a degree in physics and debated whether to pursue an academic career or work for a company. While Schrödinger weighed the positive and negative aspects of his career options, he worked as a laboratory assistant in Vienna but—by his own admission—he was not a very adept experimentalist. Schrödinger later considered himself fortunate to have pursued a career as a theoretical physicist since he did not believe that he knew how to prepare experiments properly. He did, however, make up his mind to follow his heart and become a physics teacher. Less than a year after earning his degree, Schrödinger began his climb up the academic ladder.

The eruption of the Great War in 1914 forced Schrödinger, and other young scientists in Europe, to put their careers aside temporarily and offer their services to their governments. Schrödinger was commissioned as an artillery officer in the Austrian army and spent the next 4 years on the bloody battlefields of central Europe.

After the war and the dismemberment of the Austrio-Hungarian Empire by the victorious allies, Schrödinger returned to Vienna for a brief time as an assistant to Wilhelm Wien, a distinguished physicist who had been among the earliest scientists to recognize the revolutionary nature of Einstein's theory of relativity when it was first published. Schrödinger then accepted a position as an assistant professor of physics at the University of Stuttgart where he taught both modern and classical physics, the works of Einstein and Planck having become almost universally accepted by that time in the scientific community. His theoretical work soon attracted the attention of several physicists at Breslau; they persuaded the administration to offer Schrödinger an appointment as a professor of physics. Schrödinger moved to Breslau but he was there only a short time when the University of Zurich, seeking to fill the vacancy left by the departure of Max von Laue, offered him a position as professor of physics. Although Schrödinger had already made two moves in as many years, he eagerly accepted the opportunity and moved to Switzerland. He remained in Zurich for 6 years and during that time he developed his wave equation of the electron. Aside from his work on wave mechanics, he also published papers on thermodynamics, atomic spectra, and statistical mechanics.

Schrödinger's wave equation was one of the most remarkable creations of atomic physics because it united into a single mathematical framework Louis de Broglie's wave theory of the electron and William Hamilton's expression for Newtonian dynamics. Although his wave equation was a sort of hybrid creation, it underscored the debt that modern physics owes to its classical predecessor and showed that to understand thoroughly the phenomena of the physical world, one has to be familiar with both classical and modern physics.

When Schrödinger published his wave equation, he did not reveal how he had arrived at it so that many physicists believed at first that Schrödinger had developed his mathematical formulation from scratch. Although he later published several papers revealing the exact method that had led him to the equation, the fact that he had borrowed heavily from the works of de Broglie and Hamilton did not lessen the regard with which other physicists held his work but merely showed that his own results were firmly anchored on the widely accepted works of two other important physicists.

Although Schrödinger's wave equation provided a much simpler method for understanding the behavior of the electron than the imposing mathematical techniques that had been developed by Max Born and his collaborators, Schrödinger's interpretation of the wave function differed from those of most other physicists because he did not believe that the electron has both corpuscular and undulatory characteristics (the so-called wave–particle duality). Schrödinger instead argued that the electron can be understood by focusing exclusively on its undulatory properties, a view that differed significantly from the probabilistic interpretation of quantum mechanics offered by Born.

Schrödinger's wave equation established him as one of the leading physicists of his era and led to his being awarded the Nobel Prize for physics in 1933. His work also attracted the attention of the University of Berlin in the mid-1920s; he was the unanimous choice to replace Max Planck when the latter retired in 1927. Although Schrödinger enjoyed his pleasant life in Switzerland, he saw that the University of Berlin had perhaps the greatest collection of theoretical physicists in the world at that time including Albert Einstein and Max von Laue. The opportunity to engage in frequent and informal discussions with such distinguished colleagues was too strong a lure for Schrödinger; he joined the faculty in Berlin in 1927 and remained there for the next 6 years until Hitler rose to power in 1933. Although Schrödinger was not Jewish and could have remained safely in Berlin, he was disgusted by the racial policies of the Nazi government and felt he had no other choice but to leave Germany. After spending 2 years at Oxford, Schrödinger returned to Austria in 1936 to accept a position at the University of Graz. However, his stay in Austria was short-lived as its annexation by Germany in 1938 forced him to flee across the Alps into Italy. Schrödinger made his way to the United States and then became the director of the school of theoretical physics at the Institute of Advanced Studies in Dublin, Ireland, where he remained until his retirement in 1955.

Although Schrödinger had already done his great work in quantum mechanics, he produced many important papers while in Dublin on subjects ranging from relativity and electromagnetism to gravitation and the physics of life. During this time he wrote a small book called *What Is Life?*, in which he argues that quantum principles can be used to explain biological mutations. Schrödinger's book was extremely well received and went through many printings. Although subsequent developments in fields such as molecular biology have rendered untenable some of his arguments, the book remains a classic. Several physicists who subsequently decided to pursue careers as molecular biologists credit Schrödinger's book with having influenced them to change careers.

Schrödinger left Dublin in 1955 and returned to Vienna to spend his final years. He received many honors and awards and was widely regarded as one of Austria's foremost citizens until his death in 1961. Despite the public acclaim, however, he remained a simple man who preferred walks in the country to testimonial dinners. Like Einstein, Schrödinger's brilliance was often intuitive and did not lend itself to collaboration with other physicists. Despite his reluctance to work with others, however, it is difficult to exaggerate the importance of his wave equation for the development of quantum mechanics.

According to the Schrödinger theory, one cannot speak of precise orbits for the electron in an atom because the electron is not a well-defined particle. It behaves like a wave as well as a particle, and a wave cannot move in a definite orbit. The Schrödinger equation shows that the Bohr orbits are really the most probable paths of electrons in atoms, but there is a definite finite probability for finding the electron anywhere within the atom. One can speak of orbits only in a probability sense. Actually, the Schrödinger wave equation is an equation for the probability of finding the electron at any point in space at a given instant of time if its energy is known.

While Schrödinger was busily engaged in setting up his wave equation, the physicist Werner Heisenberg was led to a similar picture of a particle but from an entirely different point of view. In carrying out some calculations on the way radiation behaves when it passes through matter, Heisenberg and the Dutch physicist Hendrik Kramers discovered that the only way they could get a result from the Bohr theory that agreed with the observations was by replacing the quantity in their equations that gives the position of the electron at any instant by an array of many quantities, called a matrix, as though there were no precise meaning to be attached to the position of an electron. They had to do the same thing with the momentum of the electron (the momentum of a particle is its mass multiplied by its velocity). In other words, according to this point of view, one cannot attach a precise meaning to the position and motion of a particle.

Heisenberg interpreted this very important result in the following way: he pointed out that classical Newtonian mechanics cannot be applied to the motion of an electron because Newtonian mechanics is based on the notion that one can measure the position and momentum (i.e., the motion) of a particle simultaneously to any desired accuracy. This is essential if Newtonian mechanics is to hold. But what is the situation if radiation exists in the form of indivisible photons? Heisenberg showed that, as a result of this quantum thoery of light, it is impossible to know simultaneously both the position and the momentum of an electron with infinite accuracy.

Suppose that, having determined the momentum of the electron, we look for it by shining light on it to determine its position. To do this with great accuracy, we must use light of very small wavelength (gamma radiation). But this means that we are bombarding the electron with very energetic photons (small wavelength means large frequency and, hence, quanta of large energy). If one of these photons hits the electron and bounces back from it into our eye, thus enabling us to see the electron and determine where it is, we no longer know its momentum because the photon, in colliding with the electron, changes the latter's momentum.

This result was expressed by Heisenberg in the form of his famous uncertainty principle, which can be stated as follows: it is impossible to measure both the position and the momentum of a particle (an electron) simultaneously with infinite accuracy. This means that the greater is the accuracy with which the position is known, the smaller is the accuracy with which we know the momentum and vice versa. It can be expressed arithmetically in the following way. If the error in our knowledge of the momentum of a particle is multiplied by the error in our knowledge of its position, then this product can never be smaller than Planck's constant divided by the number 2π.

Although, at first sight, it seemed that the Schrödinger and the Heisenberg theories were quite unrelated to each other, except that they both gave the same correct results for the behavior of an electron in an atom, it was shown by Schrödinger, Dirac, and Jordan that both are different aspects of a more general theory that is now called quantum mechanics. Dirac did one more piece of brilliant work that finally completed the theory and expressed it in the form in which we now know it and use it.

The Schrödinger equation introduces the three quantum numbers of the electron's orbit quite naturally, but it says nothing about the spin of the electron. Dirac pointed out that this is so because the Schrödinger equation is not in tune with the theory of relativity, i.e., it is not an invariant wave equation of the electron. By means of some ingenious mathematics, Dirac derived a wave equation for the electron that obeys the theory of relativity, and this relativistic wave equation correctly predicts the spin of the electron and its correct value as introduced arbitrarily by Goudsmit and Uhlenbeck.

One more thing is worth noting about the Dirac theory of the electron: it predicted the existence of a positive electron (a particle with all the properties of the electron but with a positive electrical charge) with the property that if it collides with an electron the two particles annihilate each other in a burst of radiation. In 1933, 5 years after Dirac announced his theory, these particles (positrons) were discovered in cosmic rays by Carl Anderson.

As soon as positrons were discovered in nature, physicists created them in their laboratories in two ways. On the one hand, it follows from the Dirac theory that it should be possible to create electron–positron pairs by allowing a stream of very energetic photons to flow past the nuclei of heavy atoms (the transformation of radiation into matter). To do this a source of very energetic photons was needed, and when such sources were developed, positrons were manufactured in laboratories in exact agreement with the Dirac theory. Positrons are also obtained when artificial radioactivity is induced in stable nuclei by bombarding them with very energetic protons or by subjecting them to the intense neutron beams in atomic piles. New unstable nuclei are thus created, which decay by emitting positrons.

Since a positron destroys an electron when the two meet, we may consider the former as an antielectron. Antiprotons (negatively charged protons) and antineutrons have also been discovered, and we now know that for every ordinary fundamental particle in nature an antiparticle that can destroy the ordinary particle must exist. In fact, there is good reason to suppose that somewhere in our universe there may be stars composed of antimatter having properties identical with the properties of ordinary stars.

There is a very interesting way of discussing antiparticles that reverses the usual manner of describing the way events unfold in time. It is customary to speak of time as always progressing from the past into the future, and the usual behavior of such particles as electrons, protons, and neutrons can be described completely in terms of this time pattern. Suppose, however, that we were to reverse the direction of the flow of time in the equations that describe the motions of these particles. We would then be discussing particles that are traveling from the future into the past. When we do this, the equations we obtain are precisely those that describe the antiparticles moving from the past into the future. In other words, we may consider the antiparticles either as antimatter moving according to the customary time pattern from the past into the future or as ordinary particles (electrons, protons, etc.) moving in the reverse time pattern from the future into the past.

Using the theory of the atom we have just outlined, we can interpret the fine details in the spectra of the stars, and from these details determine the chemical and physical structure of the stellar atmospheres. To see how the temperature determines the behavior of atoms in a star's atmosphere, we again trace the changes that take place in a mixture of elements similar to that in the atmosphere of a star like the sun.

If the mixture is at a very low temperature, we observe no spectrum at all, but with increasing temperature the material (initially a solid) begins to glow and we obtain a continuous spectrum. This is so because the atoms in a solid are so close together that their individual optical characteristics are lost in the gross effects (vibrations) of the entire mass. As the atoms vibrate, they interfere with each other because of their proximity. This interference results in a smearing out of the radiation into all frequencies. As the temperature rises, the material becomes gaseous and the spectra of the various molecules in the mixture become visible. These are the bands seen in the M-type stars. If the most abundant substance in this material is hydrogen, then the spectrum of atomic hydrogen begins to show itself even at fairly low temperatures. As the temperature increases still further, the collisions among the molecules become more and more violent, resulting in their dissociation into individual atoms.

Which atoms show themselves first in the spectrum of such a hot mixture of gases? The answer is, obviously, those in which the electrons are not too tightly bound; this is true of atoms like sodium, potassium, calcium, and some of the other metals. The spectra of the metallic atoms become stronger and stronger and the band spectra of the molecules begin to disappear, except for those molecules whose atoms are most strongly bound together, like the hydroxyl ion OH and titanium oxide TiO. But these molecules also disappear (they are ripped apart) as the temperature increases beyond 3000°K. With still increasing temperature the collisions among the atoms become so disruptive that the metallic atoms in the gas mixture (we may now consider this gas mixture as representing the star's atmosphere and the increasing temperature as resulting from the very hot radiation coming from its photosphere) can no longer hang on to the external, loosely bound electrons for any length of time. For most of the time at these temperatures, many of the metallic atoms are singly, doubly, or even triply ionized; we get the ionized spectra of these atoms.

If a great abundance of hydrogen is present, as there is in the atmospheres of the stars, many of the hydrogen atoms move fast enough, even at comparatively low temperatures, to have their electrons thrown from orbit to orbit during collisions, and the atomic hydrogen spectrum is visible. The intensity of this spectrum increases until reaching a maximum at about 10,000°K. By this time, most of the metals are highly ionized and their spectra consist of invisible lines in the deep ultraviolet. As the temperature increases above 10,000°K, large quantities of hydrogen become ionized so that we get no spectrum from it and the intensity of the hydrogen spectrum begins to diminish; very little is discernible that relates to the metals.

As the temperature continues to increase, the helium spectrum becomes more intense because the collisions between atoms and the photons, streaming through the atmosphere from the surface below, are energetic enough to force the two electrons in the K orbit to dance from orbit to orbit. This requires about twice as much energy as that needed to excite the hydrogen atom, so that this begins to occur with appreciable intensity at temperatures in the neighborhood of 20,000°K. At still higher temperatures one of the two electrons of the neutral helium atom is torn out so that the spectrum of ionized helium becomes more intense while the intensity of the neutral-helium spectrum diminishes.

Thus far, we have been discussing the emission spectrum of a gaseous mixture, but the same thing applies to the dark line absorption spectrum. Consider just how an absorption spectrum arises; we need a stream of continuous-spectrum radiation passing through a cool gaseous mixture. Let an atom of some sort, say hydrogen, be in the path of this radiation and let an observer be in the line of this radiation after it has passed the hydrogen atom. As we have seen, the electron in an atom does not respond to photons unless they have just the right amount of energy to bring the electron from its lowest orbit to some higher permitted orbit. But if the continuous radiation coming from below is too cool, it does not contain very many photons energetic enough to excite the hydrogen electrons, and the observer finds the spectrum of the radiation the same as it would be if there were no hydrogen atoms between him and the source of the continuous radiation.

If, however, the radiation passing through the hydrogen gas (the star's atmosphere) is quite hot, many photons have sufficient energy to force the electrons of the hydrogen atoms into the higher orbits. But we know that these electrons do not remain in the higher orbits but immediately jump down to the ground state again and reemit the photons that they absorbed. However, these photons are not reemitted in the same direction in which they were moving initially but in all directions. Thus, fewer photons of the sort that can excite hydrogen are moving toward the observer. The observer finds that the radiation is poorer in precisely those photons that the hydrogen atom emits when it is excited. This accounts for the absorption lines in the spectrum of a star, and shows us how the temperature of the radiation passing through the atmosphere (i.e., the effective temperature of the photosphere) determines the absorption lines that are observed.

If the temperature of the photosphere is high enough, the continuous radiation coming through the atmosphere of the star contains such energetic photons that they strip the optical-spectrum electrons completely away from some of the atoms and the optical spectra of these atoms do not appear in the spectrum of the star.

A few examples illustrate the points discussed in the previous paragraphs. If the temperature of the continuous radiation coming from the photosphere of the sun is about 5500°K, only one hydrogen atom in every quarter of a billion has its electron raised to the second level and is thereby capable of forming the visible hydrogen absorption lines. Only because very large numbers of hydrogen atoms are present in the solar atmosphere do we find a fairly strong hydrogen absorption spectrum in the sun.

As another example, we note that, under the same conditions as in the atmosphere of the sun, 97% of the magnesium is ionized and only 3% is neutral. In Sirius, with an effective temperature of about 10,000°K, all the titanium atoms are ionized, with 71% of them being doubly ionized (two electrons torn away) and 29% being singly ionized.

An analogy may be useful here to illustrate the manner in which an atom generates an absorption line by scattering photons. Picture a stream of footballs of all colors moving past a person who quickly counts the number of each color and tabulates this number on a graph. He finds a certain color distribution, depending on the way these footballs are emitted by the source. If another man standing on a rotating platform in the football stream between our observer and the source kicks every blue football that comes within reach of his foot, hardly any blue footballs reach the man who is counting because the kicker scatters the blue footballs in all directions. This kicking represents an absorption of blue footballs.

We have seen how the overall features of the spectrum of a

star are determined by the effective temperature, but the chemical composition of the atmosphere also plays an important role. No matter what the temperature may be, the spectral lines of any particular element will not be present unless that element is present in sufficient abundance. It is obvious that for a given set of physical conditions (e.g., temperature, pressure), the greater the abundance of a particular element is, the more intense are the absorption lines of that element. But the relationship between the intensity of an absorption line and the abundance of the element producing the line is not straightforward; it depends in a rather complicated manner on the shape of the line.

Although, theoretically, a line in the spectrum of an atom refers to radiation of just one wavelength (so-called monochromatic radiation), in practice there is no such thing; all the lines that are observed have a certain spread over a wavelength band. This is similar to what you experience when you tune in a radio set. Each station comes in sharpest and with greatest volume for a particular setting of the dial, but you can still detect the same station if your dial is set a little bit to the left or to the right of this optimum setting. This spread of a line is called broadening; it is due to many conditions. Some of the broadening arises from the way the electron jumps from one orbit to the other while it is emitting the line; this is called natural broadening. Other effects that contribute to the broadening of a line are the motions of the atoms (Doppler broadening) and squeezing together of the atoms (pressure broadening).

All these things contribute to the shape of the line and to the relationship between the intensity of a line and the abundance of the atoms producing that line. By applying proper theoretical considerations, we can calculate these influences separately and establish correct relationships between line intensities and abundances of various elements. This has been done for a few well-known stars, and in all cases it has been found that hydrogen is by far the most abundant element present.

For the sun, we find that although the lines of more than 60 of the 92 natural elements found on Earth are observed in the solar spectrum, more than 80% of the atoms in the solar atmosphere are hydrogen and more than 18% are helium. All the other elements constitute less than 1% by volume. For B-type stars we find that about 60% of the atmosphere by weight is hydrogen, and close to

40% is helium, with all of the other elements contributing a fraction of a percent.

Before we leave the subject of stellar atmospheres, one more thing is worth noting. The atmospheres of stars are very murky and hazy. On a clear day here on Earth we can see for miles through our atmosphere, but this is not so for stellar atmosphere. The reason is that the atoms in the hot atmospheres of the stars are much more opaque than those in the cool atmosphere of Earth because the violent collisions suffered by stellar atoms make them much more susceptible to ionization by visible radiation.

To understand this we note that the opacity of a gaseous medium is due primarily to a process that we call photoionization. If an energetic photon is absorbed by an electron in an atom, and the electron is torn out of the atom, two things happen: the residual atom is given a backward kick, and the ejected electron moves off with the rest of the kinetic energy supplied by the photon. It then collides with other atoms and electrons until it is captured by some stray atom. Thus, the energy that was originally present in the form of a visible photon becomes kinetic energy, distributed by collisions among a large number of particles. This type of absorption hinders photons from getting through the atmosphere, thus producing haziness.

For such a process to occur, the photons must be capable of ionizing the most abundant atoms in the star's atmosphere—the hydrogen atoms. But under ordinary conditions the electrons in hydrogen atoms are in the lowest orbit, and so much energy is required to tear them completely out of the atom that the photons in the visible part of the spectrum are unable to do this. If, however, the temperature of the atmosphere is sufficiently high, collisions among the hydrogen atoms raise the electrons to higher orbits, making these atoms amenable to ionization. The atoms in the stellar atmospheres of hot stars are in excited states and therefore more easily ionized by visible radiation than are those in cool stars.

In the sun's atmosphere, not too many hydrogen atoms are excited, as we have already noted, so that this process is not the principal cause of the haziness of the sun's atmosphere. However, this process becomes more and more effective as the temperature increases, and in the A-type stars the atmospheres are extremely opaque as a result. Another type of hydrogen-ionization process produces most of the opacity of the sun's atmosphere. The neutral hydrogen atoms in the sun's atmosphere capture an additional electron, thus becoming negatively ionized. In this negative hydrogen ion, neither electron is as tightly bound as the single electron is in the neutral hydrogen atom, so that either electron can now be torn out by passing photons. The opacity of the solar atmosphere is now known to be due principally to the negative hydrogen ions, although the metallic atoms that are also in excited states contribute to some extent.

As an example of how effective excited atoms (atoms that are about to be ionized) are in blocking out radiation, we note that it takes only a few grams of gaseous material per square centimeter of solar atmosphere to prevent most of the radiation from getting through from the photosphere below. The only reason we can peer down into the solar atmosphere as far as we do is that this atmosphere is very distended and at very low pressure. In dwarf stars, where the atmospheric pressure is much greater, the atmosphere is very shallow; in giant stars, with small atmospheric pressures, the atmosphere is extremely deep. If the opacity of Earth's atmosphere were as great as that of the stars, we could see no more than a few feet in front of us.

CHAPTER 13

Stellar Masses and Radii

Under the passing stars, foam of the sky
Live on this lonely face.

—W. B. YEATS, *The Rose of the World*

We have seen that a careful analysis of a star's spectrum reveals the chemical and physical characteristics of its atmosphere, but the spectrum tells us very little about its deep interior. It gives us only the total amount of radiation emitted per second by the star; it tells us nothing about the generation and transport of this radiation through the star.

As we pass from the atmosphere into the star's interior, such physical parameters as the temperature, pressure, and density change quite drastically. This is so because the closer we are to the center of the star, the greater the weight of gaseous material that has to be supported by the gaseous pressure at that distance from the center. Other changes also take place. In some stars the abundances of the various chemical elements change as we move toward the center because mixing does not occur; the amount of hydrogen and helium in the interior is then quite different from that in the atmosphere. We have direct evidence that this is indeed so for certain groups of stars.

The chemistry of the interior of a star differs from that of the atmosphere because of two different phenomena both of which are important for the star's structure; one of these is ionization and the other is thermonuclear fusion. We discuss ionization here and defer the discussion of fusion until later. Ionization of stellar material, which increases as we go down into the interior, changes the mean molecular weight of the stellar gas. In determining the average molecular weight of the gaseous material, we must take account of ionization because the free electrons arising from ionization must be counted just as though they were molecules even though they do not contribute measurably to the total mass of the material.

Another important aspect of the increasing ionization is its effect on the energy transported from the interior of a star to the surface. In the atmosphere this is a radiative process, with the energy being carried by electromagnetic waves. But as we penetrate into the interior and reach a region where the temperature is high enough for the hydrogen ionization to be quite well advanced, convection becomes an important means of energy transport. In the sun this hydrogen convective zone sets in a few thousand miles below the photosphere. It ends at a point in the interior where the temperature is sufficiently high for the hydrogen to be completely ionized.

In addition, as we go deeper into the star, changes occur in the opacity of the stellar gases owing to the changes in pressure, temperature, and ionization. Another important quantity that varies is the rate at which energy is manufactured by the nuclei in the star's interior. The energy-generating process is the thermonuclear fusion of hydrogen to form helium, a process sensitive to temperature and pressure changes. In view of all these changing factors, we see that constructing a model of a star's interior is a very complex problem since such a model must describe how all of these quantities change as we move from the center of the star out to the surface.

To see what is involved, we may (on a much simpler scale) compare the astrophysicist in this case to an engineer because, in solving the problem of a star's interior, the astrophysicist has a construction job on his hands somewhat similar to that which the engineer has when he constructs a bridge. The engineer is given a set of geometrical dimensions (the length, height, and width of the span), a certain kind of material to work with (e.g., steel), and a definite quantity of this material. He then applies the laws of statics to ensure that at each point in his structure all the forces acting at that point (e.g., weight of material, stresses) balance each other so that the structure is in equilibrium. The engineer knows that he has solved his problem when he has a structure that can support a specified rate of traffic. Of course, the solution given by a particular engineer is not unique because many different designs of a bridge can satisfy the given conditions.

The astronomer also starts with a set of geometrical conditions that may be quite complex for actual stars because most stars are spinning about an axis the way Earth does. These objects are thus oblate spheroids, so their geometry is not simple. However, since the amount of flattening for stars like the sun is minute, the astrophysicist simplifies his problem by assuming that a star is perfectly spherical; he therefore works with only one geometrical datum, the radius of the star. He makes the additional assumption that the conditions are the same at all points in a star that are at the same distance from the center. This is called the condition of spherical symmetry. With this simplification, all quantities describing the interior depend only on the distance from the center.

Starting with a sphere of definite size, the astrophysicist has a given amount of material (the mass of the star) to arrange within this volume, and he must do it in such a way that at each point all the forces are balanced. He is also limited as to the chemical composition of his stellar material since not all elements are equally serviceable for constructing a star. But how does the stellar engineer know that he has really constructed the star that is called for? If the sphere of matter that he has constructed radiates energy at the correct rate (i.e., has the right luminosity), then he has done his job properly.

We may illustrate these ideas by considering the sun. We have the following data: a sphere of radius 6.95×10^{10} cm, a total mass of 1.98×10^{33} g, luminosity equal to 3.72×10^{33} ergs/sec, a chemical composition of more than 73% hydrogen and about 22% helium. (The hydrogen and helium contents are all that need be known to calculate the mean molecular weight of the star's interior.) We must now arrange this total amount of material, starting with an initial assumption as to the fraction of it that is hydrogen and the fraction that is helium, within the given volume so that the correct amount of energy is released per second. This means that we must deduce from the basic physical laws a table of values of temperature, pressure, density, mass distribution, and transport of energy in the interior of the sun, showing how these quantities change as the distance from the center increases. If this table gives the correct total luminosity, then our initial choice of the chemical composition (hydrogen and helium content) is correct and our problem is solved. If we do not obtain the correct luminosity, we must start with a different choice of the hydrogen and helium content.

We have just outlined the problem that the astrophysicist faces when he investigates the internal structure of a star, and we have seen that he must take as his starting point certain data that can be obtained only observationally; the interior condition must then be made to fit these data among which are the chemical composition, radius, and mass of the star. We saw that we can learn about the chemical properties of the star from its spectrum. Now we must learn how the sizes and masses of the stars are found.

How does one measure the sizes of objects that are so far away that even with the most powerful telescopes they appear no larger than points? For the star closest to us (the sun) the problem is purely a trigonometric one because we can actually measure the apparent size of the sun's disk. The sun's disk covers about half a degree in the sky. From this and the knowledge of the distance of the sun (one astronomical unit or about 93,000,000 miles) we can calculate the diameter of the sun as follows: we take one-half of 3600 (this changes the one-half degree to seconds) and divide this by 206,265, which changes the half degree to radians. We now multiply this number by 93,000,000. The product obtained is the diameter of the sun in miles. We use the sun's diameter (about 900,000 miles) as a unit of stellar diameters.

We cannot determine the radii of other stars this way because we have no way of measuring their apparent diameters; to us they are just points of light. One cannot get around this difficulty by using greater and greater magnification with a given telescope because of the wave nature of light. If light actually moved in straight lines, a true image of a star's disk could be obtained, but owing to its wave characteristics light bends around corners slightly (diffraction) so that the image of a point source of light in a telescope is not a point but a small disk.

This disk is called the *spurious disk* and it increases in size as the objective (the front lens or mirror) of the telescope decreases. In other words, the larger the front lens of a telescope is, the smaller is this spurious disk. But even with our largest telescopes this disk for the nearby giant stars is larger than the image of the actual disk of the star formed by the telescope. This means that the size of the actual disk cannot be measured because it is hidden in the spurious disk. Increasing the magnification merely increases the size of this false disk. The flickering of a star arising from variations in Earth's atmosphere also obscures the true disk.

Since we cannot measure the diameters of stars directly, we must proceed indirectly by first determining the total luminosity and the effective temperature. From the effective temperature we know how much energy is radiated per second from each square centimeter of the star's surface; if we divide this into the luminosity, which is the energy radiated per second from the entire surface of the star, we obtain the star's surface area. We now divide the surface area by the number 4π and take the square root of the result. This gives the radius of the star in centimeters.

We may illustrate this method by comparing the luminosity and temperature of a star with those of the sun. A star like Rigel is known to be 62,000 times as luminous as the sun. If it were as hot as the sun, its surface area would therefore have to be 62,000 times as great as the sun's to radiate as much energy per second as it does. But since Rigel is a B-type star it is hotter than the sun, its temperature being about 15,000°K. Therefore, each square centimeter of its surface radiates about 80 times faster than each square centimeter on the sun. (The rate of radiation per square centimeter increases as the fourth power of the absolute temperature.) Since the difference in temperature between the sun and Rigel must account for an 80-fold increase in the luminosity, the surface area of Rigel can only be about 800 times that of the sun, so that its radius is approximately 25 times larger than the sun's.

As another example we consider the main-sequence star Sirius, which belongs to spectral class A_0 and is about 30 times as luminous as the sun. Since its surface temperature is about 10,500°K, each square centimeter of its surface radiates 15 times faster than a square centimeter on the sun. To account for the factor of 30 in the brightness, the surface area of Sirius must be twice that of the sun and its radius about $1\frac{1}{2}$ times greater than the sun's.

Using this sort of analysis for various stars we obtain the following results: for main-sequence stars, the diameters range from about 30 times that of the sun for the early B_0-type stars to 1/8 that of the sun for the red dwarfs. In other words, the main-sequence stars do not differ much in size. However, the diameters of stars along the other branches of the H–R diagram differ considerably from the sun's diameter. The supergiant and giant stars have radii ranging from about 1000 times that of the sun (the bright component of the binary system Epsilon Aurigae) to about 16 times that of the sun for a giant like Capella, which is of the same spectral class as the sun. Among the largest known stars, in addition to Epsilon Aurigae, are the red supergiants like Betelgeuse and Antares with diameters about 500 times that of the sun. On the other hand, the diameters of the white dwarfs are on the order of 1/100 that of the sun. A red supergiant must be very big to emit as much radiation as it does because it is very cool. On the other hand, a hot white dwarf can be as faint as it is only because it is small.

From these data we see that the sizes of stars cover an enormous range, from the white dwarfs that are not much larger than Earth to the red supergiants that are so large that their volumes could accommodate more than 50,000,000 stars like the sun. These results are so astounding that it is reassuring to have some means of verification that does not depend on our knowing the star's temperature and luminosity. For the supergiants, such a method was developed by American physicist Albert A. Michelson, who used the interference of light rays from different parts of a star's disk to measure the diameter. He invented an instrument, called the interferometer, to do this and he obtained results that agree very well with those obtained by the analysis outlined above.

Since the Michelson interferometer can be used only for the very large stars, other methods must be employed for the white dwarfs. Since the radii of these stars are very small and, as we shall see, their masses are not much smaller than the sun's, the gravitational fields on the surfaces of these stars are much greater than on the sun's surface. From this it follows that the radiation from these stars should exhibit the Einstein redshift, which we discussed in Chapter 5. This is indeed so, and if a white dwarf is a member of a binary system (e.g., the companion of Sirius) we can use this effect to determine its radius. Here, too, we find the results in agreement with the previous method based on the luminosity and temperature.

Another important method for measuring stellar diameters can be applied to members of a binary system in which an eclipse occurs. If the orbit of two stars that revolve about a common point lies in a plane that passes through Earth, an observer on Earth detects regular fluctuations in the light coming from this binary system. This luminosity variation arises from a periodic eclipsing of one star by the other. By measuring the duration of the eclipses (there are two eclipses in one complete period) and analyzing the light curve (the variation in the luminosity), the diameters of the two components of the binary system can be found. This method has been applied to about 50 eclipsing binaries, and the results agree with those of the other methods.

Not only must we know how big a star is before we can construct it, but we must also know its mass. Here we face a very unusual problem because it is impossible to measure the mass of an individual star directly. Its mass can be measured directly only if it is physically attached to at least one other star (a binary system). The reason for this is immediately clear if we recall that the mass of a body is a measure of its inertial reaction to a force acting on it. The only way we can measure the mass of any object is to exert a force on it and observe the acceleration that this force imparts to the object.

But how are we to observe the effect of a force on a star? In general, there is no way for us to exert a force on a star, but many stars are so close to another star that we can observe the mutual gravitational interaction between the two, which stems from their masses. In other words, we can study the stars in a binary system and from an analysis of their motions determine their masses. The procedure is based on the third (the harmonic law) of Kepler's three laws of planetary motion as corrected by Newton. Kepler discovered that the periods of the planets around the sun are related in a definite way to the mean distances of the planets from the sun. Kepler thought that this relationship was the same for all of the planets, but Newton showed that it changes slightly from planet to planet because the masses of the planets are not the same. The relationship, as shown by Newton, depends on the sum of the mass of the sun and the mass of the planet. If we carefully measure the sidereal period and the mean distance from the sun to any planet, we can apply Kepler's third law to obtain the mass of the sun (we neglect the mass of the planet), which we find to be 2×10^{33}.

The gravitational dynamics of a stellar binary system is similar to that of the motion of a planet around the sun except that the two components of the binary system are of comparable mass so that the mass of neither can be neglected. Both members of the binary revolve about a common center of mass with a period that is related to their mean separation by Kepler's third law. We can use this law to determine the total mass of a binary system by carrying out the following steps: first, measure the average angular separation in seconds of arc between the two components of the system; cube this number; divide this by the cube of the parallax of the binary system; divide this result by the square of the period of the system. This is nothing more than a set of directions for applying Kepler's third law to a binary star. The resulting number is the total mass of the binary expressed in solar mass units.

To obtain the individual masses of the two components of the binary system, we must find the common point about which both stars are revolving (the center of mass of the binary). This point can be found from the orbit of the binary if both stars of the system can be observed and their motions relative to the other background stars determined. This is so because the center-of-mass point must always move in a straight line with uniform speed regardless of what the two stars are doing, and the line connecting the two stars must pass through this point. The knowledge of the position of this point relative to that of the two stars at any moment gives us the ratio of the masses of the two stars. This information, together with the sum of the masses as found above, enables us to find the mass of each component.

The stellar masses obtained are rather remarkable, because as we go from the white-dwarf branch through the main sequence into the giant branch of the H–R diagram, the masses do not change to any great extent. The masses of most stars fall between 1/10 and 50 times the sun's mass. Some red supergiant stars have masses in the neighborhood of 100 solar masses, and one star is known with a mass equal to 400 times that of the sun, but these are very rare objects. The mass of a star is a very critical quantity that cannot be tampered with to any great degree without drastically changing the star's characteristics.

To see how small the variation in stellar masses is compared to that of other parameters, we note that the white-dwarf companion of Sirius with 95% of the sun's mass is about as large as Earth and has only the 1/400 of the sun's luminosity. On the other hand, the G-type giant Capella with about 4 times the sun's mass is 120 times more luminous than the sun and nearly 6000 times more voluminous. The supergiant Antares is 20 times more massive than the sun but its luminosity is about 10,000 times that of the sun, and it is millions of times greater in volume. From these facts we see that the mean density of stellar material changes greatly as we go from the white dwarfs to the supergiant, because the masses are practically constant whereas the volumes change enormously.

For main-sequence stars the average density ranges from about

10 g/cm^3 for the red dwarfs (M-type stars) to 0.01 g/cm^3 for the B-type stars. The mean density of the sun is 1.4 g/cm^3. Giant stars have densities on the order of 1/1000 of a gram per cubic centimeter and the supergiants have densities as low as one ten-millionth of a gram per cubic centimeter. These objects are as tenuous as a good vacuum here on Earth. But the most astounding stellar densities are those of the white dwarfs and the more recently discovered pulsars or neutron stars. The white dwarfs are so compact that an average cubic centimeter of their material has a mass of thousands and even hundreds of thousands of pounds; Van Maanen's star has a density of 10 tons per cubic inch. The densities of the neutron stars are many times larger than those of the white dwarfs; a typical neutron star has a density of 2000 trillion or 2 billion tons per cubic centimeter.

Thus far, we have been discussing the masses of binary stars because only for such stars can masses be measured directly. However, we can find the masses of many single stars by an indirect approach that uses an important relationship between the mass of a star and its luminosity. Considering that the mass of a star is so critical, we surmise that its entire structure changes with slight variations in its mass. This is evident from the fact that the luminosity of a star is very sensitive to its mass; the masses of stars change only slightly for very large changes in the luminosity.

As long ago as 1905, Hertzsprung had observed an empirical relationship between the luminosity of a star and its mass. With the large number of stellar masses and luminosities that are now available, we can construct a graph on which we plot the luminosities against the masses of the stars. We then see that the main-sequence and giant stars fall on a smooth curve. This is the mass–luminosity relationship that was first derived by Sir Arthur Eddington from theoretical considerations of the internal constitution of stars; it states that the luminosities of stars vary as the fourth power of their masses. We can use this relationship to determine the masses of stars that are not components of a binary system provided we know their absolute magnitudes (luminosities). If we know the absolute magnitude of a star, we can obtain its luminosity and then with the aid of the mass–luminosity curve find its mass.

That a mass–luminosity relationship exists for stars can be made plausible by a simple physical argument. The more massive a star is, the greater must be the internal pressure of the gaseous stellar material to support the weight of this mass. But this means that the internal temperatures must be larger to yield the larger pressures, and these larger temperatures in turn produce a more rapid release of energy, resulting in a greater luminosity.

We have already described two methods for measuring the distances of stars, and now the mass–luminosity relationship gives us a third method, which applies, however, only to binary stars. We learned how to find the total mass of a binary system, provided we know the parallax of the system. This parallax, however, has to be known quite accurately if we wish to have a fairly reliable value for the mass because the parallax has to be cubed in the procedure and any error in it is therefore tripled as an error in the mass. Suppose, however, we reverse the procedure and try to find the parallax of the binary system from the mass, assuming that we have some previous but not precise knowledge of the mass. To do this we must take the cube root of the mass, so that even if there is quite a large error in our knowledge of the mass, the error in the parallax is not too great because it is one-third of that in the mass.

But how do we know what value of the mass to start with? We use the fact that stellar masses are not very different from star to star, and assume that the total mass of the binary is twice that of the sun. If we now reverse the directions given previously for finding the mass, we get a value for the parallax of the binary system. This is obviously not correct because we started out with a guess for the mass of the system. However, with the aid of this newly found parallax we can determine a value for the absolute magnitude (also incorrect) and hence the luminosity of each member of the binary.

We now take these values of the luminosities and, using the mass–luminosity relationship, determine the masses of two stars that correspond to these luminosities. These will be different from the values (the solar mass) we started with; we can now take these new values of the masses for the two stars of our binary system and go through the whole procedure again and obtain a corrected value for the parallax. We repeat this procedure until no new value of the parallax can be obtained. This is, then, the best value for the parallax of the binary system that this procedure can give. This method, which is called the method of *dynamical parallaxes*, gives parallaxes with an accuracy of about 15%.

CHAPTER 14

The Deep Interiors of Stars

He passed the flaming bounds of space and time:
The living throne, the sapphire-blaze,
Where angels tremble while they gaze,
He saw; but blasted with excess of light,
Closed his eyes in endless night.

—THOMAS GRAY, *The Progress of Poesy*

With what we now know about the luminosities, radii, masses, and chemical compositions of stars like the sun, our investigation of their internal structures will reveal to us the conditions that must hold throughout a star's interior if its luminosity is to be consistent with its known mass and radius. To guide us in this investigation we introduce two general principles that ensure the dynamical and thermal stability of a star. The first of these imposes the condition of dynamical equilibrium at each point within the star. This means that the star is neither exploding nor collapsing, nor even, for that matter, pulsating. It does not mean, however, that the material within a star is at rest; it means that there is no outward or inward accelerated motion, nor any net outward or inward flow of material at any point. It is permissible under this condition to have a slow circulation of material (convection currents) in various zones of the star.

The second principle imposes a steady thermal state at each point. This means that despite the continuous radiation of energy from the surface of the star, the prevailing thermodynamical conditions in any neighborhood within the star are changing so slowly that the temperature at any point is constant for very long periods of time (millions of years).

With these conditions to guide us, we can now design a model

233

of the sun, or other stars like it, that specifies the run of the temperature, pressure, and density as we go from the center to the surface of the star, and that also gives the abundances of hydrogen and helium throughout the star. To go into a detailed analysis of this problem would involve us in the higher mathematics of differential equations, which we shall not do, but we can get a general insight into the procedure without using mathematics if we employ our two conditions properly.

The first condition tells us that the inward gravitational pull on an element of matter (let us say a small cube of the stellar gas) at any point must be balanced by the outward pressure of the underlying gases. This balance of gravitational forces and gas pressure is expressed in the form of an equation that relates the rate at which the gas pressure increases toward the center to the manner in which the weight of an element of matter (i.e., the acceleration of gravity) varies as we move toward the center. This equation involves three unknown quantities (quantities that must ultimately be found by solving the equations): the pressure at a point, the density at the point, and the mass of stellar material lying in the partial stellar sphere whose surface passes through the point being considered and whose center is the star's center.

We obtain a second relationship (equation) by dividing the star into a series of concentric shells, each of which differs only slightly from the two neighboring shells above and below it. This model is a good approximation if we impose the condition that the mass of material in any one of the shells differs only very slightly, and in a predictable manner, from the masses of the two neighboring shells.

These two relationships are not sufficient to enable us to solve the problem completely because three unknowns, as mentioned above, are involved in these equations, so that we must seek a third relationship. We now employ our second condition to find this third relation among our unknowns because the third relationship deals with the continuous outflow of energy from the interior to the surface of the star, and is therefore of a thermodynamical nature. This relationship defines the thermal conditions that must exist at any point to maintain this stream of energy at a constant rate.

Knowing that for the major part of its outward journey the energy moves through the star in the form of a stream of radiation, we focus our attention on a small cross section of this stream and follow it along as it moves from a given point in the star to some nearby point farther out from the center. Since the flow of this radiation is impeded by the absorbing atoms and free electrons in its path (the opacity of the stellar material), some force from below must be pushing it out. This force is determined essentially by the rate at which the temperature changes as we move outwardly (the temperature gradient at a point). The third relationship is therefore an equation stating that the temperature gradient at any point inside the star where there is a radiative transport of energy must be proportional to the outward flux of energy and to the opacity of the stellar material.

We now have three equations (stated in words above), but there are four unknowns, the temperature having been introduced in our last relationship in addition to the three we introduced previously. Moreover a new quantity, the opacity, has entered, and this varies from point to point in the star. We now seem worse off than we were before, and unless we can introduce still another relationship connecting some of the unknowns, we are unable to solve our problem. Fortunately, we can impose two other conditions that give us the required relationships.

One of these is a condition on the total flow of energy across the surface of any one of the concentric shells into which we divided our star above. This energy flow per second must equal the total energy generated per second by the mass within the sphere that is bounded by the outer surface of the shell through which the energy is flowing. The other relationship follows from the fact that the stellar material, right down to the center of the star, is in the form of a perfect gas. This enables us to relate the pressure at any point in a star to the temperature, density, and mean molecular weight of the stellar material at that point.

We have now introduced five conditions (mathematical equations) to describe the variations from point to point within the star of the following five quantities (the five unknowns of our equations): gas pressure, density, temperature, radiant flux, and the acceleration of gravity. We should therefore be able to determine the values of these five quantities throughout the star by solving the five equations. This would be so if it were not for the fact that three other quantities have been introduced that complicate things considerably. These quantities are the opacity, the mean molecular weight, and the rate of energy generation. However, these three parameters can be expressed in terms of the five unknowns at any point in the star, so that our problem can be solved.

Since as already noted, the mean molecular weight of the stellar material enters into the five relationships described above, we must determine it before we can pursue our investigations into the internal structure of stars. The extremely high temperatures in stellar interiors simplify this task because at these high temperatures collisions among atoms are so violent and the photons so energetic and numerous that most of the atoms are very nearly completely ionized. If we make this very safe assumption, the free electrons released by the complete ionization of the atoms affect the mean molecular weight considerably.

When a heavy atom is completely ionized, the average molecular weight of this atom is greatly reduced because the total weight of the atom is now distributed among all of the free electrons torn from it; these particles must be counted as separate molecules in computing the average but they do not contribute to the total weight. Thus, all heavy atoms when ionized have a mean molecular weight of about 2. However, when helium is completely ionized, its mean molecular weight is 4/3 because its atomic weight of 4 is distributed over its two electrons and nucleus (three particles). The mean molecular weight of completely ionized hydrogen is $\frac{1}{2}$. We thus see that if the stellar gas is completely ionized, as it is in the deep interior where the temperatures are high, we can express the mean molecular weight of this gaseous material in terms of the hydrogen and helium abundances. The reason for this is that we can consider each gram of stellar material to be the sum of a certain fraction of a gram of hydrogen, another fraction of a gram of helium, and a remaining fraction of a gram of everything else that has a mean molecular weight of 2. Thus, the mean molecular weight lies between $\frac{1}{2}$ (all hydrogen) and 2 (all heavy atoms).

If we keep this in mind, we find that the mean molecular weight of the internal stellar material is given by a simple arithmetic expression involving just the hydrogen and helium abundances. As soon as we know the values of these quantities or make a reasonable guess as to their values, we can write down a number for the mean molecular weight. In most of the reported investigations into the internal structure of stars like the sun, it has been assumed that the mean molecular weight is constant throughout the star. This, of course, is not a correct assumption but it is a good first approximation that can be corrected later.

As noted above, the molecular weight presents no problem once we know the hydrogen and helium abundances in the stellar gas. At first sight, it may seem that this is no help at all because we have apparently replaced one unknown (the mean molecular weight) by two unknowns (the hydrogen and helium contents), but we have actually simplified things because we have definite observational hints as to the hydrogen and helium abundances in stars. But what about the opacity? It, too, can be eliminated as a separate unknown by being expressed in terms of the temperature, density, and hydrogen and helium abundances.

A number of atomic processes contribute to the opacity of gaseous material at high temperatures. To begin with, high-temperature photons, being quite energetic, tear electrons out of atoms so that the photons' orderly transport through the star is impeded. This photoionization process is thus a source of stellar opacity. To be sure, the atoms in the interior of the stars are almost always completely ionized, but for short periods of time they do capture electrons, and these electrons are then available for the photoionization process that helps make the gas opaque. From a knowledge of the structure of the atom and the manner in which photons are absorbed by electrons that are bound to atomic nuclei, it is not too difficult to calculate how much this ionization process contributes to the opacity.

Another process is one in which an electron that is outside the atom but moving in a curved orbit (a hyperbolic orbit) that is still influenced by the nucleus of the atom absorbs a photon and changes its orbit. The rate of this absorption process can also be calculated and expressed in terms of atomic constants and the pressure, density, and temperature at the point. This gives another contribution to the opacity. A third contribution is obtained from the scattering of photons by free electrons that are not under the influence of atomic nuclei.

When we take all of these effects into account, we find a rather simple formula for the opacity: it is proportional to the density of the stellar gas at the point, multiplied by the electron density (number of electrons per unit volume) and the abundance of the heavy elements (which can be expressed in terms of the hydrogen and helium abundances), all divided by the 3.5 power of the temperature.

This is called the Kramers opacity formula because it was first derived by the Dutch physicist Kramers for a gas under ordinary conditions here on Earth.

The British astronomer Arthur Eddington pointed out that this formula cannot be applied directly to stellar material because it gives too large a value for the opacity; the formula, as Kramers derived it, applies to an atom before it absorbs a photon. Once an atom absorbs a photon and the absorbing electron is torn out of the atom, that particular atom can no longer absorb as efficiently as it did before it lost one of its electrons. Eddington described this phenomenon by saying that, in the process of absorbing, the atom is "guillotined" and should therefore not be immediately counted again as contributing to the opacity of the stellar material.

Eddington took account of this dilution of the absorbing capacity of an atom after it has first absorbed a proton by dividing Kramers's formula by a "guillotine factor." This factor does two things. It corrects the formula for the number of atoms that are incapable of absorbing at any moment because they have lost their absorbing electrons, and it also brings the formula into agreement with the wave properties of the electron. The guillotine factor can be expressed in terms of the pressure and temperature, but it is also sensitive to the relative abundances of elements with atomic weights less than 20 (e.g., oxygen, nitrogen) in the group of heavy elements.

Arthur Stanley Eddington (1882–1944) was born in Kendal, England, the son of Arthur Henry Eddington, a devout Quaker who was the headmaster at a local grammar school, and the former Sarah Ann Shout. The death of the senior Eddington in 1884 forced Sarah to move the family to Somerset, where Arthur spent his childhood years. Despite the family's tight finances, Arthur's mother managed to send him to Brynmelyn School, where he developed interests in science, literature, and mathematics. Although somewhat introverted, Arthur played cricket and football. He won a scholarship to what is now the University of Manchester in 1898 where he attended the physics classes of Arthur Schuster. In 1902 he entered Trinity College at Cambridge where he concentrated on mathematics and secured first-place honors in the mathematics examinations. He earned his degree in 1905 and spent a year as lecturer and tutor in mathematics at Cambridge before joining the staff of the Royal Observatory in 1906. Eddington gained much experience as chief assistant and made a trip to Brazil in 1912 to observe an eclipse. His fieldwork for the Royal Observatory continued to benefit him even after he returned to Cambridge University in 1913: He organized an eclipse expedition in 1919 that helped prove Einstein's general theory of relativity by showing that the path of starlight grazing the sun is bent by the gravitational field of the sun. However, Eddington himself never saw the eclipse because he was too busy changing photographic plates.

Eddington's most important work was in the field of stellar structure, though he dealt with a wide variety of problems including "the motions and distribution of the stars, the internal constitution of the stars, the role of radiation pressure, the nature of white dwarfs, the dynamics of pulsating stars and of globular clusters, the sources of stellar energy, and the physical state of interstellar matter."[1] He also helped to popularize Einstein's theory of relativity in Britain and, as a result of his own writings on the subject, made valuable contributions to relativity theory. Finally, he attempted to derive what he referred to as the fundamental constants of nature and thereby construct a grand theory to explain the dynamics of the cosmos.

Eddington's early work in stellar dynamics dealt with radiation pressure. He was influenced by Karl Schwarzschild's work on radiative equilibriums in stellar atmospheres but derived his own equations showing that a state of stellar equilibrium must take into account gas pressure, radiation pressure, and gravitation. Eddington also derived the important mass–luminosity relation showing that stellar luminosity varies directly with mass to the fourth power; thus massive stars radiate their energy much more rapidly than solar-mass stars. Eddington thus concluded that there are relatively few massive stars such as the red and blue giants because of their comparatively short life spans.

Eddington also calculated the diameters and masses of several stars and found that the white dwarf companion of Sirius has such a small diameter and, hence, such a small volume that its density averages several tons per cubic inch. Although this conclusion was initially regarded with skepticism, Eddington's calculations were confirmed by detailed examinations of the star's spectra. Almost as surprising was Eddington's wrong conclusion that to reconcile both his mass–luminosity relationship and the Russell–Hertzsprung sequence of stellar evolution, an average life span of several trillion years would have to be assumed for stars. Even the very massive stars that did not cast off their outer layers as supernovas would take at least a trillion years to burn down to the size of white dwarfs. These immensely long life spans convinced Eddington that stars must be powered by a continuous conversion of matter into radiation. Although his theory of atomic stellar energy was criticized, he was in some sense vindicated in 1938 by Hans Bethe's carbon cycle model showing conclusively that mass is converted into energy in stars.

Eddington became the first proponent of Einstein's general theory of relativity in Britain when he was forwarded a copy of Einstein's famous 1916 paper by the Dutch astronomer Wilhelm de Sitter. Eddington saw at once the importance of Einstein's work and committed himself to mastering its mathematical intricacies. His 1918 *Report on the Relativity Theory of Gravitation* to the Royal Society represented the first English version of Einstein's theory. A second edition published in 1920 contained the results of Eddington's 1919 eclipse expedition that seemed to confirm Einstein's prediction that the gravitational fields of massive objects bend the paths of light beams. Eddington's 1923 book, *The Mathematical Theory of Relativity*, was still considered by Einstein over 30 years after its publication to be the finest explanation of relativity theory in any language.

Eddington also tried unsuccessfully to reconcile quantum theory and relativity theory by determining the values of the fundamental constants in nature such as the velocity of light, Planck's constant, the mass of the electron, and the gravitational constant from basic theoretical principles. Although he did show how some of these constants are related to each other, his most memorable, if not helpful, conclusion was his statement that there are about 10^{79} protons and electrons in the universe.

Eddington wrote a series of popular books that showcased not only the revolutionary ideas of 20th century physics but also his own philosophical speculations, particularly his now-discredited selective subjectivism. His popular writings were characterized by rich metaphors, wit, and sparkling clarity. Some of his later works were of a decidedly mystical bent, reflecting both his appreciation of sensual and spiritual phenomena and his belief that spiritual truth cannot be deduced from scientific theories.

We come now to the third quantity that appears in our equations for the internal constitution of a star, and which we must express in terms of the temperature, density, and chemical composition. This is the rate of energy generation per gram of stellar material. We can picture the material in the interior of the stars as though each gram were a tiny energy factory, generating energy at a definite rate. This rate of course is not the same for every gram of material; it depends very sensitively on the conditions that hold for any particular gram. Near the center of a star, where the temperature and density are high, the rate of energy generation is much greater than out near the surface where the stellar gases are cool and tenuous.

Before we can derive a formula for the rate of energy generation we must know the nature of the generating process. The vast quantities of energy that a star emits per second tell us that no ordinary chemical process, such as the burning of fuel in an oxygen medium, can account for stellar energy. Such processes could not keep a star like the sun radiating for more than a few thousand years. We must exclude slow gravitational contraction as a source of stellar energy in most stars for the same reason. The sun could, indeed, radiate at its present rate by contracting 1/1000 of its size each year, but then the sun could not be more than 20,000,000 years old, nor could it last for more than another 30,000,000 years. These time spans are much too small to fit the observational facts.

With Einstein's discovery of the mass–energy relationship, it at once became clear that the stars are getting their energy by destroying mass on a vast scale (4,500,000 tons every second in the sun), and that they are doing so by some kind of thermonuclear process. Although this was known as long ago as 1910, not until 1938 were the structure of the nucleus and the nature of nuclear forces well enough understood to enable physicists to develop a correct formula for the energy generation in stars.

A few considerations quickly reveal which nuclei are best suited for the nuclear transformations that supply energy most rapidly under a given set of conditions. The first requirement that must be met is that of abundance; enough nuclear fuel must be present to last for many billions of years. This at once leads us to hydrogen as one of the interacting nuclei. To see what the other interacting nuclei are, we consider the end product of the thermonuclear reaction. We are led quite naturally to helium because the latter can be obtained by fusing four protons (hydrogen nuclei) and in this process about 0.032 atomic unit of mass is transformed into energy. In other words, if a thermonuclear fusion of hydrogen into helium occurs within stars, then for every 4 g of helium that is formed in this manner, 0.032 g mass is transformed into energy E according to the Einstein formula $E = mc^2$. This means that the hydrogen nuclei (protons) must interact with themselves, either directly or by way of intermediate nuclei in a series of steps that lead to helium.

That hydrogen and, in general, the very light nuclei produce the stellar energy is indicated by the electrical charges on atomic nuclei. All such nuclei are positively charged and therefore strongly repel each other, so that thermonuclear processes are quite unlikely except at very high temperatures (large speeds of the interacting nuclei). This repulsion increases very rapidly as we go to the heavier elements because of the large electrical charges on such nuclei; it is smallest for hydrogen interacting with itself or with other light nuclei. These light nuclei are thus the best candidates for thermonuclear reactions at the temperatures inside the main-sequence stars.

Before we can write down a formula for the rate of energy generation resulting from the transformation of hydrogen to helium, we have to understand something about the nature and structure of the nucleus of an atom. With the discovery of the neutron in 1933, it became clear that all nuclei are composed of neutrons and protons. Actually, in terms of modern physical concepts, we must not consider the proton and neutron as two different particles, but rather two different states or aspects of a single fundamental particle, the *nucleon*. Although the neutron and proton differ in that the former has no electrical charge, they can be transformed into each other under the proper set of conditions. Since the neutron is slightly more massive (by about the mass of three electrons) than the proton, it will, if left alone, give birth to an electron and a neutrino and become a proton. This happens to a free neutron in about 12 minutes. But the reverse is also true. If a proton, electron, and neutrino are moving fast enough relative to each other, the proton can capture the electron and neutrino and become a neutron.

The neutrons and protons are held together inside the atom by nuclear forces. Although we do not understand the exact nature of these forces, we know that they are hundreds of times greater than the electrical forces that keep electrons bound in their orbits within the atoms. Thus, the electron can be torn out of the hydrogen atom by applying about 14.5 volts (the ionization potential of hydrogen), but the removal of either a proton or a neutron from an ordinary nucleus, like that of helium or oxygen, requires the application of about 8 million volts.

We know that nuclear forces are very short-range forces relative to electrical or gravitational forces. Only when two nucleons (neutrons or protons) approach to within about 2 or 3 ten-trillionths of a centimeter do the nuclear forces between them come into play. With the discovery of *mesons* (charged neutral particles with masses about 200 to 300 times that of the electron) a clearer understanding of the nature of the nuclear forces has been obtained. Nucleons interact with each other by tossing these mesons back and forth. When a proton and neutron combine to form a *deuteron* (heavy hydrogen), the neutron and proton stay bound to each other by tossing mesons back and forth between them, with the neutron changing into a proton and back again very rapidly while the proton changes into a neutron and back to a proton again, and so on.

From a knowledge of the structure of atomic nuclei we can write down a formula that gives the rate at which energy is released for any thermonuclear reaction. We do not give a detailed analysis of this formula but just show briefly how one goes about obtaining it. There is, first of all, a geometrical factor that just represents the chance that two spheres (the two interacting nuclei taking part in the thermonuclear reaction) will collide. This factor is a bit more complicated than merely the total geometrical cross section of the two nuclei, since one must include a correction for the wave character of nuclei. This quantity must now be multiplied by a factor that takes into account the fact that the two nuclei repel one another electrically. But even after the two nuclei have interpenetrated, there is no guarantee that they will stick together to form a new nucleus; there is always the chance that they will bounce away from each other. We must therefore multiply the product we have just obtained by the probability that a new nucleus will be formed. If we multiply this result by the total number of nuclei of both kinds that are present in a unit volume of our material and also by the energy released when the new nucleus is formed, we obtain the formula we want.

If we carry out the procedure that we have just outlined, which involves a good deal of advanced mathematics and physics, we find that the rate at which energy is released per gram of stellar matter is proportional to the density of the material, the abundances of the interacting nuclei, and the temperature to some power that may range anywhere from 4 to 20 for the transformation of hydrogen to helium.

Two important sets of reactions lead to the transformation of four protons into helium and give the star its energy. Both of these sets as possible sources of stellar energy were first investigated by the physicist H. A. Bethe; he showed that thermonuclear reactions of this sort can adequately account for the luminosity of the sun and other stars.

In one of these sets the fusion of the four protons to form a helium nucleus occurs without the intermediary action of a catalytic nucleus. This set of reactions, which is known as the *proton–proton chain*, consists of the following separate reactions:

1. Two protons combine to form a deuteron, emitting a positron and *neutrino* in the process:

$$H^1 + H^1 \rightarrow H^2 + e^+ + \nu$$

2. The deuteron captures a proton to form helium-3, an isotope of helium:

$$H^2 + H^1 \rightarrow He^3$$

3. Two of the helium-3 nuclei combine to form ordinary helium and to give back two protons:

$$He^3 + He^3 \rightarrow He^4 + 2H^1$$

Before we discuss this set of reactions further, we must explain the symbols that we have used. Letters stand for the names of atoms, and the superscript gives the weight of the particular isotope of that atom. Thus, H^2 stands for hydrogen of atomic weight 2, and He^3 for helium of atomic weight 3. An isotope is an atom with the same chemical properties as another atom but with a different weight. All isotopes of a given element have the same number of protons in their nuclei but different numbers of neutrons. The symbol e^+ stands for a positron (the positively charged particle that has the same mass as the electron; it is also called the antielectron). The Greek letter ν stands for a neutrino (a neutral, spinning particle with zero rest mass), which is always emitted when a neutron changes into a proton or vice versa. Somewhat different types of proton–proton chains have been investigated more recently, and they too contribute to stellar energy but the one we have described is the most important. Using the procedure described above we can calculate the rate at which this proton–proton chain liberates energy. This can be expressed as a temperature-power law; the rate of energy release for this chain goes as the fourth power of the temperature. We see that this set of reactions is not too sensitive to the temperature and is therefore not limited to a small region near the center of the star where the temperature is a maximum, but extends out from the center.

Another set of reactions leading to the fusion of four protons to form a helium nucleus is the famous carbon cycle first introduced by Bethe. In this cycle the carbon nucleus acts as a catalytic agent, taking part in the reactions at the beginning of the cycle but reappearing with the formation of the helium nucleus. The cycle consists of the following steps:

1. A carbon nucleus captures a proton, resulting in the formation of a nitrogen isotope:

$$H^1 + C^{12} \rightarrow N^{13}$$

2. The nitrogen isotope immediately decays into a carbon isotope by emitting a positron and a neutrino:

$$N^{13} \rightarrow C^{13} + e^+ + \nu$$

3. Carbon-13 now captures the second proton, forming ordinary nitrogen:

$$C^{13} + H^1 \rightarrow N^{14}$$

4. The nitrogen nucleus now captures the third proton to form an isotope of oxygen:

$$N^{14} + H^1 \rightarrow O^{15}$$

5. The oxygen isotope immediately decays by emitting a positron and neutrino, thus becoming a nitrogen isotope:

$$O^{15} \rightarrow N^{15} + e^+ + \nu$$

6. In this final stage the nitrogen isotope captures the fourth proton, resulting in the formation of a helium nucleus and the reappearance of the carbon nucleus:

$$N^{15} + H^1 \rightarrow He^4 + C^{12}$$

Just as for the proton–proton chain, we can express the rate of energy generation by this cycle as a temperature-power law except that this process goes as the 20th power of the temperature. The amount of energy generated per second per gram by this cycle is proportional to the product of the hydrogen abundance, the density of the stellar gas, and the 20th power of the temperature. Because the carbon cycle is so sensitive to the temperature (a drop of 1% in the temperature means a drop of more than 20% in the rate of energy generation), it operates only in the neighborhood of the center of the star.

Both the proton–proton chain and the carbon cycle proceed very slowly, with the former requiring a time span of about 14 billion years under the conditions at the center of the sun. This means that a proton inside the sun lasts, on the average, about 14 billion years before being captured by another proton to form a heavy hydrogen nucleus (a deuteron). At a temperature of about 20,000,000°K (much higher than at the center of the sun) a carbon nucleus moving through a gas with a hydrogen concentration of about 80% captures a proton once every few million years.

In the interior of any particular star both cycles go on simultaneously but not at the same rate. Both carbon nuclei and protons compete in capturing some other proton, and if the carbon nucleus is successful in capturing the proton, the carbon cycle is set into motion, but if a deuteron is formed the proton–proton reaction is set off. It is fairly easy to see that the carbon cycle becomes more and more important compared to the proton–proton reaction as the central temperature increases, whereas for lower central temperatures the proton–proton reaction dominates. It has been well established that the latter is the principal source of energy for the

G-, K-, and M-type main-sequence stars, whereas the carbon cycle contributes practically all of the energy for some of the F-type stars of the main sequence and for all of the A-, B-, and O-type stars of the main sequence. For the F-type main-sequence stars that lie close to the G-type stars, the proton–proton reaction and the carbon cycle probably contribute equally. Because the giant and supergiant stars are so very luminous, we conclude that they are governed by the carbon cycle or by some nuclear reaction involving He^4 or the heavy nuclei.

CHAPTER 15

The Structure and Evolution of Stars

Silently one by one, in the infinite meadows of
heaven
Blossomed the lovely stars, the forget-me-nots
of the angels.

—HENRY WADSWORTH LONGFELLOW, *Evangeline*

Having expressed the mean molecular weight, the opacity, and the rate of energy generation in terms of our five unknown quantities and the hydrogen and helium content, we can follow the astrophysical engineer in his star-building venture. As we have already noted, he starts out with a given amount of material (the mass of the star) and the available volume for the star. He must now decide how much of this total material is hydrogen and how much is helium. Everything else is lumped together in what is left over. This gives him a numerical value for the mean molecular weight, which can be substituted directly in the equations that guide him.

But where does the astrophysicist start in constructing the star? He may either start at the very center of the star and build out to the surface or start at the surface and move in toward the center. He usually adopts the latter procedure because he knows the conditions at the surface of a star. He actually begins at the surface by taking the temperature, pressure, and density as zero there; these are the initial conditions of the problem or the boundary values. It is true that the temperature and pressure at what we ordinarily call the surface of the star (the photosphere) are not zero but they are so small compared to their values in the deep interior that the error we make in considering them to be zero is negligible.

With these starting values for the pressure, temperature, and density, and with knowledge of the total luminosity and mass of the star, our astrophysical engineer can move step by step into the interior of the star, adjusting all five quantities as he advances in accordance with the five equations that govern the variations of these quantities. These equations, as described above in words, tell him how each of his five unknown quantities changes from point to point, so that if he knows the values of these quantities at one point of the star (the surface) he can then find them for every point in the star step by step as he moves inward.

Before we discuss the results of this type of investigation, we mention one important characteristic of the interior of a star governed by the carbon cycle that may simplify the problem of its internal structure to a considerable extent. As we move into such a star's interior, the temperature increases very rapidly, and becomes so high at a point near the center that the stellar material cannot be in radiative equilibrium. This means that close to the center of a star the temperature changes so drastically over short distances that the energy is released too rapidly to be transported out fast enough by radiation. Instead huge convective currents are generated, and these currents transport the energy outwardly to a point, not far from the center, where radiation can take over again.

This convective zone sets in when the ratio of the pressure gradient (the rate at which the pressure changes as we move in and out) to the temperature gradient equals or is less than $2\frac{1}{2}$. When this occurs, we can no longer use the five equations that we described above to continue our penetration into the interior of the convective zone. These equations, however, are replaced by a single rather simple equation, the solution of which gives us all the information that we need about the convective core, right down to the center of the star. Convective cores exist in stars higher up on the main sequence than the sun (O, B, A, and F stars). The G, K, and M stars have either very small convective cores or none at all; these stars, however, have outer convective zones, whereas the hotter stars do not.

How does the stellar engineer know when he has obtained a correct solution? There is a fairly simple test: by the time he has reached the center, he must have properly accounted for all of the mass of the star and all of its luminosity. In other words, he must be sure that he has not used up all of his mass and energy before he gets to the center, nor that he has any mass or energy left over by the time he has arrived there. If this does not happen, he must start over again with a different choice of hydrogen and helium abundances, and he must keep changing these until he obtains an acceptable solution. Work of this sort done for the sun gives the following results.

The hydrogen abundance within the sun is about 75%, the helium abundance about 23%, and all of the other elements about 2%. The central temperature is very nearly 15,000,000°K and the central density is about 160 g/cm^3. There is either no convective core or a small one that contains about 4% of the mass of the sun and extends out from the center to about 1/12 of the way to the surface of the sun. The proton–proton reaction accounts for all but a percent or so of the energy that is generated, and only a third of it is produced in the core, the remaining two-thirds being generated in the surrounding envelope.

A word should be said about our assumption that the stellar material is gaseous all the way to the center and also that the molecular weight is constant throughout a main-sequence star. Since the density becomes very large as we approach the star's center (many times larger than that of the densest known solid on Earth), it may seem incorrect to assume that the stellar material is gaseous close to the center. But this is not so because of the complete ionization of the atoms near the center. Material ceases to be in the state of a perfect gas only when the molecules composing the material are very closely crowded together. For a gas consisting of unionized atoms the external electrons in their orbits prevent the atoms from getting closer than 0.000000001 cm to each other without touching. But when these atoms are ionized, they can be squeezed almost 100,000 times closer before they are touching. We are thus justified in assuming that stars are gaseous even when they are as dense as white dwarfs.

The assumption that the molecular weight is the same throughout the star is only approximately true. Near the center where hydrogen is being used up fastest there is a greater concentration of helium than near the surface of the star. Moreover, as the star passes through space, it collects hydrogen from the surrounding regions so that the hydrogen content near the surface tends to be larger than that in the deep interior. Only if thorough mixing occurred would the assumption of uniform molecular weight be correct.

As we go up the main sequence toward the O-type stars, we find that the hydrogen content decreases (the stars are radiating faster and therefore using up their hydrogen more rapidly) and the central temperature, pressure, and density increase, so that the carbon cycle becomes dominant. Unfortunately, no detailed investigations such as that for the sun have been made of the interiors of F-, A-, or B-type main-sequence stars, so that we do not know just where along the main sequence the carbon cycle takes over.

If we try to construct giant or supergiant stars according to the pattern we applied to main-sequence stars, we run into difficulty at once because these stars are very large and luminous even though their masses do not differ very much from those of main-sequence stars. The difficulty becomes clear if we consider what would happen if we tried to change the sun into a giant star. To begin with, we would have to increase its mass by about 5 times to obtain an ordinary giant and perhaps 50 times to obtain a supergiant. If we did nothing else, this would bring about an enormous increase in the central temperature, density, and pressure, But we would also have to spread out the sun with its increased mass until it extended over a volume millions of times greater than its original volume, and this would result in a large drop in the internal temperatures and densities. In fact, the temperature at the center would fall to only a million degrees or less; far too low to support a thermonuclear reaction even on a modest scale, much less on the vast scale necessary to keep a giant or supergiant star supplied with the energy it needs to exist.

From a consideration of the luminosity of these stars, we know that only the carbon cycle can give the required energy, and for this process, temperatures between 30,000,000 and 40,000,000°K are required. But how does one go about constructing such stars without running into the difficulty that we have just described? This problem was solved by the Princeton University astronomer Martin Schwarzschild, who showed that we can construct giant stars by having more than two zones in the star, with the chemical composition different in each zone.

The stellar models that have been constructed and that are in good agreement with giant and supergiant star data are those in which there is a central core composed entirely of helium, everywhere at the same temperature (the isothermal helium core). Although this core represents the ashes of the thermonuclear process that generates the energy supplied to the star, it still contributes some energy to the star because it undergoes gravitational contraction. Thus the temperature of the core continues to increase.

The main source of energy in this giant star model, however, is a thin shell surrounding the core in which the carbon cycle is transforming hydrogen into helium. This shell is rich in hydrogen and poor in helium, but as the hydrogen in the shell is used up, the carbon-cycle shell advances out toward the surface and the burnt-out helium residue in the old shell is added to the core. As the mass of the core increases in this way, it contracts faster gravitationally and grows larger.

In time, the core contracts to such an extent that its temperature reaches 100,000,000°K and a new type of thermonuclear reaction occurs. Three helium nuclei combine to form carbon, and this process releases energy fast enough to account for the luminosities of giant and supergiant stars. This triple helium reaction occurs so rapidly that it is called the helium flash. The rate at which energy is released during the helium flash is enormous but it lasts for so short a time that nothing drastic happens to the star as a whole, and after the flash, the star settles down to an orderly existence. This model accounts for giants and supergiants as stars that have evolved from main-sequence stars. We now describe briefly how stars evolve from the main sequence to the giant and supergiant branches of the H–R diagram and what happens to them after that stage. We begin by considering first the gravitational contraction of a cloud of dust and gas into a main-sequence star, which we treated briefly in Chapter 11.

Since all stars, as described in Chapter 11, are formed by gravitational contraction from clouds of gas and dust, the events of their very early lives are about the same except for the duration of their infancy, which depends on the star's mass. The more massive the star, the more rapidly it contracted from the cloud in which it was born and became a main-sequence star, and the higher up on the main sequence it landed when it began the stable part of its life. Thus a contracting cloud whose mass is only 8% of the sun's mass takes about 800 million years to reach the main sequence, whereas a contracting cloud whose mass equals the sun's mass reaches the main sequence in about 30 million years. If the initial mass of the cloud is three or more times the sun's mass, the contraction time is only a few hundred thousand years.

Because all stars contract to the main sequence in pretty much the same way, we use a star like the sun as an example. Contraction starts quite slowly but accelerates rapidly as the initial cloud collapses down to about 100 times the sun's present radius. As it does so (taking no more than a few hundred years), it releases vast quantities of gravitational energy, half of which is radiated away into space in the form of ordinary electromagnetic radiation (light and heat) and half of which remains within the protostar as heat (the internal energy of the random motions of the atoms and molecules). This internal energy raises the interior temperature of the collapsing protostar so high that the internal pressure slows the collapse down appreciably and the subsequent contraction proceeds in a slow, stable manner.

At this stage the protostar is a very luminous cool sphere and looks very much like a red giant; it has a surface temperature of about 3000°K and emits about 700 times as much energy every second as the sun does now. The chances are practically zero, however, that any of the red giants now visible are stars in their infancy, since protostars pass through this contraction so fast that hardly any are around for man to view. Moreover, if one of the typical red giants we now see were a contracting protostar, astronomers would easily have detected definite changes in its appearance during the last century.

Once the protostar contracts down to a sphere whose radius is about 60 times the sun's radius, after about 1000 years, its luminosity is about 500 times that of the present sun; its contraction then proceeds quite slowly. During this slow contraction, the convection keeps the protostar homogeneous. After about 1 million years of slow contraction, the radius of the contracting protostar is about twice that of the present sun and its luminosity is about 1.5 times as large as that of the sun. During this million-year period of contraction, the temperature in the deep interior becomes high enough for thermonuclear reactions to occur, and such light nuclei as deuterium, lithium, beryllium, and boron are transformed to helium; at the same time a good deal of the carbon is transformed into nitrogen, thus establishing the presently measured ratio of carbon to nitrogen in the sun. Contraction continues as the protostar becomes much less luminous and prepares to become a main-sequence star; after contracting for another 20 or 30 million years, it finally settles onto the main sequence and begins its life as a normal main-sequence star; it is now the initial, or zero-age, sun. The temperature at the center of the initial sun is about 10 million°K, which is high enough to trigger the proton–proton chain of thermonuclear transformations that we have already described. With the onset of thermonuclear fusion, gravitational contraction stops and all of the energy comes from the fusion of hydrogen.

The initial sun is about 4.5 billion years younger than the present sun and 38% as luminous; its radius is 4% smaller and its surface temperature about 10% lower than that of the present sun. In other words, the initial sun is a chemically homogeneous, cooler, redder, smaller, and less luminous gaseous sphere than the present sun, which, with a central temperature of about 15,000,000°K, is already chemically inhomogeneous in its central core because of the past transformation of hydrogen to helium, and which is beginning to evolve away from the main sequence.

The description of how a population-I star like the sun contracts from a cloud to its zero-age main-sequence stage applies as well, with some variations in certain specific details, to population-II stars and to population-I stars more massive than the sun. Massive stars descend to the main sequence much more rapidly than the sun and their initial main-sequence position is higher up than the sun's position; they are bluer, hotter, and more luminous than the initial sun. Their central temperatures range from 20,000,000 to 30,000,000°K, which is hot enough to keep the carbon–nitrogen cycle going.

Since the population-II stars are the oldest stars in the universe and were formed from the primordial gas following the big bang, we discuss their evolution first. Another reason for discussing the evolution of these stars before population-I stars is that the latter could never have been born if population-II stars had not existed and evolved first, for it was in the deep interiors of the very oldest population-II stars, near the ends of their lives, that the heavy elements required for population-I stars were built up by thermonuclear fusion from the primordial hydrogen and helium.

The thermonuclear fusion of hydrogen into helium proceeded extremely slowly in the deep interiors of zero-age main-sequence stars like the sun or zero-age population-II stars initially. Indeed, even now, as already noted, any single proton, on the average, may wander around for billions of years inside the sun before it is caught in the thermonuclear dance that binds it with three other protons to form a helium nucleus. The reason for this is that the probability that the first step in the proton–proton chain will occur is extremely small because it involves the creation of a positron and an antineutrino, which can occur only when one of the protons changes into a neutron. In spite of the slowness of the proton–proton chain, it is the prime thermonuclear process that accounts for the luminosities of the cool main-sequence stars. The sun is radiating at its present rate because in its deep interior about 564.5 million tons of hydrogen is fused into 560 million tons of helium every second via the proton–proton chain. To keep itself going, the sun is converting 4.5 million tons of mass into energy every second in spite of the slowness of the proton–proton chain; there are such vast quantities of hydrogen (protons) in each cubic centimeter in the deep interior of the sun that very large numbers of them are undergoing fusion at any given moment. In fact, there are about 50 trillion trillion protons in each cubic centimeter of the sun's core, where the proton–proton chain is operating. Since this core contains about 30 million trillion trillion cm^3, some 1000 trillion trillion trillion proton–proton fusion chains must be present every second inside the sun, even if any single proton meanders around for billions of years before it is captured by another proton to start its transformation to helium. Though the sun and stars like it are converting 4.5 million tons of mass into energy every second via the proton–proton chain, these stars are so massive that the process will diminish their masses by only 0.1% in 15 billion years. Hydrogen is thus a very long-lasting fuel for such stars.

Although the similarity between the sun and the early main-sequence life of a young population-II star is sufficiently great to allow comparison, as time goes on important differences develop. The precise details of the way a population-II star evolves away from the main sequence depend on the amount of He^4 that was present initially. If one accepts 25% as the concentration of He^4 in the initial main-sequence population stars, computed evolutionary models give evolutionary paths that agree fairly well with the paths deduced from the H–R diagrams of globular clusters that consist of population-II stars. As initial population-II stars, whose masses were similar to the sun's, burned their hydrogen, they moved slowly up the main sequence, becoming hotter, bluer, and more luminous. These stars began to turn off the main sequence, moving upward and slightly to the right when their luminosities increased by about a factor of 10 and their surface temperatures increased by 15–20%. The time an initial population-II star spent on the main sequence depended on its initial mass. Thus, a population-II star whose initial mass was about 0.75 solar mass spent about 12 billion years on the main sequence and then evolved quite rapidly—in less than a billion years—to the red-giant region.

The population-II star began to turn off the main sequence because the thermonuclear fusion of hydrogen in its core transformed the star from a chemically homogeneous to a chemically inhomogeneous structure. This turnoff was very slow and gradual at first but proceeded quite rapidly toward the giant branch of the H–R diagram when the star had exhausted all the hydrogen in its core, which then consisted only of He^4. The star was then in its subgiant stage, during which it generated energy in a thin shell surrounding the helium core via the proton–proton chain. No nuclear energy was generated in the helium core itself during the subgiant stage, because the temperature there was far too low to involve He^4 in any kind of nuclear burning.

The rapid increase in the amount of energy generated in the hot hydrogen-burning shell and the contracting helium core caused the star to expand (just as in the case of evolving population-I stars) and as it did so, its surface cooled. Thus, the star became larger, redder, and more luminous as it moved almost vertically upward on the H–R diagram, reaching the giant branch in about 1 billion years. At this point, owing to the continuous contraction, the temperature in the core rose to 100,000,000°K and a sudden and drastic change occurred in the core. The He^4 nuclei in the very hot core began to fuse with each other in the first step leading to the triple helium reaction and to the formation of carbon-12 (C^{12}). Although a temperature of at least 10,000,000°K is required to impart enough speed to protons to ignite the proton–proton chain, it is far too low to cause He^4 nuclei to fuse and form heavier nuclei. The reason is that although a temperature of 10,000,000 or 15,000,000°K is large enough to overcome the mutual electrostatic repulsion between two He^4 nuclei (which is larger than that between two protons) and to bring them close enough to each other for nuclear fusion to occur, it does not occur, except momentarily, because the beryllium nucleus that is formed is not stable; it breaks down immediately into the original two He^4 nuclei again. At 100,000,000°K, however, a third He^4 nucleus enters the scene to change the whole picture. At this high temperature, so many pairs of He^4 nuclei are interacting to form beryllium-8 (Be^8) that there are always many Be^8 nuclei around, even though any one of them exists for no more than a

hundred-trillionth of a second. Owing to this, to the very high temperature, and to the great abundance of He^4 nuclei, some of the
short-lived Be^8 nuclei that are formed fuse with He^4 nuclei (alpha
particles) to form the very stable C^{12} nucleus and release energy.
This, as already mentioned, is the triple helium, or triple alpha particle reaction, which may be written as follows:

$$He^4 + He^4 \rightarrow Be^8$$

$$Be^8 + He^4 \rightarrow C^{12} + radiation$$

This leads to the helium flash, which we have already described.
Once the helium flash is over, the star settles down to an orderly
burning of He^4 in the core. In the early stages of helium burning,
most of the star's energy comes from the hydrogen-burning shell
surrounding the helium core, but in time the helium-burning core
supplies most of the energy. This brief description of the helium
flash and the triple helium reaction gives a good picture of what
happened inside a population-II star about 1 billion years after it
had left the main sequence and become a red giant.

To understand better the onset of the helium flash, we must
know something about the state of the compressed helium core just
before the flash occurs. Owing to the steady gravitational contraction of the core, its temperature rises continuously until it reaches
100,000,000°K and the conditions are set for the ignition of the triple
helium reaction (the helium flash). But whether this reaction occurs
in a mildly explosive fashion or in a flash—that is, with a very great
release of energy in a very short time—depends on whether the
helium nuclei and free electrons in the core behave like particles in
a perfect gas (move about freely in such a way as to maintain a
proper relationship between pressure, density, and temperature) or
not, and this depends on the mass of the star. If this mass exceeds
1.4 solar masses the weight of the stellar material compressing the
core is so great that the core continues contracting slowly like a
perfect gas and the helium nuclei and electrons move about freely,
as in a perfect gas, so that the pressure in the core is always proportional to the temperature. In that case no flash occurs because
when the helium is ignited, the sudden increase in temperature that
is generated results in an increase in pressure, which causes the
core to expand slightly and to cool itself. This acts as a brake on

the helium burning, preventing it from running away so that this burning proceeds at a steady, unhurried pace. It is only mildly explosive when it starts.

If, however, the mass of the star is less than 1.4 solar masses, the core stops contracting when the helium nuclei are very close together and the free electrons are moving about with energy enough to support the weight of the overlying stellar material. The electrons in the core are then said to be in a degenerate state moving about as they do in a metallic conductor, and the core itself behaves something like a rigid metal sphere. Under these degenerate conditions the pressure in the core is independent of the temperature so that the pressure does not rise with increasing temperature. Owing to this degenerate condition in the helium core, the core cannot cool itself off immediately by expanding when helium burning occurs, and a runaway reaction (the helium flash) occurs. Since, even in a degenerate state, the pressure does depend on the temperature to some extent, the pressure will increase slowly after the flash has set in and the core will expand until the degeneracy is lifted. The helium burning will then proceed in a more leisurely and orderly fashion.

The population-II star considered here had an initial mass of about three-fourths the sun's mass and an initial He^4 content of 25%, but the overall picture is the same for population-II stars with initial masses ranging from about 1.2 to about 0.65 solar mass and initial helium contents ranging from 10 to 35%. The principal difference was in the time it took these stars to go from their initial main-sequence positions to the giant branch. The more massive the stars were and the greater their initial helium content, the sooner they evolved to the giant branch. A star with 10% initial helium content and 1 solar mass would have required 20 billion years to reach the red-giant branch, whereas a star with 35% helium and 0.75 solar mass would have required 12 billion years. In any case, these data, derived from theoretical evolutionary models of population-II stars, tell us that the stars in globular clusters are the oldest objects in the universe and that they had initial masses not much different from the sun's mass and helium contents of some 25%. Stars much more massive than the sun are not found in globular clusters, because such stars must have evolved so far during the cluster's life span that they are now burnt out and contribute little to the cluster's luminosity.

We have discussed the evolution of population II stars in some detail here because all elements heavier than He^4 now in the universe were formed in these stars; and the synthesis of carbon in the core of these stars was the beginning of this buildup process. As the star transformed more and more of its helium to carbon, a core of carbon developed at the center of the helium core and two things ensued: the carbon at the interface began to fuse with helium to form oxygen-16 (O^{16}), and the carbon core began to contract because no nuclear energy was generated within the carbon core itself to support the weight of the core. In time, however, the temperature within the carbon core, owing to the core's contraction, rose to hundreds of millions of degrees and the carbon nuclei began to fuse in pairs to form magnesium; at the same time, many more oxygen nuclei were being formed from carbon and helium, and these oxygen nuclei were in turn fusing with other He^4 nuclei to form neon.

Although the full details of how still-heavier nuclei were synthesized have not been worked out, a good overall picture of events during these last stages of the star's life can be painted in broad strokes. As successive cores were formed at the star's center at each new stage of the buildup of heavy elements, the star's central temperature continued to rise owing to the continued core contractions until it reached more than 1 billion°K. At these successively higher temperatures, the alpha particles were captured by the successively heavier nuclei until all of the helium in the original helium core was exhausted and nuclei of such heavy elements as silicon, sulfur, argon, calcium, and iron were built up. This heavy-element buildup produced by the capture of alpha particles did not go beyond iron, because alpha particles cannot be captured by iron nuclei. When a very energetic alpha particle collides with an iron nucleus, the latter is disrupted into less massive nuclei and other alpha particles; the buildup beyond iron does not occur. At this stage of the star's development, a drastic change occurred because all of its nuclear fuel was exhausted and there was no way for it to replenish the energy that was rapidly streaming out of its interior. As the star's internal temperature dropped, its internal pressure also dropped, and a violent gravitational collapse occurred because the star could no longer support its own weight. This collapse generated a tremendous amount of gravitational energy, which sent vast shock waves outward through the star, causing it to explode violently and to become thousands, or even millions, and billions of times as luminous as it was.

The exact details of this stellar outburst are not fully understood, but there can be no doubt that such violent events occurred and are still occurring in the lives of stars as they pass through old age. This conclusion is supported not only by the theory of the structure and evolution of stars but also by direct observational evidence presented by the eruption of supernovas, exploding stars that suddenly flare up and become millions of billions of times as luminous as they were. This happens in a very short time and is accompanied by a vast outpouring of material from the star's interior. Indeed, as in the case of the Crab Nebula in Taurus, which was observed as a supernova by the Chinese in 1054, most of the stellar material surrounding the iron-rich core of the star is blown away, revealing the very hot, dense, small core. This core star at the center of the Crab Nebula is a hot neutron star that is spinning 33 times per second and emitting intense radio pulses.

Although nuclei heavier than iron cannot be built up by alpha-particle capture, the very heavy nuclei, such as those of bismuth, gold, lead, and uranium, were built up from iron by the absorption of neutrons that are emitted in copious quantities during the gravitational collapse and subsequent explosion of a supernova. Since the metal-rich population-I stars like the sun originated from the material ejected from novas or supernovas or from material ejected in some other way from old population-II stars, the present chemical compositions of these population-I stars are accounted for; the famous planetary nebulas are examples of highly evolved population-II stars emitting material more slowly than novas.

Since the gaseous matter emitted by exploding population-II stars consisted, in addition to hydrogen and helium, of such heavier elements as carbon, oxygen, calcium, and iron, these coalesced into heavy molecules, which in turn condensed in large numbers to form particles and grains of dust in cold interstellar space, just the way water molecules in the cold upper regions of the atmosphere condense to form raindrops, hailstones, and crystals of snow and ice. Such grains of dust accelerated the formation of population-I stars in two ways: First, owing to their large masses—much larger than the masses of individual molecules—the dust particles moved about very slowly and exerted a much greater gravitational pull on each other than did individual molecules; hence, they had a greater tendency than did molecules to collect gravitationally and form the nucleus of a protostar. Second, such dust grains prevented the hot radiation from the residual core of the nova from passing through easily, so that this radiation pushed the grains closer together, thus increasing their tendency to coalesce gravitationally. These effects in dust clouds led to a relatively rapid formation of population-I protostars, as can be seen in the photographs of such clouds. In these photographs there are actual examples in the famous Herbig–Haro objects of such clouds breaking up into nodules and condensations that contract almost as we watch them. Evidence of this process of population-I star formation is seen in such clouds in the sun's neighborhood. In the constellations of Orion and Taurus, quite young population-I stars are present, which could not have coalesced from the dust and begun their main-sequence lives more than a few million years ago. In our galaxy the metal-poor population-II stars outnumber the metal-rich population-I stars, but the latter are of much greater interest than the former because they present a complete evolutionary tableau of young, middle-aged, and old population-I stars. In addition, those stars have planetary systems that can support life. Population-I stars are called metal-rich stars only because the metal abundances in their atmospheres are much larger than in the atmospheres of population-II stars and not because these metal abundances are large compared to hydrogen and helium; even in these metal-rich stars the heavy atoms constitute only 2% of the total mass.

The evolutionary track of a population-I star away from the main sequence in the H–R diagram depends on the mass of the star and is thus different for an upper and lower main-sequence star. This difference arises because the upper main-sequence stars have convective cores so that their cores remain chemically homogeneous as they evolve and the He^4 in the core is built up uniformly and quite rapidly. This is not so for the lower main-sequence stars. The H–R diagrams of such evolving stellar models show in detail the changes that occur in the structures of such stars as they depart from the main sequence. The evolutionary tracks of a number of stars with a variety of different initial masses have been calculated; they agree remarkably well with the H–R tracks of the star in clusters.

In the early stages of the upper main-sequence population-I stars, hydrogen burning proceeds quite steadily near the center for anywhere from some 10 million years for a star of 15 solar masses to about 2 billion years for a star of 1.25 solar masses. During this hydrogen-burning period, these stars grow cooler and more luminous, moving upward and to the right in the H–R diagram. The lower main-sequence population-I stars, on the other hand, spend billions of years burning hydrogen very slowly—stars like the sun remain close to the main sequence for about 7 billion years—and become hotter and more luminous as they move upward and to the left.

When all of the hydrogen is exhausted, or very nearly so, in the cores of the upper main-sequence population-I stars, these cores contract and heat up quite rapidly; this produces an overall contraction of these stars, which causes their luminosities and surface temperatures to increase quite rapidly for a period that lasts anywhere from a few hundred thousand years for the more massive upper main-sequence stars to some tens of millions of years for the less massive upper main-sequence stars. Overall contraction goes on until the temperature at the surface of the hydrogen-depleted core is high enough to ignite the CN (carbon–nitrogen) cycle in a hydrogen-rich shell surrounding the core; the star then reverses its track in the H–R diagram, becoming cooler, redder, and somewhat less luminous. This goes on until the mass of the helium core, now completely depleted of hydrogen, equals about 10% of the total mass of the star. At this point the mass of the core is so large that its rapid gravitational contraction is unavoidable. As the core contracts, its temperature rises quite rapidly and the star expands into the red-giant region, becoming redder, much larger, and much more luminous. A star whose initial mass is 5 solar masses expands until its radius becomes 74 times that of the sun's. At this point the core temperature is about 100,000,000°K and the triple helium process ($3He^4 \rightarrow C^{12}$) is ignited in the core. This helium burning contributes to the luminosity, ultimately becoming the main energy source of the star in its red-giant stage.

The evolution away from the main sequence and into the giant branch of lower main-sequence population-I stars with masses equal to, or less than, the sun's mass proceeds much more slowly and uniformly than that of the upper main-sequence stars. In fact, the evolutionary tracks of these stars are quite similar to those of the very old population-II stars. Since the lower main-sequence stars have no convective cores, they suffer no sudden overall contraction as hydrogen is being depleted. They first move leftward almost parallel to the main sequence for 7 or 8 billion years, becoming bluer, hotter, and more luminous, and then turn gradually to the right as the hydrogen in the core is slowly depleted. When hydrogen depletion is complete, the core temperature rises as the core contracts, and the star rapidly expands into the red-giant branch. At this point, helium is ignited in a flash as the core temperature reaches 100,000,000°K, and the energy thus released causes the core and the entire star to expand. Helium burning then proceeds at a steady rate until enough C^{12} has been synthesized in the core for the C^{12} + $He^4 \rightarrow O^{16}$ reaction to begin and to set the stage for the subsequent buildup of the still heavier elements as the core temperature rises to hundreds of millions, and finally billions, of degrees. The physical processes that go on during this phase of the population-I star's life are quite similar to those that go on inside population-II stars that have passed through their triple-helium-burning stage and are building up heavier elements by means of alpha-particle capture. Just like population-II stars, the massive population-I stars end their lives by becoming supernovas after all of the helium in their hot cores has been transformed into the various heavy elements up to iron. There is, however, one important difference between the buildup of heavy elements in population-II and population-I stars. Since the former start out being almost all hydrogen and helium, elements whose atomic weights are not multiples of 4, such as fluorine-19 (F^{19}) and sodium-23 (Na^{23}), have little chance of being formed. The nuclei whose atomic weights are multiples of 4 are favored because 4 is the atomic weight of He^4, which plays the dominant role in heavy-element buildup. Population-I stars are formed from a chemical mixture that already contains heavy nuclei. Since these can capture protons in addition to He^4 nuclei, the restriction to nuclei whose atomic weights are multiples of 4 is removed. Thus, the oxygen nucleus can capture a proton to become an isotope of fluorine, the neon nucleus can capture a proton to become a sodium nucleus, and so on.

Note again the important—indeed, overwhelming—role that the initial mass of the star plays in its evolution away from the main sequence. In the calculated H–R diagram of the evolutionary track of a star of 5 solar masses, the intervals of time between successive stages in the star's structure and behavior can be indicated, and similar results can be derived from the theory of stellar structure and evolution for stars of larger and smaller mass. Without going into these details we note the following: a star of 15 solar masses evolves from the main sequence to the tip of the giant branch in about 12 million years; a star of 9 solar masses evolves to the red-giant tip in about 25 million years; a star of 5 solar masses takes some 90 million years to evolve; a star of 3 solar masses takes about 300 million years; and a star as massive as the sun takes over 10 billion years to reach the red-giant tip. Roughly speaking, the time in years that a population-I star spends on the main sequence equals 10 billion multiplied by the mass of the star (in solar mass units) divided by its luminosity (in solar luminosity units).

The general description of the evolution away from the main sequence of aging population-I stars as they transform their hydrogen to helium can be applied to the sun if the data for a star of 1 solar mass are used. The evolutionary track of such a star can be plotted in the H–R diagram from the calculated data. From the data we see that the sun is a very slowly evolving star that is just beginning to leave the main sequence. Its surface temperature is still increasing, so that it is becoming bluer, larger, and more luminous as it moves almost parallel to the main sequence. It will continue changing in this way, departing more and more from the main sequence in an upward direction for about 5 billion years. It will then turn sharply to the right and continue to move to the right for about 2 billion years, becoming much larger, considerably cooler, redder, and still more luminous. After this relatively quiescent stage, it will move drastically upward as it expands into its red-giant phase and will become about 1000 times more luminous than it is now.

Since the sun is now almost 5 billion years old, it has already converted half the hydrogen in its core to helium, so that it is no longer chemically homogeneous. At this stage of the sun's development the density of matter at its center, the gaseous mixture of H^1 and He^4, is about 160 g/cm^3 (compare this density with the overall solar density of 1.44 g/cm^3) and the temperature at its center is 15,000,000°K. At present the sun is converting mass into energy at the rate of 4.5 million tons/sec, but this rate will increase as the central temperature and density of the sun rise in the next few billion years. This will in turn cause the sun to become still hotter, somewhat bluer, and more luminous. It is obvious that this change will greatly affect Earth and life on it, just as the lower surface temperature and lower luminosity of the sun about a billion years ago greatly affected the emergence of life on Earth at that time. Since the sun's luminosity was about 10% smaller a billion years ago, Earth's overall surface temperature could not have been higher than 0°C, so that advanced forms of life could not have existed. Using the same kind of argument in reverse, one can see that the hotter, more luminous sun that will dominate the solar system in a billion years will drastically alter, or entirely destroy, life on Earth. The sun's luminosity will have doubled by that time and the overall temperature of Earth will have increased by at least 50% so that some rivers and lakes will begin to boil.

In time, all main-sequence stars will become red giants after they have transformed all of the hydrogen in their core to helium, because then the gravitational energy released by their contracting pure-helium cores will force them to expand. Their core temperature will rise rapidly to over 100,000,000°K, triggering a new series of thermonuclear reactions that will start with the transformation of helium to carbon and end with the formation of iron. How long a star will remain in this red-giant heavy-nuclei-building state is not certain, but a very critical stage will be reached when all of the helium has been exhausted and the star's interior consists of some carbon, oxygen, neon, etc., but mostly of iron in successive layers going inward. The star will then be unable to maintain its gravitational equilibrium and gravitational collapse will necessarily occur. What happens after that depends on how violent the collapse is and on the behavior of superdense matter, but this behavior, which must be deduced from the general theory of relativity and nuclear physics, is still not fully known. A number of different descriptions of such matter have been proposed that are similar in their overall features but that differ in certain details, depending on what one assumes about nuclear forces at very high densities. But even without discussing these fine points, one can deduce important overall characteristics that reveal the ultimate fate of the star.

Whether all the stellar matter will be scattered into space to form a dispersed nebula after the gravitational collapse or whether most of it will settle down to form a white dwarf, a neutron star, or, ultimately, a black hole depends on what follows the gravitational collapse. If the collapse is so extremely violent that a vast amount of gravitational energy is released suddenly, the explosion that follows may, indeed, be so violent that most of the stellar material will be propelled outward at many thousands of miles per second in the form of an expanding ionic envelope. The star will then have become a supernova. The history of the residual core will then depend on its mass and density, which will determine whether it is to become a neutron star or a black hole. This type of violent explosion will occur for the massive main-sequence stars; stars like the sun will not explode but become white dwarfs gradually.

Enough of these hot dense stars have been discovered within a few light-years of the sun to indicate that they are very numerous, as is to be expected if they (white dwarfs) are the dying stages of stars. A star will become a white dwarf only if its mass is less than 1.4 solar masses; if the star's mass exceeds this value at the end of its orderly evolution, it will become a white dwarf only if it gets rid of some of its mass. If it cannot get rid of its excess mass, the star suffers gravitational instability and is forced into a catastrophic collapse that pushes it beyond the white-dwarf stage into a neutron-star or black-hole stage. Since the sun's mass is already below this critical value, the sun can and probably will become a white dwarf in an orderly manner and will never reach the iron phase of evolution but remain with its carbon core intact. If the sun or a star like it does not collapse too violently, it may go through a series of minor explosions following each other in intervals of a few hundred thousand years and thereby get rid of about 10% of its mass. It will then contract gravitationally in an orderly way until the carbon nuclei in its inert core are squeezed so close together that the electrons will move around freely inside this carbon lattice and, by their kinetic pressure, support the sun—now a white dwarf—against further collapse. The sun will exist in this state, called a degenerate electron state, for billions of years, cooling off gradually until it becomes too faint to be seen, and whatever planets are left will continue revolving around it pretty much the way they were before, but cold and dead.

Just how a star slightly more massive than the sun gets rid of its excess mass before becoming a white dwarf is revealed in a remarkable group of objects called planetary nebulas, each of which consists of a tenuous expanding shell of gas and a very hot white central star whose temperature may be as high as 100,000°K. These objects were called planetary nebulas when they were first observed and photographed because they looked like planets on photographic plates or when viewed through the early telescopes. They have nothing to do with planets, however; they are stars that have reached the end stages of their evolution and are ejecting their outer layers of mass in the form of expanding shells. If we take their expanding shells of matter into account, these objects are thousands of times larger than our entire solar system. About 1000 such objects have been observed and catalogued, but there are probably many more in our galaxy. The mass of the expanding shell in a spiral planetary nebula is about one fifth of the sun's mass and is expanding at about 30 km/sec.

To explain these objects according to the evolutionary theory we have described above, we note that when a carbon core has developed in a star only slightly more massive than the sun, the temperature in this core cannot rise much above 600,000,000°K as the core contracts. This temperature is too low for thermonuclear processes involving carbon to occur, so the core continues to contract slowly emitting only gravitational and thermal energy from its hot surface. This raises the temperature of the surrounding shell of any helium that may still be left, and helium burning continues ever more rapidly until all of the helium in the shell is exhausted. During this process the outer layers of the star absorb the radiation from the helium shell, expand, cool, and become quite red so that the star becomes a red giant again. As this expansion and cooling continue, the ionized atoms in the expanding outer layers capture electrons and become neutral. This process goes on until the expanding outer layers are so thin optically that the hot carbon core (now a white dwarf) is visible through the expanding nebulous shell.

A white dwarf does not remain a white-hot object for more than a few billion years since it must cool off the way any hot object that has no internal source of energy does. The white dwarf cannot contract (thus cannot release gravitational energy) because the great pressure of its electrons, forming a degenerate gas, prevents this from occurring. All it can do, then, is radiate away the kinetic energy of its nuclei that are moving about randomly. As this thermal energy is dissipated, the white dwarf grows ever fainter until it becomes a cold, invisible black dwarf (*not* a black hole).

If the gravitational collapse of a star that is considerably more massive than the sun after its iron stage is extremely violent, it will become either a pulsar (a neutron star), spinning extremely rapidly and sending out intense bursts of electromagnetic radiation, or a black hole, a tiny superdense dead sphere of matter from which nothing, including light, can escape. Whether it becomes a pulsar or a black hole depends on how much the core is compressed by the collapse. If the compression is not too violent, so that the density of the core does not exceed about 1000 trillion g/cm^3, the core contracts during this violent implosion, and as the core's density increases beyond white-dwarf densities—about 100 million g/cm^3—the free electrons that could support a white dwarf against collapse will be forced into the iron nuclei, where they will combine with protons to form neutrons. This will continue until the nuclei in the core contain too many neutrons to be stable. Neutrons will then drip off the nuclei in the core, slowly at first, but more and more rapidly as the compression continues, until all of the nuclei have disintegrated and only neutrons are left. Because such closely packed neutrons form a degenerate neutron gas, or a superfluid, and hence repel each other at this stage (densities on the order of 10 trillion g/cm^3), the core will be a stable neutron star with a 10-mile diameter. At first it will be pulsating very violently and rotating rapidly with a very strong embedded magnetic field, but the pulsations will die out very quickly, leaving a dense sphere spinning on its axis about 30 times per second. As the very strong magnetic field of this final pulsar stage sweeps around during each rotation, it radiates strong electromagnetic waves. At this stage the surface of the pulsar is a solid iron crust only a few hundred meters (possibly a kilometer) thick but extremely dense. Such residual stars (pulsars) were discovered in 1967 and since then they have been studied intensely.

If the implosion of the stellar core is so violent and rapid that the nuclear repulsive forces cannot stop it, the collapse will slow down but not come to a complete halt at the neutron stage. The core density will increase until the core becomes a black hole, whose properties can be deduced from the general theory of relativity.

The general theory of relativity asserts that if the mass of the collapsing core of any star exceeds 3 solar masses, nothing can stop the core from contracting down to a black hole; the inward gravitational pull will be so strong that the nucleon–nucleon repulsive forces are not strong enough to prevent the core from contracting down to a point (in theory). Although the iron core of an imploding star like the sun will certainly not meet this condition, since its mass will be smaller than that of the sun, it can still become a black hole if its density exceeds a certain critical value that is determined entirely by the mass of the core. To obtain this critical value of the density for a given mass, one divides the speed of light raised to

the sixth power by 33.6 times the product of the cube of the gravitational constant and the square of the mass. We write this as

$$\text{Critical density for a black hole} = \frac{c^6}{33.6G^3M^2}$$

where c is the speed of light, G is the gravitational constant, and
M is the mass of the core. This equation shows that the more massive a collapsing core is, the smaller its density has to be for it to
become a black hole. In fact, if a collapsing star had a mass equal
to 100 million solar masses, it would become a black hole if its
density were no greater than that of water—1 g/cm^3. If after the
initial implosion and the subsequent explosion of matter the mass
of the collapsing core is not much smaller than the sun's present
mass and if its density at any stage of its contraction exceeds 10,000
trillion g/cm^3, it will collapse down to a black hole. Its density will
acquire the necessary excess if the collapse of the core is so fast
that it is halted only momentarily, but not completely stopped, by
the nucleon–nucleon repulsive forces of the neutrons when the density is slightly less than the critical value given. The core will then
rush past its neutron-star phase and contract further, thus forcing
the density past the critical black-hole value. Everything will then
be lost and no reversal or halting of the complete collapse will be
possible; it will go on collapsing to a point (in theory).

One may ask what justifies the name *black hole* for such a
collapsing object. Consider a spherical mass like the sun from whose
surface particles are escaping; to escape, they must move with at
least the speed of escape from this surface, which is determined
entirely by the mass of the solar sphere and its radius. If the entire
sphere is now compressed so that its radius is reduced by a factor
of 4, everything on its surface becomes 16 times as heavy, the speed
of escape from its surface becomes twice as large, and so on. Suppose now that this spherical mass contracts until the speed of escape
from its surface equals the speed of light; this happens when the
radius of the contracting sphere (called the Schwarzschild radius in
honor of the German astronomer Karl Schwarzschild, who first
solved Einstein's equations) equals $2GM/c^2$, where M is the mass
of the sphere, G is the gravitational constant, and c is the speed of
light. Since no light can escape from a sphere whose radius is smaller
than this value, the sphere, on collapsing to this stage, is invisible

to any observer at a distance from it greater than the Schwarzschild radius. Not only can no light escape from such a sphere, but any light directed to it from the outside can never be reflected out; it will be trapped in the curved space-time geometry of the sphere as soon as it passes the Schwarzschild radius. In other words, such a collapsed sphere can never be seen; hence, the designation of this object as black. For this reason also, the surface of such a sphere is called the horizon of the black hole.

Now consider an astronaut in a spaceship falling freely toward the black hole; as long as he is outside the Schwarzschild radius, space and time have their usual properties, which means simply that the past and future do not depend on the direction of motion of the spaceship. He can move toward the black hole, away from it, and sideways as he pleases, or he can stop himself. But once he passes the Schwarzschild radius, he no longer has any choice as to how he can move; he will move into the black hole no matter what he does or how hard he tries to avoid it. This is so because in the geometrical region within the Schwarzschild radius, space and time are interchanged. This means that going from the past to the future in time is related to one, and only one, direction in space at a given point; that direction is in toward the black hole. An object can no more reverse its direction and go outward at that point than time can go from the future to the past. The name *black hole* is completely justified; all light passing the horizon of the black hole is absorbed by it and all objects penetrating the horizon must fall into the hole.

CHAPTER 16

Variable Stars

But he, with first a start and then a wink,
Said, "There's another star gone out, I
think."

—GEORGE GORDON, LORD BYRON, *George III*

The description of stellar evolution given in the previous chapter leads quite naturally to variable stars, because as a star evolves away from the main sequence, it may, if it is massive enough, pass through instability regions in the H–R diagram. In these regions the star's internal gravitational equilibrium is disturbed and relatively rapid structural changes occur that produce variations in the star's luminosity. All stars, of course, change slowly as a consequence of their evolutionary development, but these are not the changes we have in mind, for these slow changes occur over hundreds of millions or billions of years. Variable stars are those whose brightness variations can be observed from day to day or hour to hour.

If we define a variable star as one that shows an observable change in its characteristics, however slight, we have to include many stars as variables that are ordinarily not considered as such. Thus, the sun would have to be classed as a variable because there are constant slight variations on its surface. In addition, the periodic appearance and fading away of sunspots give a variable character to the sun. Variations have also been observed in the sun's magnetic field, and, more recently, the sun was found to be pulsating.

Since many stars probably exhibit surface spots and also variable magnetic fields, we would have very large numbers of variables if we did not limit ourselves further in classifying a star as a variable. We define a variable star here as one that shows a distinct and measurable change in its brightness. The nature of this variation is best presented by a light curve, which is a day-to-day graph showing how the apparent brightness of the star changes.

273

Two kinds of changes can occur in the apparent brightness of a star, one stemming from a structural change in the star itself, and the other from an accidental configuration of objects outside the star. In other words we have *intrinsic variables* (stars that undergo internal changes) and *apparent variables* (stars that remain unchanged internally but are periodically eclipsed by some other star). We limit ourselves here to the intrinsic variable stars.

We divide intrinsic variable stars into two groups; one contains the stars that undergo unpredictable or irregular changes, and the other contains the periodic variables. We subdivide the stars in the second group into two subgroups according to whether they take a long or a short time to go through a complete cycle of changes. A long or short time here is not sharply defined, but is relative to the midpoint between the variables having the shortest periods and those having the longest periods. Since this midpoint is about 60 or 70 days, we call stars with periods longer than 70 days long-period variables and those with periods less than 50 days short-period variables. Actually the long-period variables have periods that range from 70 to about 500 days. Most of the long-period variable stars are red and yellow giants with luminosities more than 100 times that of the sun. The amazing thing about these slowly varying stars is that their luminosities fluctuate by a factor of 100 or more during a complete cycle of variations, and any mechanism that causes such a series of changes in a star must be fairly complicated.

The most famous of the long-period variables is the star Omicron Ceti in the constellation of Cetus. In 1596 the observer David Fabricius noticed that this star was not visible to the naked eye in October although it had been visible earlier that year; it was observed again in 1603 by Bayer. In 1638, the star's brightness variations over an entire period, which is about 330 days, were observed telescopically for the first time. These variations were so remarkable that Omicron Ceti was given the name Mira, meaning "wonderful"; it is referred to most often today by that name.

During its variations, Mira ranges from an apparent magnitude of 9 to an apparent magnitude of 3, and its surface temperature changes from 1900°K to 2600°K. The changes in its surface temperature lead one to predict that its luminosity will increase by a factor of about 4 and yet the observed change is about 100-fold. These two statements are not contradictory because the factor of 100 refers to the visible radiation whereas the factor of 4 refers to the total emitted radiation. When the star is cool, most of the radiation is emitted in the infrared region of the spectrum, and as the surface temperature increases, more and more of the radiation is shifted into the visible part of the spectrum.

It is quite probable that the long-period variables are pulsating since their spectral lines suffer a periodic Doppler shift, indicating that the surface of the star is alternately approaching and receding from Earth. For Mira the velocity of the surface varies by about 10 km/sec.

An interesting feature of the long-period variables is that strong bright lines appear in their spectra just before maximum brightness is attained. These lines present quite a puzzle to the astronomer because their Doppler displacement shows that the atmospheric layer in which they originate is receding from Earth while the part of the atmosphere that generates the dark lines is approaching. Apparently, these stars are both contracting and expanding at the same time! Many of these long-period red variable stars have dense, hot companions revolving about them, which undoubtedly play an important role in influencing the behavior of these variables.

We come now to a most remarkable group of stars that have played an enormous role in extending our knowledge of the heavens and of the depths of space. These are the short-period variables, which consist of two groups: the Cepheid *variables*, named after the star Delta Cephei, which was the first one of its kind to be observed, and the RR Lyrae stars. There are two kinds of Cepheids, which are quite similar in their behavior, and for a long time astronomers did not differentiate between them. But then it was discovered that Cepheids, like ordinary stars, consist of a group of metal-rich stars (population I) and a group of metal-poor stars (population II).

The RR Lyrae stars have periods smaller than 0.8 day and even as short as 1.5 hours. These stars are also called *cluster-type variables* because they are found in quite large numbers in the globular clusters that surround our Milky Way. They are called RR Lyrae stars because one of the most conspicuous stars of this group is the star RR Lyrae.

Although the variation of the light of Delta Cephei was discovered in 1784 by John Goodricke, the great importance of Cepheid variables as yardsticks of space was not discovered until 1912. This discovery stemmed from observations that were made from 1910 to

1912 by Miss Henrietta S. Leavitt at the Harvard College Observatory. Lying at the outskirts of the Milky Way are two faint patches of stars that are known as the Magellanic star clouds (actually, dwarf galaxies) in honor of Magellan, who observed them when he circumnavigated the globe. The smaller of these clouds, at a distance of about 160,000 light-years, is rich in a great variety of stars and is therefore an excellent source of observational material. This star cloud has a particularly large abundance of Cepheid variables and Miss Leavitt investigated these stars.

The advantage of studying a particular group of stars in a very distant star cloud is that one does not have to worry about the effect of the different distances of the stars on their apparent brightnesses. Since the star cloud is very far from Earth, we may assume that all of the stars in the cloud are equally far from us and that if one star in the cloud appears to be a certain number of times brighter than some other member of the group, it actually is that many times more luminous. Miss Leavitt used this fact when she made her great discovery about the relation between the intrinsic luminosity of a Cepheid variable and its period of variation.

She discovered that the average apparent brightnesses of Cepheid variables in the smaller Magellanic star cloud are larger for the stars that vary slowly than for the stars that change rapidly. It is clear from what we have just said that this is really a statement about the relationship between the intrinsic luminosity of a Cepheid variable and its period; this constitutes the important law known as the period–luminosity relationship for Cepheid variables. It is of tremendous importance because with it we can discover how luminous a Cepheid variable is just by determining the time required for a complete cycle of changes from maximum brightness through minimum and back to maximum again.

Since the data show that the apparent magnitudes decrease (brightnesses increase) as the periods of the Cepheids increase, this must also be true of the absolute magnitudes (luminosities) because all of these stars are at the same distance from Earth. The RR Lyrae stars (cluster-type variables) do not obey a period–luminosity law. They all have about the same absolute magnitudes.

Harlow Shapley saw the great importance of the period–luminosity law, and set out at once to replace the apparent magnitudes by the correct absolute magnitudes in the graph that correlates magnitudes and periods for Cepheids. If we can determine the absolute magnitude of one of the Cepheid variables in this graph, we can then set the complete scale of the period–luminosity curve and use it to determine the absolute magnitude of any Cepheid variable. Shapley accomplished this by first assuming that the Cepheid variables in the Milky Way are the same, and therefore obey the same period–luminosity law as those in the Small Magellanic Cloud. He then determined the absolute magnitudes of some of the Cepheid variables in our galaxy by statistical methods (these stars in our own galaxy are still so far away that reliable individual parallaxes are extremely difficult to obtain) and thus established a correlation between the apparent magnitudes in the period–luminosity diagram and the absolute magnitudes.

Shapley found that a star like Delta Cephei with a period of $5\frac{1}{2}$ days has an absolute magnitude of about -1.5, and he used this value to find the correct position or scale of the period–luminosity curve on the graph. Once this was established, it was an easy matter to determine the parallax (distance) of any Cepheid variable just by measuring its period and using the graph to obtain its absolute magnitude. This was of great importance in finding the distances of the spiral nebulas like the nebula in Andromeda that contain Cepheid variables whose periods can be easily measured.

This method of measuring the vast distances of extragalactic objects opened up undreamed-of vistas to the astronomer and at last permitted determinations of the structure of the universe itself. From 1917 on, when Shapley first announced his discovery, until a few years ago, astronomers had been busily engaged in applying the method to a survey of the distribution of the closer spiral nebulas. But the important discovery by Walter Baade in 1950 that population-I and -II stars exist led to a revision of Shapley's data and a change in the period–luminosity law.

There are two period–luminosity relations: one for the classical Cepheids like Delta Cephei (population I) and the other for population-II Cepheids. The classical Cepheids are about 1.5 magnitudes more luminous, on the average, than the population-II Cepheids, so it is important to know which group one is using in determining extragalactic distances. An interesting point in connection with these two groups of stars is that the cluster-type variables are also members of population II.

Baade pointed out the importance of differentiating between the two types of Cepheids by showing that the distance of the Andromeda nebula was not the same when determined by the two period–luminosity laws. He concluded that this discrepancy was due to the error in determining the position of the classical period–luminosity curve on the graph because it was based on the uncertain measurements of the parallaxes of the Cepheids in our Milky Way. If we take the population-II Cepheid curve as correct, all the previous extragalactic distances have to be multiplied by a factor of almost 2. This also means that Shapley's absolute magnitudes for the classical Cepheids have to be corrected by decreasing them by about 1.5 magnitudes. Thus Delta Cephei must now be assigned an absolute magnitude of − 3.

The Cepheid variables are among the most luminous known stars; all of them belong to the supergiant branch and are members of spectral classes F and G. Most of the classical variables have periods in the neighborhood of 5 days. Although about 500 Cepheids have been recognized in the Milky Way, no more than a dozen are visible to the naked eye, among which are the stars Polaris (the North Star), Delta Cephei, Eta Aquilae, Zeta Geminorum, and Beta Doradus. A classical Cepheid has a range of about 1.5 magnitudes from its lowest to its highest brightness.

The cluster-type variables can be used to determine the distances of the globular clusters because they are all of about the same absolute magnitude, namely, 0.5. Since these stars are quite abundant in globular clusters, the distance of a cluster can be found simply by measuring the apparent magnitude of any one of the cluster-type variables in the cluster. From this quantity and the knowledge that the absolute magnitude of the variable is 0.5, the parallax of the variable can be found from the formula described in the chapter on stellar luminosities. This quantity is, then, the parallax of the globular cluster. This method was used quite extensively by Shapley in his investigation of globular clusters.

Spectral analysis of the light from the Cepheid variables shows that the surfaces of these stars are oscillating back and forth, which indicates that they are pulsating. Without going into details, we can easily see that regardless of the nature of the pulsations, some kind of period–luminosity relation must hold. The denser a Cepheid is, the larger the internal pressures are and therefore the more rapidly it will contract and expand once it is set pulsating by some agency or other. In other words, a dense Cepheid has a smaller period of pulsation than one less dense. But since the masses of all of the

Cepheids are about the same, the dense Cepheids are smaller than the less dense ones and therefore fainter than the latter because their surface areas are smaller. But this means that the short-period Cepheids are less luminous than the long-period ones; this is the period–luminosity law. If one applies the correct mathematical analysis to a pulsating star, the period–luminosity relation (really a period–density relation) can be stated as follows: The product of the square of the period and the average density is the same for all of the Cepheid variables.

The pulsations in Cepheid variables are caused by the trapping of energy in the zone below the photosphere where hydrogen and helium are partially ionized. The relationship between the variations in temperature and the state of ionization of the hydrogen and helium in the outer zones of a pulsating star accounts not only for its pulsations but also for the phase lag between the light curve and the velocity curve. The outer ionization zones behave like a delayed driving mechanism for the pulsations. In undergoing ionization, they absorb a certain fraction of the radiation released by thermonuclear reactions in the deep interior when the star is most highly compressed (highest internal temperature), and then release this ionization energy a quarter of a cycle later when the star is expanding most rapidly because the ionized atoms then recapture their elections. Some of this released energy keeps the star expanding and some of it contributes to the total energy radiated by the star, thus giving the star its maximum luminosity when its expansion is most rapid.

Since ionization zones are present in most stars, one may ask why most stars are not variables. The reason is that ionization zones can act as pulsation-driving mechanisms only if a very narrow range of internal conditions is fulfilled. The theory of stellar evolution has been applied to variable stars, and we now know that conditions for pulsations are fulfilled only when the evolving star lies in a very narrow, almost vertical strip (the instability strip) in the H–R diagram. This strip, which has a width of about 0.02 along the horizontal axis (the log of the surface temperature axis) or about 50°K in the surface temperature, lies somewhat to the left of the red-giant tip (the tip where the helium flash occurs for stars of 5 solar masses) and well above the main sequence. It is so narrow that a massive evolving star spends very little time in it (a few thousand years at most) when it is approaching the red-giant tip. There is thus very little chance of finding a star undergoing pulsations before it enters the helium-burning phase of its life. But after the star starts burning helium, if it is massive enough, its evolutionary track returns to the instability strip, and the star remains there for a few hundred thousand years; it can then be detected as a Cepheid. Thus, the theory of stellar evolution tells us that Cepheid variables are massive stars in their helium-burning phases, and this has been confirmed by observations. Because the evolutionary path of a massive star may pass through the instability strip in the H–R diagram more than once, a given star, if massive enough, can become a variable (for intervals ranging from a few thousand to a few hundred thousand years) at several points during its life.

Not only do stars (the Cepheids) vary in a predictable manner over a well-defined period of time but some stars change in a random, unpredictable way. In a sense the sun and probably all other stars are irregular variables because changes are continuously occurring in these objects. For example, prominent solar eruptions occur that may at times throw hundreds of millions of tons of gaseous matter a million miles into space. Although in terms of events here on Earth these are tremendous upheavals, they are quite inconsequential as far as a star is concerned. We therefore do not consider stars like the sun as irregular variables and reserve this designation for stars in which the changes are drastic enough to affect the luminosity of the star in an observable manner.

The most numerous of the irregular variables are those that undergo slow changes over fairly long periods of time, and most of these stars seem to be the cool red supergiants, such as Betelgeuse and Antares. In fact, it appears that most red supergiants have a tendency to variability. Some of these irregular variables seem half regular, perhaps possessing two or more regular pulsations superimposed on each other. Betelgeuse behaves as though a periodic pulsation of about 200 days were superimposed on an oscillation having a period of about 6 years. Spectroscopic evidence indicates that these pulsations are accompanied by an increase in the diameter of Betelgeuse of about 60% as the star goes from minimum to maximum luminosity.

Of all the variable stars that we have observed in the heavens, the most dramatic are the explosive stars, or *novas*. The explosions are produced by sudden upheavals within the stars that are so vast that huge layers of material are ejected into space at speeds of thousands of miles per second. It is unfortunate that the name "nova" was applied to these variable stars, because they are not new, as the name implies, but rather old stars that suddenly burst into great prominence. Generally speaking, these are stars so faint and insignificant among the millions of other faint stars surrounding them that there is no reason to pay any attention to them before they become novas. Then in a matter of hours they may become so bright as to outshine Venus or Jupiter at their brightest. But this state of affairs is only temporary and after a few days or weeks of expending energy at this tremendous rate, the star gradually returns either to its previous uneventful existence or to a highly compressed state (white dwarf or neutron star).

We distinguish here between two kinds of novas, ordinary novas and supernovas, because the magnitude of the explosion in the latter is thousands of times greater than in the former. When a star becomes an ordinary nova, its brightness increases by about 12 magnitudes (about 100,000 times brighter than it was) in a matter of a few hours, and it emits as much radiation during this excited stage of its life as it ordinarily would in a few thousand years. But this stage represents a mere incident in the life of the star, and has little permanent effect on it.

Enough ordinary novas have been observed to enable us to describe the series of events that accompany the explosion in considerable detail. More than 100 of these objects have been recorded in our own galaxy and this number is increasing at the rate of about 2 per year. Since, however, we can see only a small part of the Milky Way, it is safe to say that as many as 100 such explosions occur in the Milky Way each year. Since there are many billions of galaxies in the universe, there are probably hundreds of billions of ordinary novas occurring yearly throughout the universe.

When an ordinary nova explodes, the sudden release of energy causes a shell of matter near the surface of the star to rush out into space at speeds in excess of 1000 miles/sec. This material forms an expanding envelope of gas that cannot be drawn back into the star because it is moving at a speed far greater than the speed of escape. These expanding envelopes can still be seen for Nova Persei, which occurred in 1901, for the Crab Nebula (a supernova), and for Nova Aquilae in 1918. As can be expected, violent changes occur in the spectrum of the star during the nova stage, with the dark lines showing large Doppler effects and considerable broadening. Bright lines of the sort associated with hot stars appear as the star reaches its maximum brightness.

If we accept R. Kraft's hypothesis that all ordinary novas are members of very close binaries, one of which is a hot white dwarf (the star that is to become the nova) and the other a star that is just about to enter the red-giant stage, we can explain the nova outburst as stemming from the accretion of hydrogen-rich material onto the hot surface of the white dwarf. To account for the existence of a binary system with the two stars at such different stages of evolution, we need merely assume that when the binary was formed one of the stars (the one that is presently a white dwarf) was a star of a few solar masses and, owing to its large mass, evolved rapidly to its present white-dwarf stage, whereas the red companion was a star of about 1 solar mass that evolved slowly and is just now approaching its red-giant stage. The white-dwarf member of the binary must have lost most of its mass before it could become a white dwarf, and it probably did so when its outer zones expanded as the star became a red giant. This loss of mass could have occurred in various ways, but most probably as the result of a transference of mass to its companion. Such a process occurs in a region whose surface passes through the so-called Lagrangian points of the binary system. These are equilibrium points at which the gravitational field arising from the two components of the binary is such that particles situated at these points move so as to maintain fixed distances from the centers of the two binary stars. If the surface of the expanding companion extends beyond the Lagrangian points, matter rushes from this expanding surface to that of the less massive star. In this way the rapidly evolving star loses enough mass to become a white dwarf.

The stage is now set for this white dwarf to become a nova as the surface of its red companion in turn expands beyond the Lagrangian surface and hydrogen-rich material passes from it to the white dwarf. One can show that, owing to the intense gravitational field on the surface of the white dwarf, the proton–proton reaction will be triggered explosively when a large enough quantity of the hydrogen-rich material from the expanding red star hits the surface of the white dwarf. As more and more of this material falls onto the white dwarf, it compresses the underlying hydrogen, raising its temperature to a point where hydrogen begins to burn (that is, to fuse).

This occurs explosively, creating a nova; the hydrogen-accretion phenomenon may be repeated over and over, accounting for periodic novas.

Supernovas are much rarer objects than ordinary novas and probably occur no more than a dozen times every few hundred years in they Milky Way. Although about 50 of these objects have been discovered, most of them have been found in distant galaxies, and only 3 are known to have occurred in our galaxy during the last 1000 years. In November 1572, Tycho Brahe observed what appeared to be a new star created before his very eyes in the constellation of Cassiopeia. It is no wonder that he thought this was a new star and assigned the name "nova" to it. The star became so bright that it was visible to the naked eye in broad daylight for a few weeks, and at night was brighter than Venus at its greatest brilliance. It then slowly faded and 2 years later became invisible. The nova that occurred in 1604 and is now known as Kepler's nova was also a supernova and was almost as bright to the naked eye as Tycho's star.

At the present time, a faint trace of the nova observed by Tycho has been detected, and there is a faint patch of nebulosity in the direction of Kepler's nova that is assumed to be the residue of that explosion. Today we can trace these residuals of supernovas with radio and X-ray telescopes. After the explosion of a supernova the gases that form the expanding shell are set in violent turbulent motion, and they emit radio signals and X rays that can be detected here on Earth or on artificial satellites equipped with special X-ray detectors.

Perhaps the most famous supernova in our Milky Way occurred in July 1054 as recorded by Chinese astronomers. These observations are confirmed in Japanese records, and we must accept them as correct even though no one in western Europe seems to have recorded this imposing event. And a very imposing event it must have been because the residue of this supernova is the object that we now call the Crab Nebula, an expanding gaseous nebula surrounding a pulsar that is about 4100 light-years from us. Thus, the actual explosion must have occurred more than 5000 years ago, but the signal first reached Earth about 935 years ago.

From our knowledge of the luminosity of the Crab Nebula, we know that the 1054 nova must have been a supernova, and at its maximum brightness it was probably 20 times brighter to the naked eye than Sirius and about 5 or 6 times brighter than Tycho's nova. And yet there is no reference to this object in any of the European manuscripts that we are acquainted with today.

That the Crab Nebula is indeed the remnant of the great nova of 1054 can be seen from the fact that the gases in this nebula are expanding in all directions away from the center at such a rate that 35 years ago the entire mass of gas must have been concentrated in a region of space no larger than a star, and the position of this star agrees with the Chinese records. At present the nebula is expanding at a rate of about 1000 miles/sec and is almost 2 light-years in extent across its largest dimension. The total mass of the nebula is estimated to be more than 15 times that of the sun so that before its explosion this star was probably a supergiant.

We receive a large amount not only of visible radiation from the Crab Nebula but also intense radio signals and X rays, which, along with most of the visible radiation, are to be accounted for by the magnetic fields that have recently been found threading the filaments of the nebula. In a sense, then, this nebula behaves like a huge cyclotron, causing the ions and electrons in the gaseous material to move in large magnetic orbits and to radiate energy as a result of this accelerated motion. The pulsar at the center of the nebula is the source of all of its radiant energy.

As we have already stated, we must look beyond our Milky Way to the distant galaxies if we wish to discover large numbers of supernovas. If, as appears reasonable, about a dozen supernova outbursts occur in each galaxy every 300 years, considering how many billions of galaxies there are in the universe, we are led to the conclusion that one such outburst occurs every minute somewhere in the universe. Here on Earth we see very few of these because most of the galaxies outside the Milky Way are at very great distances. When a supernova occurs in one of the visible galaxies, it may rival in brightness the whole galaxy of stars in which it resides and at times even exceed the galaxy in luminosity.

In August 1885, a supernova occurred near the center of the Andromeda Nebula, and at its maximum brightness its luminosity was one-tenth that of the entire galaxy, which is composed of hundreds of billions of stars as luminous as the sun. This was an excellent example of a supernova, but unfortunately for us the light from it arrived on Earth about 20 or 30 years too soon, at a time when spectroscopy was not well enough advanced to enable observers to discover very much about the nature of these objects. This particular supernova declined in brightness by more than a factor of 10,000 in 6 months and then quickly disappeared.

Though pulsars or neutron stars are not pulsating objects like Cepheid variables, we must classify them as variables since their radiation is periodic. The first pulsar, now called CP 1919, was discovered in Vulpecula by A. Hewish, S. J. Bell, J. D. Pilkington, P. F. Scott, and T. A. Collins in the summer of 1967 with the aid of a new large radio telescope. The signals from this object consist of a series of pulses, each lasting for about a third of a second, with an interpulse interval of 1.337295 sec (a constancy accurate to one part in 10^7). Following this initial discovery, other pulsars, with properties quite similar to the Vulpecula pulsar, but over a wide range of frequencies, were found, and about 100 are known at present with periods ranging from 1/1000 sec to about 3 sec.

Although astronomers initially thought that the periods of pulsars were constant to an incredible degree, we now know that the periods do change. The first six pulsars that were discovered (which include the two supernova remnants mentioned) have now been studied with sufficient accuracy and for a long enough time to know that their periods are increasing slowly. The greatest rate of increase is found for the very-short-period pulsars NP 0532 and PRS 0.0833-45. The first of these, which has a period of 0.033 sec (30 pulses per second), has now been identified as the hot central star in the Crab Nebula, and the second one as the residual core of the supernova in Vela.

The overwhelming evidence supports the theory that pulsars are rapidly rotating neutron stars and that the energy they emit as pulses of radiation comes from the kinetic energy of rotation of these neutron stars. As they lose their rotational kinetic energy, their speeds of rotation must decrease and the periods of their radio pulses must increase. Since the youngest pulsars have lost the least amount of energy, they must be spinning fastest (shortest periods between pulses of radiation). They must also be losing energy at the greatest rate (because they have the greatest store of rotational energy) and thus be slowing down faster than the long-period pulsars. This is precisely what is found for NP 0532 and PSR 0.0833-45, which is why we believe they are the youngest known pulsars.

From the time when pulsars were first discovered, astronomers tried to identify them with optical objects. This was finally done in

1969, when it was found that the hot white central star in the Crab Nebula blinks in its visible radiation at the same rate (30 times per second) as the radio pulses from pulsar NP 0532. This pulsar was already known to lie at or very close to the center of the Crab Nebula, so the discovery was not too surprising. The central star in the Crab Nebula had been observed for many years, but its visible radiation always appeared to be quite steady when received through a telescope because it blinks so rapidly. A special technique (the equivalent of a stroboscope) had to be applied to show that the central star of the Crab sends out its visible radiation in pulses that arrive every 0.0331 sec (this is just the period of the radio pulsar in the Crab). The main optical pulse and interpulse last for 0.0014 and 0.0028 sec, respectively, and their intensities are in the ratio 3.7:1.

Quite recently, the pulsar NP 0532 in the Crab Nebula was found to emit X-ray pulses as well as radio and optical ones; all these types of pulses are emitted at the same frequency intervals. The remarkable thing about this pulsar is that it emits 100 times as much energy in the X-ray region as in the optical, and 100 times as much in the optical as in the radio region. The data show that the period of the optical pulses increased from 0.033097500 sec on February 16, 1969 to 0.033097845 on February 22, 1969.

For some time after the discovery of pulsars, there was considerable conjecture and disagreement among astronomers concerning the nature of these objects and the ways in which they emit their radiant energy. From the very beginning it was clear that pulsars must be small, compact objects because of the short duration of the pulses emitted. The size of the emitting area of the pulsar cannot exceed the duration of a single pulse multiplied by the speed of light. This means that pulsars cannot be larger than white dwarfs, and initially many astronomers suggested that pulsars are pulsating white dwarfs. However, because the period of pulsation of a typical white dwarf, if it were pulsating, would be longer than 2 sec (the square of the period is proportional to the reciprocal of the density of the white dwarf), such objects could not give the observed periods of most pulsars. Nor can white dwarfs rotate fast enough to account for the observations, if we accept rotation as the cause of the pulsar phenomenon.

We must therefore reject white dwarfs entirely and go to neutron stars to explain pulsars. A typical neutron star, with a radius of about 10 km and a mass about that of the sun, would pulsate much too fast (a period of milliseconds) to emit pulses with the observed periods. We are thus left with the rotation of neutron stars to account for pulsars, a mechanism that was first proposed by T. Gold and that now seems to account for most of the observed features of pulsars.

As we have noted in our discussion of neutron stars, these objects result from the catastrophic collapse of the core of a giant star before the star becomes a supernova. Immediately after the supernova explosion, the surface of the residual core, a neutron star, is probably very hot and may also possess rotational, vibrational, and magnetic energy. The hot surface initially emits very energetic photons (X rays) and probably neutrinos and antineutrinos, resulting in a rapid cooling that is accelerated if strong magnetic fields are present on the surface. After 1000 years (more or less), the surface of the neutron star becomes too cool to contribute much thermal energy to the pulsar phenomenon, and most of the radiated energy comes from the very rapid rotation that results from conservation of angular momentum. The angular momentum of a rotating sphere (our neutron star) of mass M and radius R is proportional to $MR^2\omega$, where ω is the angular speed of rotation. Since the angular momentum) must be conserved as the core contracts (as R gets smaller), ω must increase since M remains the same. Therefore, the speed of rotation of the core rises rapidly as it collapses just before the supernova stage. But the energy of rotation is proportional to $MR^2\omega^2$ or to $J\omega$ where J is the angular momentum of the collapsing core. Since J remains constant as the core collapses and ω increases, the kinetic energy of rotation increases very rapidly.

Calculations show that this energy is enormous for a neutron star that is spinning some 10 or 30 times per second, and certainly enough to account for the energy emitted by pulsars. From the rate at which the period of the pulsar in the Crab Nebula is increasing we deduce that it is losing energy at a rate of about 10^{38} ergs/sec (a few thousand times the sun's luminosity). This energy that the pulsar pumps into the nebulosity of the Crab probably accounts for the total radiation emitted per second by the entire Crab in the X-ray, ultraviolet, visible, infrared, and radio parts of the spectrum. The initial store of rotational energy in a neutron star can easily be about 10^{52} ergs.

The mechanism by which the neutron star radiates away its rotational energy is not entirely clear, but magnetic fields probably play a very important role. It is not unrealistic to assume that the magnetic field strengths on the surfaces of pulsars are anywhere from 10^{10} to 10^{12} gauss because the intense collapse of the core of a supernova compresses even small magnetic fields enormously and increases their intensities a trillionfold. If the magnetic poles of a neutron star do not lie on the axis of rotation of the star (just as Earth's magnetic poles do not coincide with Earth's geographic poles), the intensity of the magnetic field as seen from Earth varies as the star rotates (like a cone of light from a lighthouse) and this generates electromagnetic pulses of the right period. Moreover, the rotating magnetic field greatly accelerates charged particles emitted by the neutron star (electrons and protons). These accelerated charges then spiral around the magnetic lines of force and emit synchrotron radiation of the sort that has been observed from the Crab (in a cone of such radiation that corotates with the star but whose axis is tilted with respect to the axis of rotation of the star).

Recently, some sudden, large, but nonpermanent changes have been detected in the periods of certain pulsars. The periods of these pulsars were found to decrease quite suddenly and then to return slowly to their original values as though there had been a sudden change in the dimensions of these pulsars followed by a readjustment. These phenomena, which have been referred to as glitch "starquakes," have been explained as follows: It is believed that the surface of a neutron star is an exceedingly dense (of nuclear density) but very thin (a fraction of a meter) skin of iron that is under severe dynamical stress. Under such conditions a crack may develop in the surface of the pulsar, causing it to readjust itself as the surface of Earth does after an earthquake. This causes the period of rotation to change suddenly, but then the period returns to its initial value when equilibrium is reestablished.

Most of the pulsars (about 100) that have been discovered lie close to the plane of our galaxy but some have been found at high galactic latitudes. We believe that all known pulsars are galactic objects. In time, however, we undoubtedly will find pulsars in nearby galaxies.

Among all of the variable celestial objects, X-ray sources are among the most interesting because they are emitted from matter under very unusual physical conditions. X-ray astronomy was unknown until artificial satellites and high-altitude rockets and balloons were available; even then it was a very difficult pursuit. But with the launching of the UHURU space satellite on December 12, 1970, X-ray astronomy advanced from the rather hit-or-miss operation it was with high-altitude balloons and sounding rockets, to a precise new field of astronomy. The UHURU experiment collected far more X-ray data in its short life than had been obtained in the preceding decade by all other methods. These data reveal a variety of X-ray sources among which are:

- *Supernova remnants*, such as in the Crab Nebula and the Cygnus Loop. X rays are emitted from supernova remnants as a natural result of the violent plasma vibrations following the initial explosion of the supernova.

- *Transient sources*, like those in Lupus. The few transient sources that have been discovered are sudden intense X-ray flares in regions where none had previously been observed. At their peak intensities, these transient sources are among the strongest in the sky, and they flare up and die down the way optical novas do. These sources, which increase in intensity by a factor of 1000 or more, may disappear in a short time or last for a year or more.

- *Galactic X-ray sources* consisting of celestial objects that can be identified with known classes of stars and stellar systems. The X-ray intensity or luminosity of these sources is on the order of 10^{37} ergs/sec. Among these sources are objects such as

    1. Sco X-1. A source of X rays, radio waves, and optical radiation.

    2. Cyg X-1 (probably a black hole). A source from which X-ray pulsations are emitted in intervals of about 0.1 sec, which means that the X-ray emitting region in this variable source is extremely small (about as small as or smaller than a pulsar). Lying in the same region as this X-ray source is a variable radio source that is coincident with a 9th-magnitude BO star that is revolving around a massive invisible companion every 5.6 days. If, in accordance with the mass–luminosity relation, the bright BO star has a mass of about 20 solar masses, the mass of the invisible X-ray companion must be about 13 solar masses and it must be smaller than 0.1 light-second. Since a star of 13 solar masses is much too massive to be either a white dwarf or a neutron star, the invisible

companion of the BO star is probably a black hole. X rays are emitted by this "black hole" as streams of ions flow from the bright BO star to the dark companion.

3. Hercules X-1. A periodic X-ray source like Cyg X-1, which is eclipsed periodically. The X-ray pulse period of 1.2 sec exhibits a variable Doppler shift with a period of 1.7 days, the eclipse period. The X-ray source is coincident with the irregular optical variable HZ Hercules.

4. Can X-3. The first known eclipsing, pulsing X-ray source. The pulse period is 4.8 sec and the eclipse period is 2.1 days. Here, too, the period of the X-ray pulses shows a variable Doppler shift that is in phase with the 2.1-day period of the eclipse. This X-ray source disappears in an erratic fashion from time to time for several days.

5. Cyg X-3. A variable X-ray source of great interest because of the intense flare that occurred in its radio component or companion. Although no significant change was observed in its X-ray luminosity, its radio luminosity increased suddenly in 1972 by a factor of 1000.

•*Extragalactic X-ray sources* such as the discrete sources in the Magellanic clouds. X-ray sources have also been found in association with the quasar 3C273, the giant radio galaxy NGC 5128, the Seyfert Galaxy NGC 4151, and rich clusters of galaxies such as the Virgo cluster.

Recently, four X-ray sources were identified with globular clusters. This discovery is important because globular clusters are the oldest components of our galaxy and presumably are free of dust and gas. It is therefore difficult to see how these objects can generate X rays unless we suppose that accretion of some kind occurs on very massive but difficult-to-observe stars in these clusters. We have discussed here only the most important types of variable stars; there are other less prominent types such as flare stars, magnetic variables, and shell stars, which are discussed in standard treatises on variable stars.

We complete this chapter with a discussion of the 1987 supernova that burst into prominence in the Large Magellanic Cloud on February 23, 1987. Designated as SN1987A, it is the first nearby (within a few hundred thousand light-years) supernova observed within the last four centuries. At a distance of 164,000 light-years, it is not as close as astronomers would like, but it is still a most remarkable object and an excellent source of information that we need to check the correctness of our astrophysical models of pre- and postnova stellar structure. In particular, we can use the SN1987A data to check our theory of the formation of neutron stars and the theory of the formation of the heavy elements beyond iron.

The visual signal from SN1987A was first picked up at the Las Campanas Observatory in the Andes of northern Chile quite by accident when the operator of the telescope, Oscar Duhalde, happened to glance at the Large Magellanic Cloud and saw a 4th-magnitude star (apparent magnitude) that he had never seen before. Barely visible to the naked eye, it was the brightest object in the Large Magellanic Cloud at that time; the explosion of SN1987A had increased its brightness from a 12th-magnitude star to a 4th-magnitude star so that its luminosity had increased by about 1500-fold at that time.

With all the observatories in the southern hemisphere and astronomers alerted to the new supernova, data about its progress began to pour into various data-collecting centers from every observatory and satellite from which the Large Magellanic Cloud could be observed. The occurrence of SN1987A was an astronomical "goldmine" in many respects because of the very propitious conditions under which it developed. In the supernovas of the past, very little information about the various stages they went through during their explosions was recorded, but every stage of SN1987A's explosive history was recorded in great detail. Most of the data confirmed completely the astrophysical theories about the stellar evolution of massive stars, the buildup of heavy elements in supernova explosions, and the transformation of the cores of such stars into neutron stars.

The immediate data indicated that SN1987A was a type II and not a type I supernova, which occurs in a low-mass binary system when vast quantities of hydrogen from the larger, more massive component are transferred to the white dwarf companion. This hydrogen accretion occurs so violently that the entire white dwarf is transformed into a thermonuclear reactor and becomes a type I supernova. The rapid implosion of the iron core of SN1987A with its vast release of gravitational energy marked it as a type II supernova; this status was confirmed by two other important observations: a burst of 11 neutrinos that was detected on February 23 before the electromagnetic radiation from SN1987A reached Earth, and the detection of the radioactive decay of the cobalt-56 isotope.

The neutrinos detected in the emissions from SN1987A had just the right energy one expects from an imploding iron core in which the free electrons combine with the protons in the iron core to form neutrons and thus produce a neutron star (a pulsar). The gravitational energy released during the implosion was 3×10^{53} ergs (equivalent to one-fifth of a solar mass) and most of this energy was carried off by 10^{58} high-energy neutrinos. Since neutrinos interact only very weakly with atomic nuclei, the SN1987A neutrinos passed through the thick layers of the supernova surrounding the core without hindrance. The electromagnetic radiation from the core, however, was blocked and this radiation showing the neutron star at the core did not penetrate the dense expanding shell of SN1987A until this shell had expanded sufficiently to become optically thin enough to reveal the pulsar at the core. As of March 1988 this pulsar had not yet been observed. Cobalt-56 was detected in the expanding shell of SN1987A because this cobalt isotope decays into iron-56 with the emission of gamma rays, which were recorded. From the intensity of these rays it was deduced that the mass of the cobalt produced equaled 70 times the mass of Jupiter. The cobalt decay energy was the dominant source of the thermal energy emitted by SN1987A in its early stages; indeed, the falling light curve of the supernova exactly fits the exponential decay curve of cobalt-56, which has a half-life of 78 days.

The SN1987A supernova is unique in that it is the only supernova whose progenitor star was observed and studied before it exploded; this star was the supergiant B3 blue star Sanduleak, with a surface temperature of 15,000°K and a luminosity 20,000 times that of the sun. This star aroused some controversy as to the correctness of the physical model of a type II supernova since the precursors of such supernovas have always been accepted as old red supergiants and not young blue supergiants. But this apparent contradiction can be explained if Sanduleak—which was a blue star when it was first studied in 1969—had previously been a red supergiant and had entered a blue phase sometime in the past.

The development of SN1987A as it expanded presented astrophysicists with an unusual opportunity for confirming their theories of the evolution of massive stars and the buildup of heavy elements in their interiors as they evolve from the main sequence to their red giant and supergiant stages. The successive layers of the expanding supernova as they are revealed through the attenuating atmosphere showed the various chemical-element shells, starting with hydrogen at the top, in the order in which the elements were built up. SN1987A was thus referred to as the "cosmic onion."

CHAPTER 17

The Milky Way (Our Galaxy)

I saw Eternity the other night
Like a great ring of pure and endless light,
All calm, as it was bright,
And round beneath it, Time in hours, days,
years,
Driv'n by the spheres
Like a vast shadow moved; in which the world
And all her trains were hurled.

—HENRY VAUGHAN, *The World*

We have been treating stars as though they were isolated objects in the sky, unrelated to anything else in the universe. Actually they are members of groups that ultimately influence the history and evolution of these same stars in various ways. Every star has neighbors all around it in the sky as though it belonged to some vast ensemble of stars that seems to extend indefinitely into space. It is not very easy to get a clear idea of the nature or structure of this ensemble just by looking at the stars with our unaided eyes; modern telescopes and electronic techniques are needed to reveal the structure not only of this ensemble (the Milky Way galaxy) but also of other ensembles like it (the distant galaxies).

Sir William Herschel, the great 18th-century British astronomer, attempted to get a picture of the distribution of stars in the Milky Way by counting as many as he could see with his telescope in different parts of the sky. This very tedious process is still the only reliable method that can be used, and in the hands of 20th-century astronomers, it has been enormously helpful in understanding the distribution of matter in space. Although Herschel's telescopes were quite puny by today's standards, he developed many of the observational techniques used by astronomers today.

William Herschel (1738–1822) was born in Hannover, Germany, and spent much of his youth as an oboist in his father's traveling band.[1] After Hannover was occupied by the French army in 1757, Herschel escaped to England where he supported himself by teaching and composing music and later, as an organist in the Octagon Chapel at Bath. Herschel's single-minded devotion to his music began to wane, however, after he happened to read Smith's *Opticks*, which described in great detail astronomical phenomena and outlined rudimentary techniques for building telescopes. His interest in astronomy was encouraged in 1772 by the arrival of his sister Caroline, who remained his trusted assistant throughout his scientific career.

Herschel initially pursued astronomy as a hobby by using borrowed telescopes and components but was frustrated by their poor magnification and resolving powers. As a result, Herschel painstakingly began to grind his own mirrors. Only after he realized that his ability to peer into space varied with the light-gathering ability of the mirrors in his telescopes did he become obsessed with building telescopes of unprecedented size and mechanical precision. Herschel had to do most of the construction himself, but after several years of experimentation and mechanical improvements, he produced exceptionally fine telescopes that compared favorably with those of the Royal Observatory.

During the early years of his career, Herschel concentrated on cataloging stars by noting their positions during systematic telescope sweeps across the sky. In 1781, during a search for double stars, Herschel discovered an unusually bright object that he first thought was a comet. His report of the sighting attracted the attention of other astronomers; closer examinations of its orbit showed it to be a planet, which Herschel named after King George III although it was later renamed Uranus.

The discovery of Uranus brought Herschel the worldwide fame that had eluded him as a telescope craftsman. He was granted a royal pension and moved, taking his sister Caroline with him, to a house near Windsor Palace where his duties were few and his time his own. In 1786 he and Caroline moved to Slough to live in what became known as the Observatory House. The meagerness of his pension forced him to devote considerable time to manufacturing telescopes for sale. Yet he managed to obtain a royal grant of several thousand pounds to build a huge 40-foot telescope with a 48-inch-diameter mirror. It took 4 years to build but demonstrated its value immediately after its completion when it revealed an undiscovered satellite of Saturn. This telescope was a remarkable testament to Herschel's mechanical abilities though it was difficult to maneuver. Herschel completed three extensive reviews of the evening sky, which resulted in successively more detailed catalogs of stars. His first review in 1779 examined stars down to the 4th magnitude. His second review, which he began in August of that same year, concentrated on double stars and located the extremely bright object (Uranus) that represented the first known discovery of a planet. Herschel's second review also resulted in the first catalog of double and multiple stars that extended down to the 8th magnitude in brightness. He began his third and most ambitious review in 1781, which eventually included thousands of stars of varying magnitudes. Throughout these tedious proceedings, Caroline laboriously recorded her brother's sightings, an invaluable service that helped to ensure the survival and dissemination of his work.

Receipt of Charles Messier's catalog of nebulas in 1781 led Herschel to shift his attention to these mysterious cloudy patches of light. Whereas Messier had compiled barely 100 nebulas in his catalog, Herschel, with his 20-inch telescope, over the course of a 20-year period beginning in 1783, discovered nearly 2400 additional nebulas.

Herschel spent much of his career trying to determine the distances of stars using trigonometry to calculate the apparent displacements of stellar positions as seen from opposite ends of Earth's orbit around the sun. This method was not very successful because the observed displacements were so minute that accurate determinations could not be made. His calculations also used the then popular but erroneous assumption that all stars are equally luminous but appear different owing to their varying distances from Earth.

Herschel's concern with the distribution of the stars in the heavens probably originated during his earliest cataloging efforts. Galileo had shown that the Milky Way consists of large numbers of stars, which give it its bandlike hazy white appearance. Although Herschel had discovered over 2000 patches of light outside the Milky Way, there was little agreement as to their makeup. Some believed these nebulas to be stars while others thought them to be cloudy patches of liquid. Herschel showed that these nebulas are actually collections of stars. This discovery prompted him to hypothesize that the universe began with an even distribution of stars throughout space. He speculated that the gravitational force between the stars caused many of them to form local groupings or clusters.

Herschel's study of the Milky Way led him to suggest that it consists of a gigantic layer of stars. His search for the outer boundaries of this star system was unsuccessful owing to the limits of his instruments so his effort to map the cosmos had to be abandoned. It did, however, enable him to develop an innovative theory to explain how the nebulas might have gravitationally condensed into stars.

Although Herschel was more concerned with the structure of the cosmos, he did make detailed observations of the solar system. His observations of the sun led him to conclude incorrectly that it is essentially a planet covered by a luminous atmosphere through which only the tallest mountains protrude as sunspots. He calculated the heights of several mountains on the moon and made extensive observations of Mercury and Venus. He also studied the polar icecaps on Mars, the first two asteroids—Ceres and Pallas—to be discovered, the four known satellites of Jupiter, and the rings and moons of Saturn. Although he mistakenly supposed the rings to be solid, he did estimate their thickness to be less than several hundred miles. He also discovered two of Uranus's satellites and their retrograde orbits around the planet.

Like Herschel, modern astronomers have found that star counts alone are not sufficient to give a clear picture of the structure of the Milky Way because we are dealing not with a static distribution of stars but one that is undergoing constant change. We can therefore obtain a complete picture of our galaxy only if we also treat the motions of the stars. From an analysis of star counts and stellar motions we find that stars are quite gregarious and tend to arrange themselves in groups that range in size from small groups (e.g., binaries, triplets) to clusters, star clouds, and galaxies.

The appearance of the stars from night to night leaves one with the definite impression that their positions relative to each other are always the same. Each constellation reappears year after year with no perceptible change in the arrangements of its stars. But this is a wrong impression; the stars are at such great distances that only precise telescopic measurements of their positions year after year reveal their motions.

To investigate stellar motions we must define the "motion" of a star because, as we know from the theory of relativity, there is no such thing as absolute motion. The motions of the stars have meaning only if we first adopt a frame of reference to which we may refer these motions. Since Earth has very complex motions of its own with respect to the sun, it is not a good frame of reference; using it as a frame complicates the observed motions of the stars unnecessarily. Instead, we take the sun as our *standard of rest* for the time being. This is not the final answer to our problem because we want to know how the stars are moving within the Milky Way, and since the sun is a star, it, too, is moving within the Milky Way. We correct for the motion of the sun within the Milky Way later by shifting to another frame of reference.

We divide the motion of a star relative to the sun into two parts or components because different procedures are used for measuring these two components. We can understand the nature of the problem by considering a line segment AB as representing the distance moved by a star in one year from position A to position B (we consider here only uniform motion in a straight line i.e., unaccelerated motion). Instead of measuring this displacement directly, which, if it could be done, would give us the velocity at once in miles per year, we break it up into a displacement from A to a point C along a line at right angles to the line of sight (the line from Earth to the star), which we call the *transverse velocity*, and into a displacement from C to B along the line of sight, which we call the *radial velocity*.

If we know the transverse velocity and the radial velocity separately, we can apply the theorem of Pythagoras to the right triangle ABC, to find the actual space velocity AB. We square the transverse velocity and add to it the square of the radial velocity; the square root of this sum gives us the space velocity.

To find the transverse velocity, we measure the angle ASC (designated by the Greek letter μ) by comparing the position of the star in the sky on any night when it can be seen from the sun S with its position a year later. This angle is called the *proper motion* of the star; it is one of the most useful bits of information about a star that can be obtained. Since, on the average, this angle is larger for the nearby stars than for the distant stars, we can use this information about a star to tell us whether it is nearby or far away. This is important in determining the structure of our galaxy because we must know the distribution of stars in the neighborhood of the sun before we can investigate stellar distributions at great distances from the sun. To carry out an analysis of the sun's neighborhood, we must have a quick way of picking out among the millions of stars that appear on photographic plates those that are worthy of study because they may be nearby; proper motions permit us to do this.

Proper motions are measured by taking photographs of the sky year after year and then comparing these photographs for changes in the positions of the stars. In this game, time is on our side, for if no effect is found after one year, comparisons of plates two years apart can be made, and so on until a measurable effect is found. If necessary, we can hand the task over to our progeny for the completion of the particularly stubborn cases. Very few stars have proper motions larger than 1 sec of arc per year, but Barnard's star has a proper motion of more than 10 sec of arc per year.

If the proper motion of a star is known, its transverse velocity relative to the sun can be found from a simple algebraic formula. Divide the proper motion by the parallax of the star and multiply this quotient by the number 4.74; this is the transverse velocity in kilometers per second. Most of the naked eye stars have transverse velocities relative to the sun that lie close to an average value of about 20 km/sec.

We obtain the radial velocity of a star by measuring the Doppler displacement of its spectral lines. Most of the naked eye stars have radial velocities relative to the sun that cluster around 20 km/sec. This velocity, then, is what we may expect for the actual space velocities and we do, indeed, find that the space velocities of the stars relative to the sun are of about the same order of magnitude as the velocities of the planets around the sun. There are, however, a few exceptions comprising a group of stars that are moving with speeds of more than 60 km/sec relative to the sun, and some of them, like Barnard's star (6 light-years away), are quite close to the sun. We shall see later that this important group of high-velocity stars in our vicinity of space are population-II stars.

We can hardly get a complete picture of the entire Milky Way by looking at the nearby stars, and yet we are so situated in the Milky Way that we have no choice in the matter. There is enough dust and haze in our part of the Milky Way to obscure the stars at large distances even if we use our best telescopes; we must therefore use our ingenuity to deduce the structure of the Milky Way at great distances from data obtained from the nearby stars. We first estimate the effect that being part of the solar system has on our observations.

We have already done this to some extent by referring the motions of the stars to the sun, thus eliminating the effect of Earth's motion around the sun. But we must now take account of the sun's motion. Since no star that we have observed is standing still with respect to the sun, there is every reason to believe, even without direct evidence, that if the stellar motions were referred to some other frame of reference, the sun would also be found to be moving with respect to this new frame. Nothing that we have learned from astronomy or any other science has given us any reason to suppose that there is anything special about our sun and that it should not have the same general characteristics that the other stars have.

But if we do not take the sun as our frame of reference, what shall we take? The ideal thing would be to take either the geometrical center of the Milky Way or the center of mass of this system of stars. But it is impossible to find the latter, and until quite recently not too much was known about the former because of the aforementioned obscuration. In the early stages of this work (and even today), stellar motions were measured with respect to the center of mass or the geometrical center of the nearby stars themselves.

When we do this we find that the stars are not moving about at random but appear to be drifting in a definite direction as though the whole solar system were moving through them. This is evidence that the sun is moving in the opposite direction relative to these nearby stars. This effect has been measured using the proper motions as well as the radial velocities of the stars, and we now know that relative to the neighboring stars the sun is moving with a speed of about 20 km/sec toward the constellation of Hercules.

We see from this that the motion of Earth is very complicated, for in addition to spinning on its axis and revolving around the sun, it is moving in a straight line toward another group of stars. But this is by no means the final word, for all we have discovered is our motion with respect to a few stars in our neighborhood that do not even account for a thousandth of the Milky Way. This entire group of nearby stars with the sun included may itself be moving very rapidly with respect to the entire galaxy.

That this actually does occur has been known since 1904 when the Dutch astronomer Kapteyn showed that even after the sun's motion with respect to the nearby stars is taken into account, these

stars still show a definite pattern in their motions. This lack of ran-
domness in the motions of the stars has been investigated by the
greatest astronomers since then, and only recently has the answer
been found in the structure and motion of the Milky Way as a whole.
The astronomers who have dealt with this problem found that the
stars in our neighborhood have a tendency to stream more along
one direction than along any other direction. This preferred direction of stellar motions or star streaming, as it is called, is along a
line that is parallel to the Milky Way (it actually lies in the Milky
Way) and points from Earth toward the constellation of Sagittarius.
We shall see later that this is the direction to the center of our galaxy.

An interesting point in connection with this direction from the
sun to Sagittarius is the behavior of the high-velocity stars mentioned previously. With only a few exceptions, these stars appear
to be moving away from Sagittarius in a general direction that points
toward the constellations of Orion and Carina. Although it looks as
though these high-velocity stars are trying to escape from the Milky
Way, we shall see that this is only an apparent effect arising from
the same things that cause star streaming and the solar motion.

Thus far, we have been wandering around in our own backyard,
so to speak, as far as our galaxy is concerned, and whatever conclusions we have drawn about the motions of the stars and the sun
may be greatly changed when we look beyond the stars in our immediate neighborhood. If we wish to get a complete picture of stellar
motions, we must investigate the distant stars, the great majority
of which cannot be seen with the naked eye but which show up on
photographic plates after a long exposure in one of the large telescopes. Even with the aid of excellent photographic emulsions and
these telescopes, however, we still cannot penetrate very deeply
into the great star clouds that comprise our Milky Way because
space is filled with dust that obscures the distant stars. We are in
the same plight as the forester who tries to discover the extent and
distribution of the trees in a woods on a foggy day.

In spite of these handicaps, we must try to discover the structure of the Milky Way as best we can, and our best procedure is
still to count the stars in various directions in the sky. This task is
so enormous, however, that no one astronomer can hope to do very
much in a lifetime, and it even dwarfs the facilities of a large observatory. For this reason the work is carried on as a joint task, on
an international scale, by almost all the observatories in the world.

Astronomical research knows no national boundaries and no limits in time as each observer collects his data year after year either to use in his own construction of a model of the Milky Way or to pass on to the next generation of explorers of the universe.

To construct a reasonably good model of our galaxy, an astronomer must have all kinds of information about millions of stars. Each of the characteristics of individual stars is important in helping us form a true picture of the Milky Way. Parallaxes, proper motions, colors, magnitudes, spectral classes, and other similar data for thousands of stars are constantly being examined, sifted out, correlated, and built into a magnificent structure; but the demand is for ever more data even though the main features of the structure are now fairly well known.

At first sight, the task of counting the stars and analyzing their distribution seems an easy one because it appears to the naked eye that the stars are distributed uniformly all around us in the sky and that we are at the very center of this spherical distribution. The first hint that our initial impressions may be wrong comes from the faint luminous band (visible only on a clear moonless night in the country) that looks like a narrow cloudy band in the sky. Its position is marked by such bright constellations as Cygnus, Sagittarius, and Aquila, and we call it the Milky Way. Our unaided eyes can tell us very little about the nature of this band of light, but even a pair of binoculars reveals that it is not a nebulous structure but consists of millions of stars that become more and more numerous as we look toward the center of the band, piling up into vast star clouds in which the individual stars lose their identities among the many billions that are present.

To the casual observer it may seem that this structure is some kind of stellar distribution beyond our galaxy to which we do not belong. But this is not so; we and all the stars in our neighborhood are part of the Milky Way. This means that the stars are not distributed uniformly in the sky and that our first impression was faulty because our naked eyes gave us a prejudiced view of the heavens.

Without a telescope, the sampling we work with in making our deductions is greatly weighted in favor of the very bright stars like Rigel, Betelgeuse, and Antares, as well as of the very close stars. In either case we get a distorted picture of our star system. To see how this distortion arises, picture a very nearsighted person standing in a crowd of people celebrating the arrival of the New Year at

Forty-second Street and Times Square. If he is not wearing his glasses, he sees the people distributed uniformly all about him (since without glasses he cannot see very far beyond half a dozen people in any one direction) and he concludes that he is at the very center of the crowd. This wrong conclusion is based on a poor sampling of the people (the nearby ones) that surround him.

With his glasses he sees that the people are not distributed uniformly in all directions but instead form a narrow band that extends north–south along Broadway. They thin out when he looks east or west. The same thing is true of the stars. When we free ourselves of the prejudices that the naked-eye stars force upon us, we see that our sun is a member of a rather narrow band or disk of stars (the Milky Way) that extends along a line from Earth to Sagittarius.

Unfortunately, owing to the interstellar haze or dust, we cannot determine the extent of this band of stars by looking right along the line of the greatest concentration of stars (the line toward Sagittarius). For every thousand light-years that light travels along this direction from the distant stars, the images on our photographic plates lose almost half a magnitude of brightness so that one-third of the initial light is lost. This means that if a star is at a distance of 6000 light-years in the direction of Sagittarius, we receive no more than 9% of the light that we would receive if there were no dust between us and the star. Except for the very brightest stars (which are so few in number that they have little bearing on the structure of the entire system), visibility in the plane of the Milky Way is low, a mere 6000 light-years.

Fortunately, various devices can be used to penetrate this galactic fog, the most important of which is the radio telescope. But there are certain other things that we can do to penetrate the dust clouds with our telescopes. The very existence of the dust gives us an important clue as to where our solar system is, relative to the plane of the Milky Way.

Because we are surrounded by dust that lies in the plane of the Milky Way, we must be situated in this plane and not above or below it. If we look at right angles to the plane, we can see distant galaxies millions of light-years away, so that there can be no dust except in the plane. If we look parallel to the plane, we see none of these distant galaxies because of the dust. Since the dust characterizes the outer spiral arms of a galaxy like our Milky Way, we know that we cannot be at or near the center of our galaxy, but in the outer regions and probably in a spiral arm. A study of the dust reveals much about the way in which certain stars are born and about the distribution of chemical elements in interstellar space.

In addition to learning all we can from the dust, we must still carry out as thorough a count of the stars as we possibly can. But it is fruitless to do this out to very great distances from the sun because, as we have seen, our counts are greatly distorted by the large numbers of bright stars that are visible long after the very numerous faint stars are blotted out. Instead of trying to count all of the stars, we cut out a small region of space, say a sphere of about 20 or 30 light-years in diameter about the sun, and get a complete census of the stars within it, which is not an impossible task and is well on the way to completion at the present time. Assuming this region to be a fair sample of the distribution of stars in the arms of the Milky Way, we get a reasonably good picture of the structure of the spiral arms. We do not find any of the very bright stars or the unusual variable stars in our little nook of space, but we can add these stars to our count when we extend our findings to the Milky Way as a whole by counting the very bright stars and the variable stars separately.

To see how this works, we note first that very luminous stars are extremely rare, for to find another star similar to and as luminous as the sun we need only go out to a distance of 4.5 light-years (the star Alpha Centauri); but we do not find a supergiant such as Rigel, thousands of times more luminous than the sun, until we go out about 700 light-years. In other words, we must survey millions of times as much space to find a very luminous star as we do to find another star like the sun. In drawing our conclusion about the distribution of stars from the results of our local census, we must keep such facts in mind.

From such star counts in the neighborhood of the sun, we learn something quite different from what our naked eyes reveal. The naked-eye stars are quite the exception and not the rule as far as the average stars in the Milky Way go. We see the aristocrats and bright lights among the stars when we look at the sky at night. The most common variety of star within a distance of 16 light-years from the sun is the faint red dwarf with a luminosity less than one-thousandth that of the sun. Indeed, of the approximately 50 stars within a 16-light-year radius of the sun, more than 20 are of absolute magnitude 11.8 (about 700 times less luminous than the sun) or greater and only 3 are more luminous than the sun.

Our knowledge about our own vicinity of space can now be extended to the distant regions of our galaxy, but we must be careful not to portray the entire Milky Way in our own image, for there are important differences between us and the regions of the galaxy near the center. Certain conclusions, however, may be safely drawn. Thus, we know that the most numerous stars in the Milky Way are the faint red ones, with the white dwarfs running a close second. Our best estimate is that stars of absolute magnitude +15 (about 10,000 times less luminous than the sun) are the most numerous in the Milky Way; stars more luminous than the sun represent only a small percentage of the total number.

Another interesting feature is that the individual star like the sun is not as common as might at first be supposed. At least ten of the nearby stars we have been investigating are binary or triple systems and this is true of stars in general. It may very well be that more than half of the stars in the Milky Way are compound systems.

Not all that we have deduced from star counts in the sun's neighborhood is true for the galaxy as a whole. We find, for example, that of the very luminous stars within 1000 light-years of the sun, many are blue-white giants or supergiants (like Rigel), which are not present in the central core of the Milky Way. And what of the clouds of dust and gas that envelop us? Do they extend throughout the Milky Way or are they characteristic of our part of the galaxy only? Here, too, we find a difference between our region of the Milky Way and the center. There is very little gas and no dust in the core of the galaxy.

Having drained the star counts in our immediate neighborhood of all their information, we must try to extend the counts to greater and greater distances from the sun even though the dust hinders us greatly. Astronomers for some years have been counting all the stars along the Milky Way down to the 15th apparent magnitude. This gigantic project is nearing its end and a great deal of information can even now be obtained from it. One of the most striking features of such counts of the distant stars is the way they confirm the existence of the interstellar dust. If we assume that the stars are distributed uniformly to great distances (a few thousand light-years) in all directions from the sun in the same manner as they are distributed near us, the total number of stars counted should increase in a predictable manner. But we find, as we go to the faint, distant stars, fewer than the predicted number. This means that we cannot see some of the faint ones because of the dust.

An interesting and important deduction from counts of the more distant stars is that our sun may be a member of a local stellar cluster consisting of perhaps a few hundred stars. Some evidence of this is found in the distribution of the bright naked-eye stars, especially the B type, that can be observed at very great distances. Most of these have been catalogued to a distance of 3000 light-years and they show a definite thinning out in all directions as though we were at the center of a condensation. This, however, is not entirely borne out by the distribution of the other stars, and there may be other explanations for this phenomenon.

If we are a member of a local group of stars, we enjoy a distinction that is not shared by many stars in our galaxy as is indicated by the few such clusters in the Milky Way. These clusters, which we named *galactic or open clusters* in a previous chapter to distinguish them from the *globular clusters*, are composed of anywhere from 100 to 1000 stars and have diameters of about 18 light-years, although they are only roughly spherical. About 500 such clusters have been identified in the visible part of the Milky Way and there may be ten times as many more throughout the entire galaxy. They consist of population-I stars, gas, and dust.

Such open clusters as the Pleiades, the Praesepe, the Hyades, Ursa Major, and the double cluster in Perseus are the familiar naked-eye objects that make our sky so beautiful. A telescope reveals many more; the study of their distribution aids us greatly in understanding the structure of the Milky Way. These objects hug the broad band of the Milky Way and, since they are not as subject to extinction by dust as ordinary stars, they reveal the Milky Way to much greater distances than the 6000 light-years to which we are limited by individual stars. These clusters have been observed to distances of 15,000 light-years.

The open clusters provide us with direct evidence for the presence of dust in our galaxy. As we go out to the more distant of these objects, they appear to grow fainter faster than they should as indicated by their apparent sizes. In other words, a cluster that appears half as large as another because it is twice as far away sends us less than one-quarter as much light. This larger-than-expected drop in brightness can only stem from some absorption of light because of the presence of dust. Here then is additional evidence for the presence of interstellar absorbing material.

All the stellar members of a particular local cluster can be picked out from among the many surrounding stars because they all have the same proper motion. In other words, the members of the cluster move through space together. This means that they must all have had a common origin, as already noted, and the different rates of evolution of stars in the same cluster are determined by their different masses as described in Chapter 15.

These open clusters show the same general characteristics in their distribution as do the individual population-I stars in our galaxy; they show a definite increase in their number as we go toward the Milky Way, with the greatest concentration in the direction of Sagittarius.

All observations show that we are in a flattened system of stars that extends for at least 15,000 light-years in the direction of the constellation of Sagittarius (the direction of the center of this system), but we cannot tell just from these observations how far beyond 15,000 light-years the center is located. There are other methods, however, for finding the center of the Milky Way, which we discuss presently.

Star counts out to a few thousand light-years show that the stellar densities drop in some directions and remain unchanged in other directions. The general conclusions that have been drawn from the most recent data on the variations in stellar densities and in star counts are that in our part of the galaxy the stars are distributed along spiral arms and that our solar system is situated on the inside of one of these arms. This spiral arm, the Orion arm, lies between the Perseus arm, which is still farther away from the center, and the Sagittarius arm, which is the first spiral arm that one meets as one goes from the sun to the center of the galaxy.

The first indications that not all the matter in the Milky Way is crowded into the stars come from the huge clouds of gas and dust that are so close to the very hot stars (the B-type stars) that they glow either by reflected light or the way a neon sign glows. The intense ultraviolet radiation from these hot stars excites the atoms in these bright gaseous nebulas and causes the atoms to glow. Such nebulosities are found associated with the B stars in Orion and the Pleiades. But in addition to these bright nebulas, we also can observe dark gaseous clouds that completely obscure the stars in a particular region of the sky. These dark and bright gaseous nebulas are distributed throughout the Milky Way and are found in ever greater numbers as we move toward the central regions of the galaxy.

The gaseous nebulas are examples of an unusually great concentration of the usually very rarefied dust and gas into rather small dense clouds, extending over a few light-years. The density of this matter is exceedingly low by our standards here on Earth, but very high as far as interstellar space goes. An example is the Orion nebula, which has a diameter of about 25 light-years and is accompanied by several hundred hot B-type stars that supply it with ultraviolet radiation. The total mass of this extensive gas and dust cloud is about 10,000 times that of the sun, so that its density is less than one millionth of a billionth of the air that we breathe. In spite of this, each cubic centimeter of such nebulas contains about a million atoms.

There is still other evidence, provided by the stars, for the existence of interstellar material. In the chapter on the characteristics of the stars, we saw that the spectral class, the surface temperature, and the color are all intimately related. You cannot change one of these without affecting the others. But suppose you observe a star whose color is redder than it should be according to its spectrum and its temperature. You must conclude that this reddening is not intrinsic but is caused by the dust particles that lie between us and the star. When light passes through a dusty medium, the red component of the light gets through more easily than the blue components because the wavelength of the red light is longer than that of the blue light and very tiny dust particles are less effective in blocking the long waves than in blocking the shorter waves. That this reddening of the stars is due to the dust in space is demonstrated by the fact that the more distant stars are reddened more than the nearby stars. From an analysis of the amount of reddening, astronomers can determine the size, density, and nature of the dust particles.

The spectroscope also gives us some information concerning the interstellar gas clouds; in fact, it was in the observation of the spectra of the stars that astronomers first learned about the existence of atoms and molecules between the stars. As long ago as 1904, Hartmann discovered the absorption lines of ionized calcium in the spectrum of the double star Delta Orionis. This in itself was not surprising, but it was surprising that these lines remained fixed and did not show the Doppler effect characteristic of all the other lines in the spectrum of this binary system. This means that the ionized calcium atoms that produce these lines are not inside the double star but are moving about freely in interstellar space. More recently, the interstellar lines of neutral sodium, neutral calcium, potassium, titanium, and iron have been observed.

But one may not conclude from the absence of the optical spectral lines of other atoms or the bands of molecules that the atoms just listed are the only ones between the stars and that there are no interstellar molecules. An atom like hydrogen or helium is not affected by most of the visible light, and the ultraviolet radiation that does affect them cannot get through the ozone of our atmosphere, so that we can get very little information about the abundance of these atoms in interstellar space from the spectroscope. But we have every reason to believe that hydrogen is the most abundant of all the atoms in interstellar space and helium is next in importance. We also know that many kinds of molecules, ranging from simple diatomic to complex organic molecules, cannot be detected in interstellar space with the ordinary spectroscope, although the spectral lines of certain simple molecules such as CH, CN, and CO have been observed in interstellar space.

We can, to some extent, determine the abundance of atomic hydrogen in interstellar space by photographing the faint light emitted by ionized hydrogen atoms (protons) as they capture electrons in extended regions of the Milky Way, but most of our recent information about the distribution of hydrogen in our galaxy has come from the application of radio waves to astronomy, the most recent and perhaps the most fertile branch of this mother of the sciences. Radio astronomy refers to the information about our universe that we receive in the form of radio waves, which are longer than ordinary light by a factor of 10,000 or more. We can use radio waves to peer through the dust because these waves are not affected by the small dust particles (many thousands of times smaller than the radio waves) that press all around us in our region of the Milky Way. These waves pass around these dust particles the way a large water wave on the ocean bends around a breakwater or a small rowboat. By analyzing the radio signals that come to us from our galaxy, we can penetrate the dust clouds that surround us and look into the center of the galaxy and beyond. Radio astronomy and infrared astronomy also reveal the interstellar molecules that cannot be detected with optical spectroscopes.

Radio astronomy was born in 1931, when the radio engineer Karl Jansky (1905–1950) discovered that the constant background of faint static in a special radio receiver was coming from the Milky Way. His interest in astronomy was casual as he had been trained as an engineer at the University of Wisconsin before joining the Bell Telephone Laboratory in 1928. Jansky dabbled in various research projects for 3 years until turning to the problem of static interference, which was of considerable concern to his employer because it reduced the clarity of ship-to-shore radiotelephone transmissions.

Jansky investigated many sources of static such as thunderstorms and electrical wires and, in the course of his work, discovered a persistent but very weak type of static interference. Unable to pinpoint a localized source for the static, he was temporarily at a loss as to how to proceed. "[The static] came from overhead and moved steadily. At first, it seemed to Jansky, it moved with the sun. However, it gained slightly on the sun, to the extent of four minutes a day. But this is just the amount by which the vault of the stars gains on the sun. Consequently, the source must lie beyond the solar system."[2] The realization that he was not dealing with an earthbound source of interference prompted Jansky to perform a number of calculations, which proved to his satisfaction that the static was coming from the constellation of Sagittarius, in the direction of the center of the Milky Way.

Although Jansky's discovery attracted worldwide attention when it was announced in December 1932, few scientists bothered to study these cosmic radio waves in great detail. This neglect was unfortunate because it delayed the day when astronomers began to use these waves to peer through interstellar dust and clouds at regions of the galaxy normally invisible to ordinary light-gathering telescopes. Jansky himself did little to stimulate radio astronomy because he soon abandoned his investigations of this galactic static and returned to more mundane engineering research projects. Not until World War II demanded the development and rapid perfection of radar did our microwave technology progress to a stage that made possible the development of what is now known as radio astronomy. Although Jansky's own participation in this work was halfhearted at best, he was honored as its discoverer by having the unit of radio-wave emission named after him.[3]

Radio astronomy has advanced enormously since Jansky made his great discovery, and today there are many huge radio telescopes (parabolic antennae) hundreds of feet in diameter, with vast electronic receiving systems, throughout the world. With these devices, astronomers obtain extensive radio spectra for large regions of the Milky Way and beyond. This radio emission spectrum, which consists of a continuous background and discrete lines, comes from all regions of the Milky Way and from localized sources (so-called radio stars and radio galaxies) within and outside the Milky Way.

The most dramatic discovery outside of quasars and pulsars that has yet been made with these new astronomical techniques has been of great importance in revealing the distribution of gas and dust in our galaxy, and in displaying the structure of the galaxy itself. What we found almost impossible to do with visible telescopes, namely, to determine the structure of the Milky Way, we can now do with great ease with radio telescopes, and the very gas and dust that hindered us before are now the sources of most of our information. This discovery is one of those remarkable scientific achievements that stem from brilliant analyses and applications of previously established theories.

Fundamental particles, like electrons and protons, are spinning about an axis, which means that in addition to having an electric charge they also behave like little magnets. You know from experience that when you have two magnets, you can line them up either with the two like poles together or with the two unlike poles together. The latter arrangement is a much more stable one than the former because unlike poles attract each other and like poles repel.

When a proton (an ionized hydrogen atom) captures an electron, the spins of the two particles either point in the same direction (the spin of the proton and the spin of the electron are in the same sense; they are both spinning clockwise or counterclockwise so that unlike magnetic poles are together) or in opposite directions (like magnetic poles are together) (note that the like and unlike pole alignments of the electron and proton are opposite to their spin alignments, owing to their opposite electric charges—this fact is incorrectly presented in astronomy books). Since a parallel-spin capture of an electron by a proton is three times as probable as an anti-parallel-spin capture, 75% of the ionized hydrogen atoms in between the stars will capture free electrons with spins parallel to the proton spins and 25% will capture anti-parallel-spin electrons. But the hydrogen atom with its electron spin parallel to the proton spin is more stable (the electron is more tightly bound to the proton) than the atom with the two spins antiparallel.

This means that the electron in the hydrogen atom with its spin antiparallel to that of the proton will ultimately flop over so that its spin is parallel to the proton spin. When this happens, a burst of energy is emitted (a photon of very low frequency) and it reaches us as a radio signal. This hydrogen radio line with a wavelength of 21 cm was first predicted theoretically in 1944 by the Dutch astronomer H. C. van de Hulst. It was observed in radio telescopes at Harvard and then throughout the world in the early 1950s.

The 21-cm line comes to us with considerable intensity from all parts of the Milky Way; by carefully plotting the variation of the intensity of this line in different regions of the sky, we can get a fairly reliable picture of the distribution of gas (hydrogen) and dust throughout the galaxy. This 21-cm line shows us that there are large amounts of neutral hydrogen in interstellar space.

Since one-third as many hydrogen atoms have antiparallel-spin electrons as have parallel-spin electrons, we should ordinarily not expect to receive much 21-cm radiation because it takes on the average about 11,000,000 years for an electron in an unmolested hydrogen atom to flop over to a parallel-spin position. But there are so many such atoms in interstellar space that the occasional emission of a 21-cm photon from any one of them adds up to the observed intensity of the 21-cm line. The collisions among the neutral hydrogen atoms that occur every few hundred years keep the numbers of hydrogen atoms with parallel and anti-parallel electron spins constant or in thermal equilibrium.

From the data that have now been collected, we know that the dust in interstellar space, which constitutes only about 1% of the interstellar material, is almost entirely responsible for the dimming of the stars. The size of a dust grain is about 0.00001 cm (about the size of the wavelength of visible light), and only one such grain, on the average, is present in each 10,000,000,000 cm^3 of space. Many of these grains are probably elongated, needlelike structures that are lined up parallel to the magnetic fields that surround them. They behave like nonmetallic charged particles, their temperatures ranging from 10°K to 50°K. Although we are uncertain about the chemical nature of these dust particles, we can be quite sure that they are composed mostly of hydrogen compounds.

Although the dust between the stars is almost entirely responsible for obscuring the stars, most of the interstellar matter consists
of clouds of atomic and molecular gases, composed principally of
hydrogen atoms and molecules. About 90% of the gaseous interstellar material is hydrogen, distributed in such a way that there is
on the average about one hydrogen atom per cubic centimeter of
space. Of course, in some regions the density of the gas cloud may
be 100 times as large as this and in other regions as many times
smaller. This means that in our region of the galaxy, matter is divided about equally, half of it in the form of stars and the other half
in the form of gas and dust.

Another category of objects recently found in interstellar space
are large organic molecules, which emit a variety of spectral lines
and bands (absorption and emission) in the infrared and radio parts
of the spectrum. Very intensive work has been done in this field
since 1969, and molecular spectroscopy in the infrared and radio
regions of the spectrum has become a major astronomical tool. At
present, 137 molecular spectral bands and lines have been observed;
they are emitted by some 35 terrestrially identified organic molecules, among which are hydrogen cyanide (HCN), formaldehyde
(H_2CO), cyanoacetylene (HC_3N), and methyl alcohol (CH_3OH).
Some of the organic molecules that have been identified from their
spectral bands and lines, such as CH_3CHO and CH_3C_2H, contain
as many as seven atoms, and there is reason to believe that still
larger molecules are present in the interstellar medium. Since both
the formic acid molecule (HCOOH) and the methanimine molecule
(CH_2NH) have been detected, the simple amino acid glycine
(NH_2CH_2COOH) ought also be present in interstellar space because
it is easily formed from the other two molecules. The presence of
amino acids on meteorites confirms the synthesis of amino acids in
interstellar space, which has important biological consequences. Of
still greater biological significance is the recent work of F. Johnson,
who has matched 16 laboratory spectral lines of the very large organic molecule $MgC_{46}H_{30}N_6$ (one of the porphyrins, an important
constituent of blood) with the corresponding diffuse interstellar
lines, within an experimental error of 2 Å.

This distribution of galactic material into stars and into gas and
dust occurs only in the outer regions of our galaxy; the central
nucleus of the Milky Way is apparently dust-free and contains only
stars and small quantities of gas.

Most of the hydrogen in the interstellar gas clouds is un-ionized; only rarely is an electron torn away from a proton. But when this does happen, the proton spends a long time wandering around rather footloose before it captures another electron. In these regions of un-ionized hydrogen, the atoms in the gas move about rather sluggishly, with a speed equivalent to a temperature of about 100°K.

In regions close to the very hot O-type stars, most of the hydrogen is ionized. One can deduce that surrounding an O-type star, there is a spherical zone with a radius of about 100 light-years in which the hydrogen is ionized. In this zone the ionized hydrogen atoms move with speeds that correspond to a temperature of 10,000°K. The intense radiation from the O-type stars sets up enormous turbulent motions in the gas clouds and, in turn, causes great masses of material to intermix, coalesce, and ultimately contract into stars. This process has actually been observed, so that we can say today that in our dusty regions of the Milky Way, stars are being born before our very eyes.

These new stars are among the hottest, the most luminous, the youngest, and the most quickly aging stars that we know, for they are emitting energy at such a rate that they cannot have a lifetime of more than 10,000,000 years. This leads to the very important conclusion that the dusty spiral arms of the Milky Way consist of young, hot, blue-white giants, which are still being born from the gases and dust, and of other population-I stars like the sun and the red dwarfs.

The center of the Milky Way consists of very old population-II stars, which were formed from the primordial hydrogen and helium at very early stages of the development of our galaxy. They are redder than the population-I stars, and no white, blue-white, or blue-white giants are found among them. Population-II stars probably account for the bulk of the stars of our galaxy and certainly produce most of the gravitational fields that keep the system together and determine the motions of the outer population-I stars like the sun.

With the rapid advances that have occurred in radio and optical astronomy, we can now draw a fairly complete picture of the Milky Way. It is a flattened disklike structure consisting of about 200,000,000,000 stars and stretching for about 100,000 light-years from one point on the rim of this disk to another diametrically opposite point. The structure is principally that of a bulging central nucleus of population-II stars, with practically no gas or dust between them, surrounded by spiral arms. The central region is an oblate spheroid with a diameter of about 25,000 light-years; the stars within it are uniformly distributed with a number density that is probably hundreds of thousands of times greater than the number density of the stars in the regions surrounding the sun. If Earth were attached to some star at the center of the galaxy, we probably would never have real night because the large number of stars in the sky would furnish as much light as 1000 full moons.

That the nucleus of our galaxy is flattened indicates that the nucleus is rotating about its center, and from the uniformity of the stellar distribution within it, it is probably rotating like a solid object. The diameter of the nucleus along the axis of rotation of the disk is about 8000 light-years.

The central region of our galaxy is surrounded by a very flat disk of gas, dust, and population-I stars, interspersed here and there with some population-II intruders such as the high-velocity stars mentioned previously. This outer disk is no more than 1,000 light-years thick, and is revolving around the central region not like a solid wheel but rather the way the planets in our solar system revolve about the sun (in accordance with Kepler's laws). This planetary type of rotation, with the stars near the central nucleus moving more rapidly in their orbits than the stars at greater distances, produces a shearing action of the stars against the gas and dust, so that spiral arms are formed. Three such spiral arms have been recognized and traced out on the basis of the information contained in the 21-cm line of hydrogen.

Our sun lies at the inner rim of one of the spiral arms at a distance of about 30,000 light-years from the center of the Milky Way. The number density of stars in this part of the galaxy is so small that it would be quite invisible to an observer at a distance of 100,000 light-years. Using a good telescope, an observer at that distance would detect our region of space as a very sparsely populated portion of the galaxy characterized by a few O- and B-type giants of apparent magnitude ranging from 12 to 15.

The spiral-arm stars that lie at about 16,000 light-years from the center of the galaxy are revolving with a circular velocity of 260 km/sec, and as one goes out toward the sun, this circular speed decreases. In the neighborhood of the sun, the stars (the sun included) are revolving with a speed of about 250 km/sec, and the sun itself is revolving about the galactic center with this speed and moving in the general direction of Cygnus. This rotational motion generates the effects that are observed in the motions of the stars in the neighborhood of the sun (e.g., star streaming, high-velocity stars).

This overall picture accounts for some of the features of stellar motions that we described earlier in this chapter. Since the sun is rushing around the galaxy at about 150 miles/sec (it makes a complete revolution every 250 million years), certain stars that are moving very slowly around the galaxy appear to be moving rapidly because of the sun's motion. Thus, suppose that a population-II star had enough speed to leave the nucleus of the galaxy and reach our region of space. By the time it reaches us, it would have lost so much kinetic energy in its uphill fight against the gravitational field of the core, that it would be moving very slowly. Since such stars are being left behind by the sun, they appear to us to be moving away at great speeds, and all in the same direction (opposite to that of the sun's motion). This accounts for the asymmetry in the motion of the high-velocity stars in the neighborhood of the sun. They are slowly moving population-II interlopers from the core of the galaxy. Stars in our region of the Milky Way cannot be moving much faster than 220 miles/sec in their journey around the galaxy, for if they were, they would escape completely from the Milky Way and we would not see them. Thus, the high-velocity stars cannot be moving in the same direction as the sun.

The rotational motion of the Milky Way was first demonstrated in 1927 by the Dutch astronomer J. H. Oort, who analyzed the radial velocities of the nearby stars in various parts of the sky. In Oort's analysis of the radial velocities of nearby stars, the sky is divided into four quadrants. One first draws a line from the sun toward the center of the Milky Way and then draws a circular arc to represent the orbit of the sun in its motion around the galactic center. If we now draw a circle in the plane of the Milky Way with the sun as the center, we get four quadrants, I, II, III, and IV. The stars in quadrants I and II are closer to the galactic center than the sun and stars in III and IV are farther than the sun. Oort assumed that the stars are moving around the center of the galaxy in accordance with Kepler's laws so that the stars in quadrants I and II are, on the average, moving more rapidly than the sun (being closer to the center) and the stars in the other two quadrants are moving more slowly.

In other words, the stars in I and III should, on the average, appear to be approaching us and those in II and IV should appear to be receding.

Oort examined the radial velocities of O- and B-type stars, of Cepheid variables, and of certain stars with extended nebulous envelopes (called planetary nebulas because they look like planets when first seen in a telescope) and found that a rotation of this sort is indeed present and that the center of the Milky Way is in the direction of Sagittarius. From this analysis he deduced the speed of galactic revolution of the stars in the neighborhood of the sun. From the speed of rotation of the sun about the galactic center one can calculate the mass of the central nucleus if the distance of the sun from the center of the galaxy is known. One applies Newton's law of gravity treating the core of the galaxy and the sun as two bodies attracting each other gravitationally. It should be noted that to find the actual speed of the sun in its revolution about the galaxy, the distance to the center is also required. Oort's procedure gives only the angular speed of the sun.

After this epoch-making discovery, many other observations confirmed Oort's results, and today radio astronomy has added the final touches to the picture. Using the 21-cm line, we can trace out the spiral arms of the Milky Way and show how the hydrogen in these arms is revolving around the galactic center. Oort himself started this work with the 21-cm line; he showed that the Doppler displacements in the 21-cm line are exactly of the sort that must be present if the spiral arms are rotating in the manner described above.

Oort was the first in his family to become an astronomer but he did grow up in a family of scholars: his grandfather was a professor of Hebrew at the University of Leiden and his father was a medical doctor. Oort himself excelled in his studies at the University of Groningen and then spent 2 years as a graduate student at Yale University. He first became interested in astronomy when he heard of Kapteyn's work on the structure of the Milky Way; Kapteyn's conclusion that the solar system is located at the center of the galaxy intrigued Oort so much that he geared much of his own career to trying to determine the true location of the sun and its planetary satellites. His earliest work involved measuring the velocities of stars in various parts of the sky; Oort's discovery that the stars in various parts of the sky do not move in a manner that indicates that the sun is at the center of the galaxy was the basis for his later conclusion that the sun is actually near the rim of the galaxy. Extensive astronomical observations showed that the visual evidence in favor of Kapteyn's hypothesis was misleading because much of the galactic system examined by Kapteyn is obscured by interstellar dust. Oort also did extensive investigations of the distribution of stars in the galaxy and found that they become less plentiful as one moves away from the center, suggesting that most of the matter in the galaxy is concentrated in a central hub.

Oort joined the Leiden Observatory in 1924 and became a professor of astronomy at Leiden in 1935. He helped to establish a center for radio astronomy in Holland soon after the end of World War II because he recognized the potential importance of Jansky's discovery of galactic radio waves. Along with several other astronomers, Oort mapped the distribution of hydrogen gas in the galaxy. He also studied the rotational dynamics of the galaxy and showed how colliding clouds of gas in interstellar space tend to clump together gravitationally over time. He became the director of the Leiden Observatory in 1945 and served as the general secretary and the president of the International Astronomical Union. Oort also investigated comets and found that they come from a vast revolving shell of comets beyond the outer edge of the solar system. Some of these comets are occasionally slowed down by the gravitational field of the nearby stars and fall into orbits that carry them into the inner regions of the solar system. Oort suggested that these comets are the remnants of diffused matter left over from the formation of the solar system—matter that gradually clumped together as "dirty snowballs."

If you live where there are no city lights, smoke, or dust to obscure the stars, you may detect a small, faint patch of light in the direction of the brightest star in the constellation of Hercules. This object is not a star but a cluster of stars at a distance of almost 40,000 light-years. That it is not another one of the open clusters like the Pleiades, is evident even with a small telescope, which reveals that it is an object of an entirely different sort. Its most amazing properties are its almost perfect sphericity and the tremendous concentration of stars toward its center.

The first surprising thing about these globular clusters (more than 100 of them have been recognized) is their distribution. Since we can see the globular cluster in Hercules even though it is at a very great distance from us, we conclude that it is not in the plane of the Milky Way, and this is indeed true of all of the known globular clusters, for if they were, the dust would hide them. Some of them do, however, lie near the galactic plane and show the same reddening effect, caused by absorbing dust, as do the stars in the Milky Way. The globular clusters form a halo or shell around the core of the Milky Way, which is why most of them are concentrated on one side of the sky in the direction of Sagittarius and Scorpio. In fact, they are distributed symmetrically with respect to the Milky Way in the region of Sagittarius as though they were on the surface of a sphere with its center in Sagittarius.

Before it was known that the sun is not at the center of the Milky Way, this one-sided distribution of the globular clusters was extremely puzzling, but now it fits our picture of the galaxy and our position in it very well. This was first clearly recognized by Harlow Shapley who proposed the idea that the globular clusters form a spherical superstructure surrounding, and concentric with, the core of the galaxy. This is supported by two facts: the stars in the globular clusters are similar to those in the center of the Milky Way, i.e., population-II stars; and the globular clusters have the same apparent high space velocity as do the high-velocity stars, described previously. This means that the globular clusters as a group are not revolving around the Milky Way (the individual clusters are) so that there is a large relative velocity between them and the sun; they appear to be moving very rapidly only because of our own motion around the center of the galaxy.

When Shapley properly interpreted the distribution of the globular clusters, he saw that he could use it to find the distance of the galactic center from the sun provided he had some way of determining the distances of the globular clusters. If he could determine their distances, he could then, from their distribution, determine the position and the distance of the center of the entire globular cluster system, and this would then be the distance of the center of the galaxy. His reasoning was quite sound, and it led him to the first correct estimate of the true size of the Milky Way, which is many times larger than had been estimated previously. To do this, Shapley had to find the distances of the globular clusters, which he did by using certain characteristic properties of globular clusters, the most important of which is the presence in these clusters of very-short-period Cepheid variables (the RR Lyrae variables). Almost 1500 such variables have been found in globular clusters, all with periods of less than a day. These cluster-type variables all have about the same absolute magnitude; the distance of a cluster can be found by comparing the absolute magnitude of a variable in the cluster with its apparent magnitude. Shapley checked his findings by noting that since most of the globular clusters are of about the same size, their relative distances are inversely proportional to their relative sizes. He applied an additional check on the relative distances by comparing the relative apparent brightness of the brightest stars in each of the clusters.

Using such procedures, Shapley constructed a correct model of the globular cluster system and from it determined the distance and the position of the center of the Milky Way. His methods (with improved accuracy) are still applied; the distance we use to calculate the mass of the core of the Milky Way is thus obtained.

Harlow Shapley (1885–1972) was born in Carthage, Missouri, the son of a farmer and part-time schoolteacher.[4] He had little formal education as a boy, leaving school after the fifth grade to help on the family farm. As a teenager, he worked as a junior reporter for several years before deciding that he wanted to go to college. Because he had neglected his secondary education, Harlow attended a preparatory institution for a year and then enrolled in the University of Missouri in 1907, intending to become a journalist. Since the school of journalism was not yet open, Harlow searched for something that would interest him in the meantime and finally settled on astronomy. Three years later he graduated with a degree in mathematics and physics. He received his master's degree the following year and applied for an astronomy fellowship at Princeton University, which he received in 1911.

After Shapley arrived at Princeton, he began collaborating with Henry Norris Russell to determine the sizes of eclipsing binary stars and the shapes of their orbital paths. Shapley also suggested that Cepheid variables are not double stars, as had been assumed by many astronomers, but instead single stars whose brightnesses vary with their oscillations in size. Although Shapley's conclusions were not generally accepted, they were supported by Arthur Eddington's calculations. After Shapley finished his dissertation at Princeton, he joined George Hale's staff at the Mount Wilson Observatory in Pasadena, California. His recent marriage, coupled with the suggestion by Harvard astronomer Solon Bailey that he use the 60-inch Mount Wilson telescope to study globular clusters—dense systems

containing tens of thousands of stars—provided Shapley with both
the emotional support and professional orientation he needed to help
him establish himself as a leading astronomer.

Shapley studied Cepheid variables in globular clusters and devised a way to measure the distances to such stars based on their
brightnesses and periods. These calculations led Shapley to conclude that the structure of the Milky Way is very different from that
which had been previously accepted by astronomers. The Dutch
astronomer Kapteyn, whose work was instrumental in the earliest
efforts to understand the structure of the Milky Way, believed that
the sun lies near the center of the Milky Way and that the galaxy
is about 10,000 light-years in diameter. After very detailed and laborious observations, Shapley estimated that the Milky Way is
about 300,000 light-years in diameter and that the sun itself is about
60,000 light-years from the center of the galaxy. Although Shapley's
figures are larger than the currently accepted figures, his model of
the Milky Way was staggering in its vastness and attracted the attention of scientists throughout the world.

In the years after World War I, astronomers were engaged in
a lively debate over whether the Milky Way is a solitary star system
or whether there are other comparable "island universes" scattered
throughout space. Much of the controversy arose because the telescopes then available were not large enough to determine whether
newly discovered spiral nebulas are satellites of the Milky Way or
independent galaxies. Shapley calculated that the Andromeda nebula is over 1 million light-years away and implied that it was probably a separate "island universe." However he was not very sure
of his own conclusion because the equipment then available could
not adequately determine the sizes of such distant objects. Shapley
and astronomer Herber Curtis engaged in a spirited debate in 1920
about the dimensions of the Milky Way and the possibility of other
"island universes"; there was no clear winner as each man successfully made a number of points in support of his views. However,
the publicity given to this debate sparked the interest of other astronomers in galactic astronomy.

In 1921 Shapley left California to become the director of the
Harvard Observatory. For the next few years he used Harvard's
facilities, especially its observation station in the southern hemisphere, to study the Magellanic Clouds, which he estimated to be
about 100,000 light-years away from the Milky Way. He also began

an extensive correspondence with Pasadena astronomer Edwin Hubble, who had made a name for himself through his investigations of "extragalactic nebulas." Although Shapley and Hubble respected each other's work, both took great pride in the work of their respective institutions and encouraged the growing competition between eastern and western astronomers in the United States. In any event, Shapley spent much of the 1920s cataloging tens of thousands of galaxies in the northern and southern skies. These surveys revealed the uneven distribution of galaxies in the observable universe, a finding that seemed to be at odds with Hubble's belief that the galaxies are uniformly distributed throughout space.

At the same time, Shapley attracted many talented astronomers and students to the Harvard Observatory. He also helped to establish at Harvard what became one of the finest graduate astronomy programs in the United States. Shapley's ability to deal with both university and government bureaucracies was useful in the 1930s when the rise of Hitler in Germany forced many prominent Jewish scientists to seek refuge elsewhere. Shapley managed to cut through much of the red tape and made it possible for many of these talented individuals to emigrate to the United States.

Shapley received many awards and was elected to the National Academy of Sciences in 1924. He received more than a dozen honorary doctorates and served as president of the American Astronomical Society and the American Academy of Arts and Sciences. A popular speaker, he frequently participated in conferences about science and philosophy. His interest in international affairs made him a natural choice to be one of the American representatives charged with drawing up the charter of the United Nations Educational, Scientific and Cultural Organization (UNESCO). A proponent of closer contacts between Soviet and American intellectuals, Shapley visited the Soviet Union in 1946 when the Cold War was threatening to erupt into a world war. His outspoken views about the need for international cooperation at a time when there was growing concern about the possible Soviet infiltration of the American government caused him to be subpoenaed by the House Committee on Un-American Activities. In 1950 he was attacked by Senator Joseph McCarthy but was later cleared of charges of wrongdoing by the United States Senate. He spent the last 20 years of his life lecturing and traveling and he continued to devote much of his time to the work of various scientific organizations.

As already noted, globular clusters are almost perfect spheres, but there is a slight flattening indicating that they are slowly rotating about an axis. A cluster is held together by the mutual gravitational attractions among its stars just the way a sphere of hot molecules like the sun is held together. And just as the density of atoms and molecules in the sun increases toward the center, so the density of stars increases toward the center of a globular cluster. We cannot say at the present time how many stars there are in a distant globular cluster because we detect only the brightest and those near the edge of the cluster. If we try to capture the faint ones by a long exposure, the center of the photographic plate is completely burned out. Estimates of the total number of stars in one of these globes range from 50,000 to over 1,000,000; most of them certainly contain several hundred thousand stars.

The globular cluster M3 in Hercules has a diameter of about 100 light-years, and the central dense region, containing tens of thousands of stars, is 30 light-years across. We might compare this with our previous discussion of the number of stars within a radius of 16 light-years of the sun to see the differences in stellar number density.

We have already noted that stars evolve differently depending on their mass and so the state of development or evolution of a star in a globular cluster depends on its initial mass; if its mass is about 1.5 times that of the sun, it will move off the main sequence when about 12% of its hydrogen has been converted to helium. It will then begin to swell, grow redder, and become more luminous as it moves into the giant branch. This behavior is strikingly confirmed by the H–R diagrams of globular clusters. In an H–R diagram of the globular cluster M3, all the stars above spectral class F have left the main sequence, while the G-type and cooler stars are still on the main sequence. This distribution holds for all other globular clusters that have been studied. Since a simple calculation shows that a star slightly more massive than the sun will use up 12% of its hydrogen after 5 billion years, this means that 5 billion years must be the minimum age of globular clusters. These clusters are actually about 10 billion years old. The stars in the cluster that are less massive than 1.5 solar masses are still on the main sequence because they have not used up 12% of their hydrogen; the very massive stars are well up in the giant branch because they used up much of their hydrogen early in their lives.

When the hydrogen of more massive stars is almost completely exhausted, they begin to descend down toward the main sequence again, moving toward the left and spending part of their lives as cluster-type variables (RR Lyrae stars). They probably end their lives as white dwarfs after a series of novalike explosions.

Since the globular clusters are composed of population-II stars, the oldest stars in the galaxy, the age of these clusters gives us a fairly good idea of the age of the galaxy, which is between 12 and 15 billion years old. From the fact that the stars of the globular clusters leave the main sequence at spectral class F, we have seen that these objects are at least 5 billion years old. We may conclude that this is the smallest value that we may take for the age of the Milky Way. A better figure for this minimum age based on the evolution of stellar models is 7 billion years, but since the theory of the evolution of stars is still somewhat uncertain this figure is subject to change. In any event, we can be sure that the globular clusters are among the oldest objects associated with the Milky Way.

Important information about the total mass of our galaxy and its true dimensions can be obtained from an analysis of the Milky Way rotation curves as one moves from the galactic center out to the galaxy's edge. Taking only the visible matter of the galaxy into account, we find that the velocity curves indicate that more mass is present in our galaxy than can be accounted for by the visible matter in the form of stars, interstellar gas and dust, and dark and bright diffuse nebulas. The amount of such "hidden" or "dark" matter (the so-called "missing mass") relative to the observable matter in the galaxy increases if one takes into account the halo of the galaxy. This halo is a sphere with a radius of about 200,000 light-years in which the disk and the central bulge of our galaxy are embedded. The ratio of hidden matter to visible matter in our galaxy is about 100 to 1.

CHAPTER 18

The Galaxies

Away! the moor is dark beneath the moon,
Rapid clouds have drunk the last pale beam of
even:
Away! the gathering winds will call the
darkness soon,
And profoundest midnight shroud the serene
lights of heaven.

—P. B. SHELLEY, *Stanzas—April 1814*

In the 18th century, before spectroscopy was introduced into astronomy, the galaxies (spiral nebulas) were a source of considerable speculation, and with the exception of Thomas Wright, the English instrument maker, and Immanuel Kant, the German philosopher, most astronomers were convinced that these objects are part of our system of stars. Owing to their fuzzy appearance, when first viewed in telescopes the distant galaxies were thought to be nebulas like the gaseous nebulas in the Milky Way. That is why they were called spiral nebulas when many of them were found to have spiral arms. We know today that the spiral nebulas are vast agglomerations of stars, most of them at great distances beyond the Milky Way, but we learned this only within recent times. That these objects are not associated with our galaxy is evident from the way they are distributed with respect to its plane as seen from Earth. Instead of finding an increasing number of these objects as we look in the direction of the Milky Way (which holds for the stars), we discover that they thin out; we can see no spirals within 20° of each side of the plane of the Milky Way. But in all other directions these objects are distributed quite uniformly.

At one time this curious distribution was very puzzling, but now we can explain it, knowing of the obscuration in the disk of the Milky Way. The spiral nebulas surround our galaxy in all directions, but those that lie in the direction of the plane of the Milky Way are blocked from view by the galactic dust.

To one astronomer in particular, the presence of the spirals and globular clusters in the heavens was a nuisance. He was Charles Messier, the French comet hunter, who found that he was always mistaking a globular cluster or a spiral nebula for one of his beloved comets. Since the clusters and the nebulas are relatively fixed objects in the sky and the comets are not, he decided that he would rid himself and all future comet hunters of these annoying decoys by cataloguing them with an appropriate word of warning. In his final catalogue, which was published in 1784, he listed 103 bright clusters and nebulas.

It is ironic that Messier's comet hunting activities have long been forgotten, but his catalogue and designation of the clusters and the spiral nebulas are still used. If you run across a celestial object designated as *Messier 13, Messier 3*, etc., you will know that it is a cluster or a galaxy.

Although the nature of the spiral nebulas was not known with certainty until large telescopes and spectroscopes were introduced, a few bold and imaginative thinkers suggested quite early that these objects are probably huge aggregates of stars like our Milky Way. Thomas Wright argued in 1750 that nature is never content with a single specimen of anything and that, just as there are many suns in our Milky Way, there must be many Milky Ways like ours throughout space. He cited the few visible galaxies as examples.

About 5 years later, Immanuel Kant took up the same theme and argued on the basis of his concept of the uniformity of nature that, in his words,

> It is much more natural and reasonable to assume that a nebula is not a unique and solitary sun, but a system of numerous suns, which appear crowded, because of their distance, into a space so limited that their light, which would be imperceptible were each of them isolated, suffices, owing to their enormous numbers, to give a pale and uniform luster. Their analogy with our own system of stars; their form, which is precisely what it should be according to our theory; the faintness of their light, which denotes an infinite distance; all are in admirable accord and lead us to consider these elliptical spots as systems of the same order as our own—in a word, to be Milky Ways similar to the one whose constitution we have examined.

The first important developments in the research into the nature of nebulas stemmed from the work of Sir William Herschel and his son, Sir John Herschel. They catalogued a few thousand galaxies, and Sir William Herschel suggested that all nebulas can be resolved into stars with the aid of sufficiently large telescopes. Although telescopes revealed the spiral structures of these external systems to the early investigators, who made excellent drawings of these objects, their true nature was established only after photographic plates and the spectroscopes were used in conjunction with telescopes. It is easy, with modern photographic techniques, to resolve the nearby spirals such as Andromeda and the Magellanic star clouds and show the individual stars that compose these systems. For the very distant systems that cannot be resolved on a photographic plate, the spectroscope tells us that we are dealing with groups of stars, for we do not obtain a bright-line emission spectrum such as we get from the gaseous nebulas in the Milky Way, but an absorption-type spectrum that we associate with the stars in our galaxy.

It is unfortunate that the name *nebula* has been applied to these objects since they are not nebulous structures in the sense that this term is used when we are describing the gaseous nebulas in the Milky Way. However, the term *spiral nebula* is so commonplace that there is hardly any possibility of getting rid of it. But we must keep in mind that it refers, not to a single nebulous cloud of gas, but to an aggregate of stars similar to the Milky Way.

Up until about 1950, astronomers estimated the distance of the Andromeda nebula to be about 750,000 light-years, a figure based on the period–luminosity law of the classical Cepheid variables. But we have already noted that there are two different populations of Cepheids, and one has to be sure that the correct period—luminosity law is used. This was not known until the galaxy expert, Baade, decided that if the Andromeda nebula were no more than 750,000 light-years away, it should be easy, with the 200-inch telescope, to photograph the very-short-period cluster variables (the RR Lyrae stars), since these variables were then thought to have an absolute magnitude of zero.

However, instead of showing up on the photographic plate, as zero-magnitude stars should at the supposed distance of the nebula, they were missing. This was so puzzling to Baade that he decided to reinvestigate the question of the period–luminosity law for classical Cepheid variables. He decided on good evidence that the absolute magnitude of the cluster-type variables could be trusted, but that the period–luminosity law was suspect. With this as a basis, he concluded that the reason the zero-magnitude RR Lyrae stars did not show up on his photographic plates was that the Andromeda nebula is much farther away than had previously been thought. In fact, the distance of this and all other galaxies had to be doubled to make things correct. Today we know that the absolute magnitude of the RR Lyrae stars is 0.5 but Baade's argument still holds.

The Andromeda nebula is a flattened disklike structure with a bright bulging nucleus and spiral arms that extend out in all directions in the plane of the disk to a distance of at least 60,000 light-years from the center. Except for its size (it is about twice as large), Andromeda is similar to our own stellar system. The nucleus is free of dust and gas and is densely filled with population-II stars (red and yellow giants and main-sequence stars) in agreement with the model of this galaxy developed by Baade, who showed that the individual stars in the central region of Andromeda can be photographed (resolved) on red-sensitive photographic plates. If one uses ordinary plates, the nucleus cannot be resolved and one obtains a fused image of all of the stars. Closer inspection shows that the population-II stars are not confined to the nucleus but extend all the way out beyond the dusty spirals themselves, thinning out gradually from the center. In fact, these stars form the main substance of this galaxy, which contains about 400 billion stars. In this respect, Andromeda is again similar to our Milky Way, which is a disklike substratum of population-II stars in which are embedded the spiral arms with their dust, gas, and population-I stars. Baade compared the Andromeda nebula to a cake, with the population-II stars being the cake itself, and the spiral arms the frosting on the cake.

One can trace out the spiral arms with their bright gaseous emission nebulas, their blue-white (O- and B-type) supergiants, their dust and gas, their Cepheid variables, and all the other features of the stellar distribution in the neighborhood of our sun superimposed on the basic structure of the population-II stars. Four spiral arms are on one side of the nucleus and five on the other side, forming tightly wound loops around the central region. In one of the inner arms of the nebula, the densely packed blue-white supergiants thin out toward the central region and drop off quite suddenly, leaving only a dusty path into the nucleus itself. Surrounding the main body of the Andromeda nebula and forming a spherical distribution concentric with it are about 500 globular clusters, which have the same general characteristics as those around the Milky Way: huge dust-free and gas-free globes of population-II stars. Andromeda abounds in variable stars of all kinds, with Cepheids present in great numbers. There are scores of ordinary novas and every now and then a supernova explodes with awe-inspiring violence.

The motion of the stars in Andromeda is similar to those of the stars in our galaxy. Near the outskirts of Andromeda the motions are about the same as those of the stars in the neighborhood of the sun, so this nebula is rotating, with the stars in the outer regions moving around the center the way the planets move about the sun. The speed of rotation of the nebula increases from the center out until a maximum speed of about 150 miles/sec is reached at a distance of 18,000 light-years from the center. From that point on, the speed of rotation decreases because the outer regions do not rotate like part of a solid disk. By applying Newton's law of gravity to the motions of the outer stars about the central region, we find that Andromeda's mass is about 400 billion times the mass of the sun, which is about twice the mass of the Milky Way.

The Milky Way and the Andromeda spiral are the two best-known spiral galaxies, but there are various types of spirals as well as nonspirals. The first impression one gets from a superficial examination of the regions of space outside of our galaxy is that most of the galaxies have a spiral structure, but we must be cautious about accepting this conclusion because it is based on a misleading selection of the brightest of them. We run the same danger here that we did in surveying the stars in the Milky Way; we tend to become blinded by the great luminaries in the galactic population and thus to overlook the dwarf galaxies.

Actually there is a variety of shapes and a great spread in the luminosities of the galaxies. As an example we note that accompanying the Andromeda nebula are two small satellite galaxies that are dwarfs compared to Andromeda. Two such companion galaxies are attached to our own galaxy. In 1519, the great explorer Magellan, while sailing through the southern seas, observed what appeared at first to be two small clouds in the night sky; there are in fact huge star clouds outside, but not too far from the Milky Way.

Since one of these clouds is larger than the other, they are called the *Large Magellanic Cloud* and the *Small Magellanic Cloud*. Although these irregular galaxies (they do not show a definite spiral structure though there does appear to be some evidence of a barlike structure) are much smaller than our galaxy or Andromeda, they are above average in luminosity. We must keep in mind that a general survey of the sky does not give a correct picture because the very faint galaxies are overlooked. We now know that there are numerous dwarf galaxies in space.

Just as we find a range of luminosities, we find quite a variety of structures, ranging all the way from the irregulars, such as the Magellanic star clouds, to the well-defined spirals like Messier 81 and Messier 101. Edwin Hubble divided the galaxies into four groups according to their spiral structures or lack of such structures. These are the irregular galaxies, the spirals, the barred spirals, and the ellipticals. Hubble suggested that the elliptical galaxies are related to the spirals and the barred spirals through an evolutionary development shown in his famous "tuning fork" diagram. Although this figure seems to show a definite relationship of an evolutionary kind among the various types of galaxies, there is a great deal of difficulty in this particular evolutionary model.

Edwin Hubble was born on November 20, 1889, in Marshfield, Missouri.[1] His father, John Hubble, was a lawyer who encouraged Edwin from an early age to think logically and to approach problems by breaking them down into manageable components. Edwin's mother was very supportive of her son's many interests; she often told him that he should never hesitate to ask questions and evaluate the answers he received with a critical eye. Edwin was fortunate that both parents understood the value of education and that gains could be realized from the pursuit of knowledge itself even if it sometimes did not have a particularly "practical" application.

Edwin's family moved to Chicago when he was a teenager. While attending high school in Chicago, his intellectual and athletic talents began to surface. He earned excellent grades in his subjects and won a scholarship to the University of Chicago. Although Hubble entered Chicago with the intent of becoming a lawyer, he soon began to doubt his career choice after listening to the lectures of astronomer George Hale. Hale instilled in Hubble an appreciation of the majesty of the universe; Hubble was particularly impressed by the way in which the entire structure seemed to be governed by the relatively few principles of celestial dynamics. To Hubble, this profound simplicity was awe-inspiring and it evoked in him such a devotion to astronomy that he began to attend as many lectures and courses as he could manage and to read various astronomy books in his spare time. He put his interest in the law aside for the time being and majored in astronomy and mathematics. Hubble was also a fine boxer and his athletic prowess attracted the attention of several professional trainers who believed he would be an excellent professional boxer. Although Hubble considered an athletic career, he eventually decided that he was more interested in astronomy than boxing.

In 1910 Hubble won a Rhodes Scholarship enabling him to go to Oxford for 2 years of study. When he arrived at Oxford, he began studying mathematics, believing that it would be helpful to his career as an astronomer. However, he found the mathematical studies to be too detailed and rigorous for his liking so he switched to law. Two years later he received his B.A. in jurisprudence and returned to the United States.

Still uncertain as to whether he should pursue law or astronomy as a profession, Hubble decided to give the law a try, having enjoyed his legal studies at Oxford. He was admitted to the bar and opened an office in Louisville, Kentucky. He soon found that the day-to-day routine of running an office and protecting the legal interests of his clients did not appeal to him; certainly it was quite far removed from the leisurely studies of the common law that he had done while in England. In any event, he closed his office after less than a year in 1914 and went back to the University of Chicago to do graduate work in astronomy. He received his Ph.D. in 1917 for his photographic investigations of nebulas.

While Hubble was at Chicago, he became reacquainted with the visiting Hale, who was then working at the Mount Wilson Observatory in California. Hale saw that Hubble had exceptional abilities as an astronomer even at this early stage of his career and offered him a job at Mount Wilson. Hubble would have left for California immediately had the United States not entered World War I. Hubble enlisted in the army and served in France until the Armistice was declared in 1918. He remained in Europe for an additional year as part of the Allied occupational force in Germany before returning to the United States in 1919.

Hubble spent the early years of his career at Mount Wilson investigating galactic nebulas; his researches led him to suggest a classification system for galactic and nongalactic nebulas. He also discovered several planetary nebulas and variable stars. He made his first important discovery when he detected a Cepheid variable in the M31 nebula, Andromeda. As such variable stars had been used by other astronomers such as Harlow Shapley to determine galactic distances, Hubble's discovery was an important first step in determining that the Andromeda star system is not a part of the Milky Way but a separate "island universe." The uniqueness of the Milky Way galaxy was argued in 1920 during a famous debate between Heber Curtis and Shapley, the former arguing that there are other galaxies outside the Milky Way and the latter denying it. The outcome of the debate was inconclusive, in part because sufficient observational evidence had not yet been collected. During the next 4 years, Hubble detected more than 30 Cepheids in M31 and, through a series of calculations, estimated that star system to be some 900,000 light-years distant. Because most astronomers agreed that the Milky Way is only 100,000 light-years in diameter, Hubble concluded that M31 must be a separate galaxy. The report of Hubble's discovery in 1924 resolved forever the debate over the existence of other galaxies.

Pursuing his research into the structure of galaxies, Hubble introduced a comprehensive galactic classification scheme, which he published in the *Astrophysical Journal* in 1925. He had developed this scheme during his extensive astronomical studies of galaxies; he found that nearly all such stellar systems rotate around a central hub and that they can be distinguished on the basis of either spiral or elliptical characteristics. Hubble called the few galaxies that do not rotate (such as formless clouds of stars like the Magellanic clouds) irregular galaxies. He found that the structures of both the spiral and the elliptical galaxies undergo similar changes with age and that the former can be further subdivided into normal and barred spirals. Hubble's scheme was not initially conceived as a map of the evolutionary stages of galaxies but was based solely on observations.

By using Cepheid variable stars as galactic measuring rods, Hubble then systematically measured the distances of the most distant objects in the known universe. By measuring the distances of the brightest stars in galaxies and even clusters of galaxies, Hubble extended the boundaries of the observable universe out to some 250 million light-years. In the next few years he measured the distances of nearly 20 galaxies; this information, coupled with the discovery in the early 1920s that the distant nebulas are receding from the Milky Way, led Hubble to propose what is now known as Hubble's law. This law simply states that the rate of recession of the galaxies increases linearly with distance. Hubble's discovery showed that the static universe assumed by Einstein is not correct; Hubble's work supported the de Sitter model which predicts that the light of distant galaxies is redshifted, indicating that these are receding from us. With the aid of several colleagues at Mount Wilson, Hubble showed that his law of recession can be extended to distances exceeding 100 million light-years. As a result, he then concluded that the relationship between the velocities and distances of galaxies holds to the limits of the observable universe. This work demonstrated that the universe is not static but expanding, its galaxies receding from each other at speeds that are proportional to their distances from each other.

Hubble's discovery not only prompted many astronomers to investigate the cosmological features of the general theory of relativity but also rekindled interest in the cosmological models that had been proposed earlier by A. Friedmann and G. Lemaître. Hubble's law also enabled astronomers to calculate the age of the observable universe by taking the ratio of velocity to distance (called Hubble's constant) and inverting it to obtain a time that is taken as the age of the universe. The original estimate of this age was about 2 billion years, but with subsequent determinations of the dimensions of the universe that number has been revised upward to about 15 to 18 billion years.

An important by-product of Hubble's exhaustive astronomical studies was his conclusion that the universe is homogeneous and isotropic: although galactic systems tend to cluster into local groups, all matter is fairly evenly distributed throughout the universe. This work coupled with the absence of evidence for a more hierarchical system such as that proposed by H. L. V. Charlier led astronomers to conclude that the galaxies are the basic building blocks of the universe.

The entry of the United States into World War II led Hubble to put aside his research programs and to volunteer for the army. The United States War Department politely declined the offer but suggested that Hubble would be more useful as the director of the Supersonic Wind Tunnel Laboratory at the Aberdeen Proving Ground in Maryland. Hubble eagerly accepted this offer and remained at that post until 1946. His outstanding contributions to the government's research programs at Aberdeen were recognized when Hubble was awarded the Medal of Merit in 1946.

After Hubble returned to California, he spent much of his time helping to design and oversee the construction of the 200-inch telescope to be placed at Mount Palomar. Hubble was instrumental in planning a detailed sky survey, which was completed in the late 1940s. When the 200-inch telescope was finally installed in 1949, Hubble was given the honor of being the first to look through it. Among the first and most important contributions of the new telescope was the discovery by Walter Baade that the universe is actually much larger than had originally been estimated owing to the correction of the period–luminosity law of Cepheid variables and the unanticipated faintness of many of the stars that Hubble had originally used to measure galactic distances. Baade's discovery did not detract from the brilliance of Hubble's efforts to determine the structure of the universe but simply showed that it needed to be more finely tuned.

In the last few years before his death on September 28, 1953, Hubble spent much of his time directing research projects at Mount Palomar. Although he devoted an increasing portion of his career to administrative duties, there was probably little else that Hubble could have accomplished in his lifetime had he chosen to continue his researches on a full-time basis. Hubble had shown that there are billions of galaxies in the universe and that the universe itself is expanding; his work transformed the way in which humanity views its place in the cosmos as profoundly as did the Copernican heliocentric theory. In any event, Hubble's work represented the high point of 20th-century observational astronomy and formed the basis for most subsequent cosmological investigations.

Although tens of billions of galaxies are visible in our telescopes, no more than a few thousand are close enough to exhibit a definite structure. These range all the way from the irregular, spotty-looking objects to the smooth, spiral-free, elliptical types. However, the division of the galaxies into groups, according to how well their spiral structures are defined, may not be as significant a clue to their evolution as is the presence or absence of dust and rotation.

The irregular systems are composed almost entirely of population-I stars, and are rich in bright gaseous nebulas, dust, gas, and variable stars. Although these objects show no definite structure, they are more related to the barred spirals than to any of the other galaxies. In some cases, as, for example, the Magellanic star clouds, there is a faint suggestion of a barlike structure.

The barred spirals, which account for about one-third of all the spirals, are characterized by a central bar that is dust-free, gas-free, and rich in population-II stars. A thin spiral arm originates from each end of this bar and forms a large encircling loop that may make a complete turn around the central structure. The arms are composed of dust and population-I stars.

Our own galaxy and the Andromeda nebula are both spiral structures. In numerous instances, astronomers have photographed spirals that are so tilted that the planes of their disks pass right through Earth, so that we view these objects edgewise and the structure of the spiral arms is not evident. The central nuclei, however, bulge out well above the flat disks and the halos of globular clusters are easily discerned. In all such photographs the layer of obscuring dust that cuts right across the center of the galaxy is easily visible.

The Andromeda nebula and our Milky Way are intermediate spirals with the arms wound fairly tightly around the central regions; of the external spirals that have been studied, the structures range from those in which the central core extends all the way out with the spiral arms forming thin almost circular wisps, to those in which the core is highly concentrated, with the well-defined arms accounting for the great bulk of the galaxy.

For a long time, the elliptical galaxies were among the most puzzling objects in the heavens, for although they exhibit a symmetrical, ellipsoidal structure, the stars of which they are composed do not show up on standard photographic plates. This was quite surprising since some of them are close enough for the individual stars to have been visible on photographic plates taken with the 100-inch telescope. Moreover, it was evident from spectroscopic studies of these elliptical nebulas that they are composed of stars. The solution to this riddle was finally found by Baade, who obtained individual star images of two elliptical nebulas by using photographic plates that are sensitive to red light. He thus showed that the elliptical nebulas are composed of population-II stars and, of course, are gas-free and dust-free. Although many of the elliptical galaxies are rather small faint objects, some of them are among the most luminous stellar systems that we know and rank in size and luminosity with the Milky Way and Andromeda.

Hubble's evolutionary scheme (galaxies evolving from ellipticals to spirals) is supported to some extent by the theory of formation of the heavy elements in the interiors of the old population-II stars as previously described. Applying this theory of heavy-element formation, we can then say that the elliptical galaxies become spirals when the population-II stars in the ellipticals eject enough material from which second-generation stars like our sun are born, and from which the spiral arms can be formed. According to this picture, the elliptical nebulas are agglomerations of almost pure hydrogen and helium stars (the population-II stars) that have not yet advanced far enough in their evolutionary development to have "cooked" very many of the heavy elements. The spiral structures would then be the oldest galaxies in which enough dust has been expelled by the primordial population-II stars to form large numbers of second- and third-generation population-I stars.

If this were a correct evolutionary theory, we would expect to find, as we go out to greater and greater distances, that the number of elliptical galaxies increases more rapidly than does the number of spirals. This follows because when we look at the very distant galaxies we see them as they were in their youth. At the present time there is no evidence that the number of elliptical galaxies increases with respect to the others as we explore the very distant regions of space.

If we assume that the evolutionary scheme is one in which the galaxies evolve from the irregular structures through the spirals into the elliptical types, we run into a more serious difficulty than that of accounting for the dust. It is easy to see that ultimately, as more and more stars of population I are formed from the dust in a spiral galaxy, the galaxy becomes dust-free and gas-free and looks somewhat like an elliptical type; but it is not easy to see how a flat disklike structure of a spiral type can evolve into the almost spherelike structure of the globular type. This could happen only if the spiral galaxies lost their rotational motion, and there is no way for that to happen without violating the law of conservation of angular momentum. (See the discussion of Kepler's second law of planetary motion.)

To overcome these difficulties, Oort proposed a theory of galaxy evolution such that the elliptical and spiral galaxies evolved together from the same gaseous material that constituted our universe in its early stages. Since the density of this expanding gas was not everywhere the same, there were regions in which the mutual gravitational attractions of different parts of a very large irregular cloud caused this cloud to contract, get warmer, and radiate heat (energy). This caused further contraction in the cloud, and in time stars were formed and the cloud became more and more regular in its structure.

At this stage, further development depended on the rotation of the cloud. If it was spinning rapidly, the gas cloud contracted rapidly into one plane (the plane of the Milky Way, for example, as in our galaxy) leaving the stars that were first formed as a kind of symmetrical halo (population II stars and the globular clusters). In the later stages, hotter stars (population I) and spiral arms were formed from the gas and dust in the thin compressed disk. If the cloud was not spinning rapidly in its early stages, there was no flattening, and no concentration of gas and dust into a thin layer. In such a system, the mutual gravitational attraction of the population-II stars led to a globular structure.

Before considering the clustering of galaxies, we discuss briefly certain types of unusual galaxies which are playing an ever-greater role in astronomy. The most important of these are radio galaxies, Seyfert galaxies, N-type galaxies, and quasars. Taken together, we call these active galaxies.

We define a radio galaxy as any strong extragalactic source of radio waves. We emphasize *strong* here because although all normal galaxies emit some radio waves, the radio emission from normal galaxies is very weak compared to that from radio galaxies. If the emission of radio waves from all galaxies were no greater than that from normal galaxies like our own, the most distant objects known would never have been observed. These objects, many billions of light-years away, have been observed because they emit radio energy so intensely that our radio telescopes detect this energy even though it has been traveling toward us for billions of years and hence is greatly attenuated when it strikes Earth. That these galaxies were emitting such vast amounts of energy billions of years ago (we do not know exactly what they are doing now) tells us two things: (1) they were extremely active when they were young; and (2) they cannot be nearly as active now as they were in the past because they could not possibly have enough fuel to sustain such activity for billions of years.

More than 100 extragalactic radio sources have been identified as individual galaxies. These can be divided into two groups:

(1) Those like our galaxy, which are normal spirals and which, in addition to the 21-cm hydrogen line, emit a continuous spectrum of radio energy from their disks, their nuclei, and their coronas. The total radio energy emitted, on the average, by these galaxies is on the order of 10^{38} ergs/sec, but for some galaxies it may be as high as 10^{41} ergs/sec, which is nonetheless negligible compared to the total emission of these galaxies in the visible part of the spectrum.

(2) The peculiar radio galaxies; these have very unusual structures and are very powerful sources of radio waves. The energy they emit in the radio part of the spectrum exceeds 10^{44} ergs/sec and is as large as or greater than what they emit in the visible part of the spectrum. The galaxy NGC 1068 emits 100 times as much radio energy as does a normal spiral, and the elliptical galaxy M87 in the Virgo cluster is 1000 times as luminous in its radio spectrum as normal spirals.

The peculiar radio galaxies have complex structures and emit their radio energy in very unusual ways. Some emit this energy from a single concentrated source, whereas others emit it from a concentrated core and an extended halo. In still others the emission occurs from two extended sources, one on each side of the galaxy, which is generally a normal elliptical system. Each of the two radio sources, which generally lie on a line passing through the center of the elliptical galaxy, is about 1 million light-years from the center of the galaxy. If these radio sources originated in the core of the galaxy, the explosive events must have been of incredible intensity to have projected the radio sources to such vast distances on either side of the core. The nature of such explosive events in galactic cores is not known, but such possibilities as a chain reaction of supernovas, gravitational collapse of large quantities of matter, annihilation of matter by antimatter, and large numbers of stellar collisions suggest themselves.

That violent events occur in the cores of certain galaxies is clearly evident in the galaxies discovered in 1944 by Carl Seyfert. As one passes from the outer regions of Seyfert galaxies to their small luminous nuclei, there is a marked increase in brightness. The spectra of these dozen or more galaxies consist of strong, broad emission lines, which means that the cores of the galaxies contain concentrated regions of hot gases. The large widths of the spectral lines, which are caused by random Doppler effects, indicate that the hot gases are expanding rapidly, with speeds as high as thousands of kilometers per second. Seyfert galaxies are often strong radio emitters, and they all are intense sources of infrared radiation. The luminosities of these galaxies are about the same as those of normal galaxies, but when the infrared radiation is added to the visible radiation they are about 100 times as luminous as the normal galaxies.

Certain galaxies such as NGC 5128 and M82, which has a very peculiar shape, are very strong radio emitters and show evidence of highly explosive events in their cores. These are not Seyfert galaxies but constitute an unusual group that have no special characteristics other than their explosiveness and unusual shapes. These galaxies are generally far more luminous than normal galaxies such as the Milky Way.

A group of very luminous galaxies known as N-type galaxies (where N denotes nucleus) have characteristics that lie between those of the Seyfert galaxies and those of the quasars, which we discuss later. They have very small, bright nuclei (they appear starlike on photographic plates) immersed in faint wispy, nebulous clouds.

Because Seyfert galaxies, N-type galaxies, and quasars have certain similarities, various astronomers have suggested that the cores of Seyfert galaxies and N galaxies are members of a continuous sequence of objects that begin as ordinary galaxies and end up as quasi-stellar objects or quasars. According to this theory, the Seyfert galaxies and N galaxies are intermediate stages in the evolution of galaxies that lie between ordinary galaxies and quasi-stellar objects. The regions outside the cores of nearby Seyfert galaxies look rather like the spiral arms of normal galaxies and their apparent sizes and speeds of recession at their observed distances are as expected. At greater distances, however, only the nuclei of the Seyfert galaxies can be observed (N-type galaxies), and these objects then appear more like stars than like galaxies. This, coupled with their intense luminosities in the visible, infrared, and radio parts of the electromagnetic spectrum, suggests that Seyfert galaxies are related to quasars.

More than 300 galaxy-like superluminous objects called quasars have been discovered. All are at vast distances, if their large redshifts mean that they are moving away from us at the speed of expansion of the universe, in accordance with Hubble's law of cosmological recession, and all show the following similarities:

- They are starlike objects on photographic plates and emit radio waves; the luminosity of a typical quasar is about 100 times as great as that of a normal galaxy.
- Their radiation varies in intensity with periods that range from a day or less to a few years. This means that they are much smaller than ordinary galaxies because the region from which the radiation of a quasar is emitted cannot be larger than the speed of light multiplied by variation time or period of variation of the quasar.
- They emit an excess of ultraviolet radiation as compared to ordinary stars or galaxies and great quantities of infrared radiation that does not fit the usual infrared pattern of cool stars.
- Broad emission lines are present in their spectra, and in some cases absorption lines are also present. The emission lines appear to arise in regions where the electron temperatures range from 10,000 to 100,000°K. In addition to the Balmer lines of hydrogen, the lines of ionized O, Ne, and Mg are also present. Both the emission and absorption lines in the spectrum of a quasar exhibit very large Doppler red shifts (no blueshifts have ever been found). These redshifts represent speeds of recession up to 90% of the speed of light.
- The absorption lines are much narrower than the emission lines, which are similar to those in the spectra of hot gaseous nebulas such as the planetary nebulas.
- If Hubble's law of recession applies to all quasars, the two quasars with the largest known redshifts discovered early in 1973 are the most luminous concentrations of matter known. They are both at a distance of about 10 billion light-years and so must be incredibly luminous to be visible from Earth.
- A comparison of the intensities of the radiation emitted in various parts of the spectrum by quasars with the intensities given by the Planck radiation formula shows that quasars do not radiate like blackbodies; the quasar radiation is nonthermal. No single temperature can be assigned to a quasar to make its radiation agree with Planck's formula for such a temperature.

We have already noted that stars tend to form gravitational groups and finally to collect into galaxies. The same is true of galaxies themselves; they occur in large clusters consisting of hundreds and even thousands of individual galaxies.

The Local Group, which consists of about 20 individuals spread out to a distance of more than 2 million light-years from our galaxy, is a good example of a small cluster of galaxies. Included in the Local Group are the two Magellanic clouds, the Fornax and the Sculptor systems, Andromeda and its satellites, the spiral in Triangulum (M33), and a few other smaller structures in addition to our Milky Way. The distances to these neighboring galaxies can still be measured easily because we can pick out the Cepheid variables in these galaxies and obtain their distances from the period–luminosity relationship.

At a distance of about 62,000,000 light-years we find the large Virgo cluster, which has a diameter of approximately 8,000,000 light-years and contains about 2500 galaxies, 75% of which appear to be spirals and the remainder elliptical galaxies. At such large distances, we can no longer use the period–luminosity law because the Cepheids cannot be observed. However, we can still recognize the individual supergiants in these galaxies, and by comparing the apparent brightnesses of these stars with their known luminosities we can find the distances of the galaxies. This method has been applied to a few hundred individual galaxies that lie outside our local group. Out to a distance of 3 billion light-years we have identified about 30 galactic clusters, among which are the Perseus cluster with 400 members, the Coma cluster with 800 galaxies, and the Hercules cluster with 300 members.

We can pick out all the galaxies that belong to a particular cluster by comparing the motions of the individual galaxies with the motion of the cluster of galaxies as a whole. These clusters are much too far away for us to measure their proper motions (i.e., motion across the line of sight), but we can measure the radial velocities (velocity toward or away from us) of the individual galaxies in a cluster and the radial velocity of the cluster as a whole using the Doppler effect.

The Doppler effect shows that the distant clusters of galaxies are receding from us with increasing speeds as we go out to greater distances. We can determine which individual galaxies belong to a cluster by noting that, although each member galaxy of a cluster is moving haphazardly within the cluster, the speed of recession of the entire cluster as a whole is what one expects for an object at the distance of the cluster. Thus, the Doppler redshift shows that the Virgo cluster is receding from us at 1200 km/sec, but the individual galaxies within the cluster are moving about with respect to the center of the cluster at about 500 km/sec.

We determine the total mass of a cluster of galaxies from the distribution of the speeds of the individual members of the cluster and the size of the cluster. To do this we note that a group of bodies can be held together by their own mutual gravitational attractions only if the negative of the total potential energy of the group equals twice the kinetic energy. This is known as the virial theorem; it must apply if gravitational equilibrium exists. From this theorem one can show that if a collection of N bodies (galaxies in our case) are in equilibrium and occupy a sphere of radius R, then the total mass of the N bodies must equal the product of the square of the random speed (in the Virgo cluster, 500 km/sec) and the radius of the sphere divided by the gravitational constant. When we calculate this quantity for the Virgo cluster of galaxies, we find that the total mass of this cluster must be at least 30 times as large as the sum of the masses of all 2500 galaxies that we observe. In other words, most of the mass that keeps this cluster together gravitationally is invisible. All attempts at finding this so-called "missing mass" have failed. This is true of other clusters of galaxies as well so that most of the mass that should be in the universe is not present in the form of self-luminous matter (e.g., stars).

As we investigate the regions of space in our universe at greater and greater distances, the number of clusters increases quite rapidly, and we are presented with one of the most important and interesting problems in cosmology (theory of the structure of the universe). Since the distant galaxies and galactic clusters as seen by us at this moment do not look the way they really are now but rather the way they did when they emitted the light that is at present striking our photographic plates, we do not get an instantaneous picture of the universe as it was at any one moment, but instead a series of flashbacks to earlier and earlier times as we go out to greater and greater distances. This enables us to say something about the evolution of our universe if we know the way in which the number of clusters of galaxies increases with distance.

If the distribution of the very distant galaxies is different from the distribution of the nearby ones, we must conclude that our universe has not been static and unchanging for hundreds of millions of years but has been evolving; but if the distributions are the same at all distances, our universe may well be a static unchanging one. At present, the evidence definitely supports the evolutionary theory and not the steady-state theory.

We have seen that radio astronomy helps us plot the structure of the Milky Way, and now we shall see that the same technique teaches us a great deal about the galaxies. We have already noted the existence of intense radio sources at great distances beyond our galaxy and indicated that these sources are galaxies in some kind of agitated state resulting from some vast cataclysm. Radio telescopes enable us to see such galaxies far beyond the distances now available to the optical telescopes because, at these great distances, the Doppler shift is so large that much of the light from the galaxies is too red to have much of an effect on our photographic plates. But even though the radio signals show the same Doppler effect as light does, these signals can still be detected by our radio telescopes because the Doppler effect does not shift them beyond the wavelength range of radio telescopes.

During the past few years, extensive surveys of the sky have been made with giant radio telescopes and more than 2000 individual radio sources have been located, distributed uniformly over the heavens. Although some of them such as the Crab Nebula and the faint gaseous nebulosities in Cassiopeia belong to our galaxy, the observational data can be best explained if one assumes that these "radio stars" are extragalactic sources that extend out to billions of light-years. Since the concentration of these radio galaxies increases quite rapidly as we go out to larger distances, we conclude that the universe was more compact billions of years ago than it is now since the distant objects show us the universe as it was at an earlier age.

To understand why some galaxies are intense radio sources while others that are plainly visible in an ordinary telescope do not send us intense radio signals, we consider one of these sources in more detail. One of the most intense radio sources known seems to lie in a completely empty region of the sky, at least as viewed with ordinary telescopes. But extraordinary events require extraordinary methods of analysis, so Baade in 1951 made a very careful survey of that part of the sky with the 200-inch telescope and photographed a truly remarkable phenomenon. On his photographic plate, in the region of the constellation of Cygnus, from which the intense radio signals were coming, he found two galaxies super-imposed in such a way that one could only conclude that these objects were in a collision. It is easy to see that this is not a mere accidental overlapping of two images of objects that are really at great distances from each other but on the same straight line from Earth. The nuclei and nebular arms are so distorted as to show that each galaxy is greatly affected by the gravitational field of the other.

This radio source, in which the two interpenetrating galaxies can be seen, is known as Cygnus A and its radio luminosity is enormous. It emits 10^{45} ergs/sec as radio waves, which is millions of times as much as our galaxy radiates. Some radio galaxies are undoubtedly colliding galaxies, their radio energy resulting from these collisions. In Cygnus A, the 21-cm line of hydrogen shows precisely the same Doppler shift that is present in the visible part of the spectrum, and from this we know that Cygnus A is about 1 billion light-years away from us. This is relatively close for radio telescopes, for we can detect radio sources as intense as Cygnus A out to distances about ten times greater than this, far beyond the reach of the 200-inch Mount Palomar optical telescope.

Earlier we described many strange and fascinating phenomena that are occurring in the universe, but the events we have just been discussing—the vast upheavals that give rise to intense radio emission, be they explosions in the cores of galaxies or collisions of galaxies—are certainly the most awesome and grandest of all. Compared to such explosions or to the collision of two galaxies that rush into each other at speeds of thousands of miles per second, the explosion of an ordinary nova is a mere whimper, and even the outburst of a supernova is but a trivial incident.

When two galaxies collide, collisions between the individual stars in each galaxy do not occur, but the clouds of dust and gas between the stars in each galaxy do collide and are set into violently turbulent motions. The individual atoms and molecules in these interstellar gas clouds move about with speeds that are equivalent to temperatures of millions and even hundreds of millions of degrees. This rapid molecular motion excites the atoms and molecules, and causes them to emit the radio noise that we described above.

If this analysis of a collision between two galaxies is correct, we should expect to find that after a collision has occurred the galaxies are not as dusty as they were before the collision. Observational evidence and theoretical calculations seem to support this.

In this regard we must have some way of estimating the frequency of galactic collisions. If by a collision we mean a direct head-on encounter in which the two central regions of the galaxies are no more than a few hundred or even thousand light-years apart, then these are very rare events, indeed. But the gases within the colliding galaxies can be greatly excited and stimulated into emitting radio signals even though only the edges of the galaxies touch. Collisions of this sort are not rare, and we should expect to find at least ten such encounters occurring right now among the 2 billion galaxies that lie in the region of space that extends out from the sun to about a half billion light-years.

The frequency of collisions increases considerably in regions where galaxies are tightly packed, as they are in galaxy clusters. Thus, at the center of the Coma cluster, which contains hundreds of galaxies within a region of less than 3 million light-years, two or three collisions should be occurring right now. Baade and Lyman Spitzer investigated the collisions among galaxies in clusters and showed that collisions should be frequent enough so that during the last few billion years most of the galaxies in these tightly packed clusters should have been shaken free of dust, and therefore should now contain no spiral arms. This is borne out by the observations, which show that most of the galaxies in compact clusters are shaped like spiral galaxies but contain no dust and show no spiral arm structure. In loose clusters, on the other hand, where collisions are infrequent, the galaxies are the typical spiral structures with dust and arms.

One might argue that all the elliptical galaxies are the result of collisions among spirals, but the arithmetic of galactic collisions seems to contradict this. There are far too many elliptical galaxies to be accounted for by the known rate of collisions.

We have already noted that as we go farther and farther out to more distant galaxies, we find that they show a larger and larger Doppler shift toward the red end of the spectrum. If we accept this evidence as a true indication of the motions of these galaxies, and thus far there has been no other acceptable explanation, we must conclude that all galaxies are rushing away from us with speeds that increase with increasing distances. This does not mean that every single galaxy follows this pattern because, as we have seen, galaxies are members of large clusters, and the law of recession applies to the clusters as a whole. Within a cluster the individual galaxies have a random motion.

What are we to conclude from this spectroscopic evidence that the galaxies are receding from us? The first thing we must keep in mind is that this does not mean that we occupy a kind of central point from which all galaxies are rushing away. There is nothing special about the Milky Way that makes other galaxies want to avoid it; we would observe the same effect if we were stationed in any other galaxy. In other words, all galaxies are receding from one another as though they were part of a huge explosion that occurred billions of years ago. This has given rise to the theory of the *expanding universe*.

One explanation of the expanding universe is that originally all the matter that is now present in the form of galaxies and all of space itself were concentrated in a single highly condensed tiny sphere that exploded (the big bang) and started to expand. It is easy to show from our knowledge of the present speed of expansion of the distant galaxies that this explosion must have happened about 15 or 16 billion years ago. This figure agrees very nicely with other estimates of the age of the universe based upon the evolution of stars and stellar clusters, and the formation of heavy elements inside the red giants and supergiants. Other theories have also been advanced to account for the origin of the galaxies and their apparent recession, but at the present time, enough evidence has been presented for us to decide definitely in favor of the big bang theory. The observations of the Doppler shift in the spectra of the most distant galaxies that can be observed with the 200-inch telescope, taken together with phenomena we shall discuss in the next chapter, favor an expanding universe that is undergoing a continuous evolution as though the expansion had started explosively at a definite moment in the past from a very compact state.

But we are just beginning to penetrate the regions of space from which the answers to important cosmological questions must come because the farther out into space we look, the farther back we can go in time. Ultimately we shall see objects the way they were when our universe was but a few million years old, and these observations will tell us a great deal about the birth and evolution of galaxies. As we start from the most distant galaxies and head back to the Milky Way, we shall have a chronological picture, from youth to old age, of the galactic systems that comprise the universe.

Before leaving the distant galaxies, we describe the most recent discoveries concerning the large-scale distribution of large clusters of galaxies at great distances. Although we cannot measure the distances of these clusters directly, we can use Hubble's law of the recession of galaxies (discussed in detail in the next chapter) to deduce these distances. Hubble's law states that the speeds of recession of these clusters of galaxies (measured by their Doppler redshifts) are proportional to their distances. Assuming this law to be correct, one can deduce the cluster distances from their redshifts. Astronomers have done just this and found that, as one goes out to ever greater distances (ever larger redshifts) in space, the galaxy clusters are not distributed uniformly but instead are strung out along curved filaments forming vast lacy patterns on the surfaces of enormous contiguous bubbles (150 million light-years in diameter) that appear to be empty.

Most astronomers have used these findings to argue that the matter in the universe is not distributed uniformly but in strings and that the universe is bubbly—consisting mostly of voids. But this argument is not warranted because all we can say about the bubbles is that they contain no luminous matter. However, these bubbles may well be where the dark matter resides; if this is so, then the distribution of mass in the universe is uniform since the dark matter outweighs the visible matter by about 100-fold. These "bubbles" would then be gravitational spheres on the surfaces of which the clusters of galaxies are revolving slowly around the centers of the "bubbles" in accordance with the law of gravity. This model would be an accurate representation of the cosmos if the dark matter were present in the form of mini-black holes, each of mass 10^{-5} g, which one of the authors (L. Motz) has postulated and studied in great detail in a series of papers.

CHAPTER 19

Cosmology
(The Structure and Evolution of the
Universe)

Though earth and man were gone
And suns and universes ceased to be,
And Thou wert left alone,
Every existence would exist in Thee.

—EMILY BRONTE, *Last Lines*

Until relatively recently in the history of civilization, we believed tenaciously that we occupy a special position in the scheme of things, and the early cosmologies (theories of the nature of the universe) that were developed before the time of Copernicus enforced that belief. If, as we then believed, mankind is important to the proper running of the universe, what was more natural than to conclude that Earth is at the very center of the universe and that all the other heavenly bodies were placed in the sky for the glorification of God and man? That all the naked-eye evidence seemed to agree with such a concept strengthened the hold of these early theories and made it exceedingly difficult and even dangerous to advocate anything that threatened the belief in this favored position of Earth.

The first great blow to our ego came with the final victory of the Copernican concept, which removed us from the all-important central location and put us on one of the lesser planets circling the sun. But we were not even to be left with the consolation that the sun is the dominant central body of the universe. The sun is but a relatively unimportant star among billions of others that make up the Milky Way; and far from being at the center, our solar system is near the edge of the Milky Way. Nor can our sun claim the dis-

351

tinction of being a charter member of our galaxy (population-II stars), for the sun is a second- or third-generation star, born from the dust and gases thrown off by the original stars.

We cannot even say that ours is the only star with a planet that has life on it. Astronomical evidence is preponderantly against the notion that our planet is the only abode of life in the universe. We can divide the main-sequence stars into two broad groups: one composed of the A, B, and O stars, which are spinning fairly rapidly about an axis the way Earth is, and the other composed of the F, G, K, and M stars, which are spinning quite slowly. The sun, which is a good example of the second group, spins very slowly about an axis, requiring 25 to 30 days to make one complete rotation.

The difference in the speed of rotation between these two groups of stars is really only apparent because some of the total rotation of the second group of stars is distributed over their planetary systems, which we cannot see. In other words, the stars in the first group are spinning rapidly because they are not surrounded by planets, so that all the rotational motions are concentrated in the stars themselves. The stars in the second group have the same total amount of rotational motion as the stars in the first group except that it is shared with the planets that accompany them. If this interpretation of the difference in the speeds of rotation is correct, and all the evidence strongly favors it, then there must be many millions and even billions of planetary systems like our own in the Milky Way. This makes it highly probable that life, as we know it, exists in millions of places throughout our galaxy, for we know that nature repeats its pattern over and over again wherever the right conditions are present. If billions of planetary systems exist in the Milky Way, then many have just the right conditions for life even though the probability for just these right conditions to occur in any one place may be quite small.

If we must content ourselves with being attached to an unimportant star, perhaps we may take some comfort from the unusual qualities of our galaxy. But even this crumb of satisfaction is denied us because we now know that although our galaxy appears to be larger than the run-of-the-mill galaxies, there are many galaxies, like the Andromeda spiral, that are considerably larger and more massive. In other words, there is nothing about our planet, our sun, or our galaxy that is outstanding or noteworthy.

We approach the cosmological problem with these things in

mind; we cannot assume that there is any special quality about our galaxy, the local group of galaxies, or the region of space in the neighborhood of these galaxies that differentiates it from any other region of space.

If the greatness of a theory is measured not only in terms of how well it explains the known features of the universe, but also in terms of the new things that it predicts and the new avenues of research that it opens up, then the general theory of relativity stands at the very peak of human achievement. Not only did this theory explain the Newtonian discrepancy in the motion of Mercury, and predict the bending of the path of light and the slowing down of the vibrations of atoms in gravitational fields, but it also produced a modern cosmological theory.

Shortly after Einstein developed his general theory of relativity, he saw its great importance for cosmology. His reasoning was simple: if massive bodies in a given region of space alter the geometry in that region from flat Euclidean geometry to curved non-Euclidean geometry, then the notion that our universe is an infinite expanse of flat space in which some islands of matter are floating about must be abandoned since the general geometrical properties of the universe must depend on the distribution of masses. In any case, the existence of an infinite Euclidean universe leads to insurmountable mathematical difficulties in general relativity theory.

Since there is no good reason, aside from venerable tradition, for imposing the Euclidean "straitjacket" on our universe, Einstein decided to free himself and the universe, once and for all, of the preconceived notion that space is necessarily Euclidean. He felt that the local hills and wrinkles in space arising from the presence of stars and galaxies produce an overall curvature of space-time that causes space to curve back on itself and close up. To test this idea, he assumed a simplified model of the universe in which matter is not present in the form of little isolated lumps (the stars), but rather as a thin fog spread out uniformly and filling all space. To simplify the problem still further, Einstein assumed that the universe is static, so that the density of matter is always the same and the distances between points are not changing in any systematic way.

To solve the problem under these conditions, Einstein had to alter his fundamental gravitational equations somewhat, adding another exceedingly small term that is equivalent to assuming that there is a negative small pressure everywhere in space. (This tiny term, which is called the cosmological constant, is far too small to affect ordinary gravitational phenomena, but it has important cosmological effects.) Under these conditions, he showed that one can obtain a solution of the problem in the form of a spherical space of constant positive curvature (like the surface of Earth); in other words, a closed space.

Following Einstein's initial work, the Dutch astronomer de Sitter found that the Einsteinian universe would remain finite even if almost all the mass were removed from it, but if one did this the universe would seem to behave as though the very distant bits of matter still left were running away from us. He showed that this apparent running-away effect could be traced to the presence of the cosmological term that Einstein had introduced into his equations. From the way in which this term enters into the equations, it is clear that it tends to oppose the effect of gravity; in other words, it behaves like a repulsion. The remarkable thing about this term is that it has the same effect on all bodies, regardless of their masses, and becomes larger, and therefore more important, with increasing distance. This is a queer kind of force, indeed, and one that physicists and astronomers had never met before.

De Sitter showed that if the density of matter in the universe becomes vanishingly small, so that its gravitational cohesion can be disregarded, the cosmical repulsion takes over, and the universe behaves as though it were expanding. At any rate, the light from the very distant objects in the universe behaves as though it were emitted from objects that are receding from us.

In De Sitter's paper, published in 1917, the germ of the idea of an expanding universe was born. De Sitter suggested at that time that it might be worthwhile for astronomers to study the light from the distant galaxies to see whether it is shifted toward the red end of the spectrum.

By 1912, V. M. Slipher, the American astronomer, had measured the Doppler shift for 40 spiral galaxies and had found that in all but 4 cases (these 4 are the nearest), these objects are receding from us (Doppler shift toward the red). Hubble later extended this work, using the 100-inch telescope at the Mount Wilson Observatory, and made the very important discovery, by measuring the distances of Slipher's galaxies, that not only are the distant spiral nebulas receding, but the speed of recession increases with distance. This later became known as Hubble's law. In fact, Hubble made the bold proposal that the universe is, indeed, expanding. This hypothesis was fully borne out by the extensive work of Humason at Mount Wilson, which shows that out to distances of 40 million light-years the increase in the recessional velocity of the spirals is strictly proportional to their distances (Hubble's law). By 1935 Humason had extended his investigations to distances more than 30 times that of the Virgo cluster and had measured the velocities of recession of 150 galaxies and clusters of galaxies. He found that the speeds of recession increase in proportion to the distances out to the very remotest objects known at that time, i.e., out to a distance of about 1.5 billion light-years.

We must say a few words here about the special methods of determining the distances of these remote galaxies. Of course, it is impossible to use the properties of individual stars to determine the distances because these galaxies are so far away that the individual stars cannot be recognized on a photographic plate. However, one can assume that, on the average, these stellar systems are about equally luminous and equally large. We can then compare the distances of any two of these objects by comparing their apparent sizes or their apparent brightnesses. Thus, if one galaxy appears a quarter as bright as another one, we conclude that it is twice as far away from us as the latter. Although this method does not always give the correct answer, we should still get good statistical results if we work with a large number of galaxies.

Since the Doppler shifts for galaxies out to these great distances show that the speed of recession is strictly proportional to the distance, the Doppler effect itself (in the form of Hubble's law) is used to determine the distances of the remotest galaxies. One might refuse to accept this as a universal law because, as we saw, a few galaxies near the Milky Way do not show a displacement of the spectrum toward the red. These galaxies, however, are not to be taken as exceptions to the law for two reasons. To begin with, the speeds of recession for the nearby galaxies are quite small so that they might very well be offset by the motion of the sun itself around the center of the Milky Way. If the sun's speed of 175 miles/sec around the nucleus of our own galaxy is taken into account, most of the nearby galaxies are found to be receding from the center of our galaxy with speeds that agree with Hubble's law. The other point is that Hubble's law applies to the cluster as a whole and not to each individual galaxy in the cluster. The individual galaxies have speeds of recession that have a normal distribution about the correct speed as given by Hubble's law, some of them moving away more rapidly than the cluster as a whole and others moving more slowly. Since some of the nearby galaxies belong to the local cluster, they do not have individual speeds of recession, but move along with the local group itself.

By 1936 Humason had reached the limits of the 100-inch telescope but could still find no deviations from Hubble's law. This law can be stated as follows: *the speed of recession of a cluster of galaxies increases by about 55 km/sec for each increase of 3.26 million light-years in the distance of the cluster.* The number, 55 km/sec per million parsecs, is known as the Hubble constant; this value, which stems from the most recent work of Allan Sandage, points to an age of about 14 billion years for the universe.

As a few examples of Hubble's law, using recent data we note that the Virgo cluster at a distance of about 70 million light-years is receding with a speed of 1200 km/sec; the Ursa Major cluster at a distance of 960 million light-years has a speed of recession of about 15,000 km/sec; and the Corona Borealis at a distance of 1.4 billion light-years is running away at a speed of 21,500 km/sec. With the 100-inch telescope, Humason could not go beyond these distances. Because of this limitation and because of World War II, the redshift program was not resumed until 1951 when the 200-inch telescope on Palomar became available. With this telescope and with spectroscopes of much greater sensitivity, Humason made measurements with an accuracy of 0.5%, and recorded velocities of recession of up to 60,000 km/sec. These results are of tremendous interest to cosmologists, because these rapidly receding objects are about 4 billion light-years away from us, so that we see them now just the way they were when our universe was 4 billion years younger. The evidence from these and even more distant galaxies should tell us whether we are in a static unchanging universe or whether our universe is evolving from some earlier form. If the universe was different more than a billion years ago, the evidence for it should be contained in the redshifts.

The evidence obtained from the 18 remotest galaxies measured by Humason, though still uncertain, shows that the velocity of recession does not continue to increase in direct proportion to the distance of the galaxies but rather increases more rapidly than this. In other words, the universe was expanding more rapidly when it was younger (more than 4 billion years ago) than it is now. We consider the significance of this a bit later.

With the discovery of the redshifts in the spectra of the distant galaxies, the old notions about the structure and origin of the universe had to be changed. The guide to the new developments is contained in the general theory of relativity. We have already seen how Einstein and de Sitter formulated the problem and arrived at solutions that indicate that the universe is expanding; however, certain features in their treatment are not entirely satisfactory. The imposed restrictions—that the density of matter in the universe and the distances between points always remain the same—were much too severe and had led both Einstein and de Sitter to overlook a simple solution that automatically gives an expanding universe.

In 1922, the Russian mathematician Aleksander Friedmann reexamined Einstein's model of the universe and discovered that he could obtain the observed recession of the galaxies without introducing a negative pressure or cosmic repulsion term. All he had to do was keep the original general relativity equations unchanged but allow the energy–matter density in the universe and the curvature of space to change with time. This enabled him to construct a model of the universe that has a finite (nonzero) density of matter everywhere and that is expanding. Friedmann's work went unnoticed while Lemaître in Belgium, Robertson and Tolman in the United States, and Eddington in Great Britain carried on independent investigations into theories of expanding universes. All of these researches led to the same picture of an expanding universe that originally started from some supercondensed state of matter. This primordial origin appeared to be a natural consequence of the theory and the observations. If the distant galaxies are running away from us at the observed rates, they must have been much closer to each other billions of years ago and the universe must have been smaller.

The point is that these galaxies are not just receding from us; they are running away from each other and from every point in space. In other words, every observer in every galaxy sees things just the way we do. This means that the bits of matter in the vast sea of space that we call our universe behave as though they were fragments of some vast explosion that occurred billions of years ago in some incredibly dense globule of matter. The result of this explosion is the dispersal of the galaxies that we are now viewing; it is going on at such a rate that all the distances between the galaxies double every 18,000,000,000 years. According to this picture, our cluster of galaxies is being deserted by its neighboring clusters; we must then contemplate a time when our galaxy, with a few others of the Local Group, will be left alone in space if this expansion continues.

If any one person can be credited with conceiving the big bang model of the universe, it is Abbé Georges Lemaître.[1] He was born in Charleroi, Belgium, on July 17, 1894, the eldest of four children.

The Lemaîtres were devout Catholics and they passed their solemn respect for the Church and its doctrines down to their children. Georges received his early schooling at the local parish and then attended the Jesuit high school at Charleroi where he studied Greek and Latin. Georges was not an enthusiastic student of languages but he did show some aptitude for mathematics. By the time he finished high school, he had decided that he wanted to be both a mathematician and a priest. Bankruptcy of the family business, however, forced him to gear his education to more practical ends. In 1910 Georges entered the Jesuit preparatory school in Brussels to prepare for the entrance examinations at the College of Engineering at Louvain. After spending a year studying mathematics and Latin, Georges passed the examinations and began his studies in 1911. He received his A.B. in engineering in 1913 and then began his professional training as a mining engineer.

Lemaître's engineering career was interrupted by the German invasion of Belgium in August 1914. He immediately volunteered for active duty and was assigned to a unit digging trenches to fortify the defensive positions of the Belgian army. The sheer power of the German onslaught forced the Belgian army to retreat and Lemaître found himself caught in the cross fire for several days until he was able to make his way back to the Belgian lines. After the initial German sweep through Belgium became bogged down, the opposing sides began to wage an indecisive but bloody war of attrition. With the battle lines stabilized, Lemaître's routine degenerated into weeks of inactivity interspersed with short bloody battles. He spent much of his free time in the trenches reading scientific books to keep up with the advances that were being made in physics, especially relativity theory and quantum theory. Although Lemaître spent the entire war in uniform, he never obtained a commission as an officer. That this oversight occurred was all the more remarkable given the numerous decorations Lemaître received for bravery; it was a snub that he never forgot.

After the Armistice was declared in 1918, Lemaître returned to the University of Louvain to resume his studies in mathematics. Within 18 months he had completed all of his course work as well as his doctoral dissertation in mathematics. Acting on his earlier desire to be both a scientist and a priest, Lemaître then enrolled in the Maison Saint Rombaut seminary. After 3 years of study, he was ordained a priest in 1923, and applied for and secured a grant from the Belgian government to study astronomy at Cambridge University under Arthur Eddington. Lemaître spent a year in England and became acquainted with several outstanding physicists including Ernest Rutherford. He then went to the United States for a year and worked as a research fellow at the Harvard College Observatory. While in the United States, Lemaître not only met several leading astronomers (including Harlow Shapley and Henry Norris Russell), but also attended several international conferences. After his fellowship at Harvard ended, he enrolled in the graduate physics program at the Massachusetts Institute of Technology. His dissertation dealt with some features of Karl Schwarzschild's solution of Einstein's gravitational field equations.

In 1925 Lemaître returned to Louvain and began working as an associate professor in the mathematics department there. The university's provost gave him some leisure time to pursue his interests in physics and astronomy. Lemaître's teaching duties included a graduate course in the history of physics and mathematics; this survey of the evolution of ideas about the universe led him to examine critically the de Sitter model of the universe. He then published a paper that argued that the observed recessional velocities of extragalactic nebulas are due to the expansion of space itself; but his conclusions were ignored until he sent a copy of his paper to Eddington, who reviewed it favorably in the 1931 *Monthly Notices* of the Royal Astronomical Society. Eddington also forwarded Lemaître's paper to Shapley and de Sitter; the sheer weight of its well-crafted arguments persuaded most astronomers of its validity in a short time.

Although Lemaître's model of the expanding universe had several deficiencies, such as a time scale that was initially much smaller than the known age of Earth and an inability to explain the clustering of galaxies, it did provide an explanation of the recession of the galaxies. In a 1931 lecture to the British Mathematical Society, Eddington argued that the second law of thermodynamics can be viewed as a measure of the degree of order in the universe; this suggestion led Lemaître to conclude that the universe was originally a superdense clump of matter that exploded. The assumption that the universe originally existed with a radius equal to zero enabled Lemaître to offer a physical counterpart to the mathematical singularities predicted by Einstein's general theory of relativity, even though Einstein himself never was very enthusiastic about the analogy. In 1932 Eddington gave a lecture on the expanding universe at the International Astronomical Union; in it he demonstrated the central importance of Lemaître's work to modern cosmology and convinced most remaining skeptics about the importance of the big bang model. The following year Lemaître gave two lectures on the expanding universe and cosmic rays (which he suggested were the remnants of the primordial explosion) at the California Institute of Technology to an audience that included the leading cosmologists and physicists of that time. He was also featured in an article in the *New York Times Magazine*.

Lemaître was elected to the Royal Academy of Belgium and received the Francqui Prize in 1934. The latter honor brought with it a check equivalent to about $200,000, which helped to ease his financial worries considerably. He returned to the United States and spent some time at Harvard and Princeton's Institute for Advanced Studies doing research on cosmic rays and general relativity theory. He also created many algorithms that established him as a major figure among mathematicians in the emerging field of scientific computing. In 1938 he taught cosmology as a visiting professor at the University of Notre Dame; during this time Arthur Compton and Carl Anderson's researches into the nature and origin of cosmic rays indicated that the latter are not the fossils of the primeval explosion as had been proposed by Lemaître and others. Not until 1965, when Arno Penzias and Robert W. Wilson detected the faint echo of the primeval explosion in the form of low-temperature microwave radiation uniformly distributed throughout space, was the big bang theory accepted.

Although Lemaître returned to Louvain in 1938 and occupied an endowed chair, life at the university was becoming more disrupted by the repeated military crises along the German border. His greatest fears were realized with the spring invasion of the Low

Countries by Germany in 1940, the beginning of 4 years of military occupation. The horrors of the war were brought home when the Germans burned down the university library; the faculty had already been decimated by the drafting of many of its members into the Belgian army. Although there was no money available for research and most international scientific contacts were cut off, the university continued to function throughout the war even through the later years saw food and fuel supplies drop to bare subsistence levels. In 1944 Lemaître's house was bombed during an Allied air raid; he was treated for shock and injuries at the local hospital. Although he suffered no visible long-term injuries, the effects of the war exacted a great toll on his health and his generally optimistic personality.

After Belgium was liberated in 1944, Lemaître moved to Brussels to live with his mother; he commuted to Louvain several times a week to fulfill his academic responsibilities. During this time he also renewed many scientific contacts that had been severed during the war and prepared a collection of his popular lectures on cosmology, which was published in 1946 and translated into several languages. The commercial success of his book led to numerous lecture invitations in Belgium and France. In 1947 he was awarded the Insignia of Commander of the Order of Leopold II, the highest honor that could be conferred on a civilian citizen, for his bravery and selfless service to the University of Louvain during the war. Lemaître was also the recipient of several honorary doctorates and scientific medals and was appointed president of the Pontifical Academy of Sciences in 1960.

After the death of his mother in 1956, Lemaître sold the family home and moved back to Louvain. He continued to teach twice a week but his research interests shifted from cosmology to celestial mechanics; his big bang theory was also increasingly criticized by some cosmologists and church leaders for its alleged metaphysical and mythical features. Because the background radiation of the big bang had not yet been detected, Lemaître turned his attention to the more fruitful area of scientific computing and urged the president of the University of Louvain to purchase an electronic computer. When Lemaître's lobbying efforts failed, he bought a Burroughs E101 and donated it to the university. He taught himself machine and computer languages over the next few years and led the efforts to establish a computer center at Louvain. Less than a year after the detection of the microwave background radiation that firmly established the big bang theory as the leading cosmological model, and after repeated illnesses and a heart attack, Lemaître died on June 20, 1966.

According to the Lemaître picture, the universe began as a highly compressed hot primeval atom about 8 or 9 billion years ago. As it expanded, it cooled and condensed into galaxies, stars, and planets. In the Eddington picture, our present universe is in a kind of midway state between a pure Einstein static universe and an empty expanding de Sitter universe. It began, according to Eddington, as a relatively small static sphere having a radius of 1,068,000,000 light-years, and according to this theory it will go on expanding until the galaxies are infinitely dispersed. In the Eddington theory, the cosmic repulsion term is retained and plays a very important role in relating the structure of the universe as a whole to the structure of the fundamental particles inside the atom. A necessary consequence of this picture is that the universe is finite and curved back upon itself; however, this does not mean that we can receive light from and see every part of it. The expansion is so rapid that a signal from the very remotest objects is never quite able to reach us.

The expanding universe theory leads to one very important conclusion to which we have already referred: our universe is evolving from some earlier form and therefore we may point to some instant in the past and say, "This is when it all began." In other words, an expanding universe implies a kind of beginning and an end, unless our universe is a pulsating one, alternating between expansions and contractions that have been going on forever so that there is no beginning or end. If we adopt this view, we are now in the expanding phase of the pulsation of the universe, which will go on for some 60 billion years and then give way to a contraction. This means that the supercondensed state will again be reached, to be followed by another expansion, with history repeating itself every 80 billion years or so. According to this theory, you are doomed eternally to reread these lines every 80 billion years if each expansion phase repeats itself in exactly the same way.

The concept of an evolving expanding universe that began about 18 billion years ago was almost universally accepted, until some 25 years ago when a group of outstanding cosmologists advanced the notion that our universe is really standing still. Although it appears to be running away with itself, this is only a deceptive aspect of a universe that had no beginning and will have no end and extends infinitely in all directions. This is the steady-state theory that was proposed by F. Hoyle, H. Bondi, and T. Gold. Although this theory is vastly different from the expanding universe theories, it also has its foundations in the general theory of relativity.

Hoyle arrived at his steady-state theory by introducing into the Einstein equations certain time-dependent stress terms that generate automatic self-creation of matter in empty space. In this theory, all the conditions at any point in the universe are maintained the same for all time by the creation of matter in such a way as to compensate for the matter that escapes in the form of the receding galaxies. The matter is created in the form of hydrogen at a rate of one proton per 1000 cm^3 of space every billion years. This process is much too weak to be observed, but we shall see that there are other tests of this theory. Bondi and Gold arrived at the same ideas by introducing their "perfect cosmological principle," which states that the universe must be the same as viewed by all observers at all times. This is a generalization of Einstein's cosmological principle that the universe must be the same everywhere but not at all times.

How are we to decide between these two views of cosmology and determine whether we are evolving or just moving on a treadmill of time? We shall see that the observational evidence strongly favors an evolving universe and opposes the steady-state theory.

We have already mentioned that the distributions of distant radio galaxies and quasars support the evolutionary theory (the big bang theory) of the universe much more strongly than they do the steady-state theory. Radio telescopes reveal that the number of radio sources and quasars in a given volume of space increases with distance. This observation means that space must have been more crowded a long time ago (the time when the radio signals that we now receive were sent out) than it is now, so that an expansion and thinning out of galaxies must have occurred since then, which contradicts the steady-state theory.

Perhaps the most powerful argument against the steady-state model of the universe is the isotropic (same in all directions) 2.7°K background cosmic radiation, which was first predicted by Gamow in 1948. It was discovered in 1965 by A. A. Penzias and R. W. Wilson (who had been tracking down the source of 75-cm background radiation) after it had been suggested by R. H. Dicke and his co-workers, who were unaware of Gamow's prediction. This radiation has been studied in great detail since then, to determine whether it is blackbody radiation or not. If one plots the logarithm of the observed intensity of this microwave radiation against the frequency, the curve obtained agrees very well with Planck's blackbody radiation curve for 2.7°K.

If we accept this deduction, we can draw some important conclusions about the origin of the universe, or at least about its cosmological state billions of years ago, because this radiation must have originated when the universe was formed. It can be shown from basic thermodynamical principles that the blackbody nature of the radiation is not altered if one either slowly compresses or expands the region of space occupied by the radiation. In other words, the blackbody radiation in an expanding universe remains blackbody radiation but its temperature (i.e., its wavelength of maximum intensity) changes owing to the Doppler effect. That means that the blackbody radiation that we observe now must have been blackbody radiation at a much higher temperature when the universe was much younger and hence much more compact. Accepting this conclusion, we can now obtain a picture of the origin of the universe and of this blackbody radiation from the following considerations.

As we trace the universe back in time (i.e., reverse the expansion we now observe), we note that as it gets smaller and smaller, the background blackbody radiation is squeezed together and gets hotter and hotter (i.e., bluer and bluer owing to the Doppler effect). This process can be traced back until the universe was concentrated into a very small region of extremely high density. If we take this as the initial state of the universe, we have the big bang theory of its origin, as first analyzed in detail by Gamow. According to Gamow's big bang theory the temperature of the universe about 1 second after the initial explosion (the big bang) exceeded 10^{10} °K, so that the universe was filled with very hot blackbody radiation and neutrons. He estimated that the expansion of the universe from this state up to the present did not alter the blackbody character of the radiation but only reduced its temperature (owing to the continual redshift arising from the Doppler shift) to about 6°K. This is in good agreement with the 2.7°K radiation now observed. Since the steady-state theory, with its eternal low density of matter, cannot possibly account for the origin of the background thermal cosmic radiation, we must reject this theory in favor of the big bang theory.

The redshift data from distant galaxies also show that the present universe differs from one billions of years ago. This evidence clearly supports an evolving universe, but whether it definitely favors a pulsating universe rather than one that will go on expanding forever is still uncertain. A few billion years ago, the galaxies were dispersing faster than they are now; in other words, the expansion of the universe is slowing down. If the rate at which the universe is slowing down is fast enough, then in time the expansion will stop and the universe will begin to contract until a state of great compression recurs. This will be followed by another expansion phase and so on, and we have a pulsating universe. If the universe is pulsating, the deceleration parameter (the quantity that tells how rapidly Hubble's constant is decreasing) must be larger than a certain minimum value. Over the years the measured values of this constant have ranged from well above this minimum value to below it. Since the evidence is not all in yet, we still cannot say definitely whether the universe will stop expanding in time and begin to collapse or will go on expanding forever.

Ultimately, the redshift data will reveal whether the universe is closed and pulsating or open and expanding forever. This question is intimately related to the space-time geometry of the universe and to the nature of the curvature of space. From the study of the geometry of surfaces in three-dimensional space we know that a surface can have zero curvature (a flat surface, such as a plane in Euclidean geometry), positive curvature (an ellipsoidal surface, such as a sphere in Riemannian geometry), or negative curvature (a hyperbolic surface, such as the surface of a saddle in Lobachevskian geometry). The last two types of geometry are the non-Euclidean geometries, and one of these applies to our universe. Just as a surface can have different kinds of geometries, so, too, can our four-dimensional universe as a whole because of the presence of massive bodies; we must see which geometry observational evidence favors. The Einstein–Friedmann equations applied to the universe, show that the correct choice among the three types of geometries can be based on the measurement of two different quantities: the deceleration parameter, which we have already discussed, or the density of matter throughout space.

To see how the overall density of matter in the universe comes into play, we assume that all the matter in the universe is spread out into a thin fog so that we can assign the same density to every point in space. We then find, according to Einstein's equations, that the geometry (or the sign of the curvature of space) in our universe depends on how large the density, in grams per cubic centimeter, is compared to the number that is obtained by taking three times the square of Hubble's constant of galactic recession (expressed in centimeters per second per centimeter) and dividing this number by a certain multiple of the Newtonian constant of gravitation. This number (called the critical density) with which the density of smeared-out matter must be compared is very small—just about equal to 1 divided by 100 thousand trillion trillion.

If the density of matter *exactly* equals this small number, the universe has no curvature and obeys Euclidean geometry; it is a flat space-time manifold that extends to infinity in all directions, and the galaxies are dispersing into this infinite void. If the density of matter is larger than this critical density, the curvature of space is positive and the universe is closed (i.e., finite but boundless); it is a spherical manifold that has been expanding from an initial point-like state. Since a state of positive curvature can never pass over into a state of zero or negative curvature, the universe in this particular case cannot continue to expand forever but must reach a point of maximum expansion (maximum radius) and then begin to contract again. After contracting to zero size, another expansion occurs and so on forever. The mass in the universe is then large enough so that the gravitational pull acting among all of the galaxies finally brings them to rest and causes them to fall in toward each other again.

The curvature of the universe must be negative (space hyperbolic) if the density is smaller than the critical number defined above. If this is so the universe must have started initially from an infinitely diffused state, contracted to a highly condensed state (a point), and then passed into the present state of infinite expansion. If the curvature of space is negative, the geometry is similar to that on a saddle surface, and the expansion will continue forever until the radius of the universe becomes infinite and the geometry becomes Euclidean.

To decide among these three cases, we have to know the value of the density with reasonable accuracy, and this is precisely what we do not know. If we accept the value of the density as given by the visible stars, gas, and dust in all the visible galaxies, the hyperbolic universe (negative curvature) is favored since the density of known matter is smaller than the critical density by a factor of about 50, but the actual density throughout all of space may be larger than this by a factor of 100 because of the presence of dark matter. Recent evidence indicates that there is some dust and gas between the galaxies, and the intensity of radio signals and cosmic rays in our own galaxy indicates that it is surrounded by a vast corona of gas and dust, and this may be true of all spiral galaxies. Furthermore, great quantities of invisible matter may be present in the form of mini-black holes, each having a mass on the order of 10^{-5} g. If this is true, only one such particle for every 10^{17} protons can supply the mass required to close the universe. We see from this that we can use the density data to exclude a hyperbolic universe, but we can never be sure on the basis of the observable density alone that the universe is not of positive curvature and therefore finite and closed. The observed density of matter can never be reduced; the discovery of new material in space can only revise it upward. As more and more matter is discovered, the case for positive curvature becomes ever stronger.

If we accept the closed, positive-curvature universe as the correct one, we can estimate the present size of such a universe by using the value of the density in the universe as we find it today and comparing it with the density in an original Einsteinian static universe. From this we deduce that a closed universe, expanding the way ours is, should be about five or ten times larger (in radius) than a static universe of the same mass. Since a static universe, having a mass of about 10 billion trillion suns (i.e., 100,000,000,000 galaxies such as ours), has a radius of 1 billion light-years, we must conclude that our present universe has a radius of 15 billion light-years. We can therefore see only one-tenth of the way around the universe to the backs of our heads.

The expanding-universe theories point to some instant in the past, about 15 or 18 billion years ago, when all the matter in the universe and all of space were concentrated in a single tiny supernucleus of incredible density. This globule of matter exploded (the big bang) and we are now living as remnants of that vast explosion, the evidence for which is found in the background thermal cosmic radiation. That various parts of the universe have evolved from some initial chaotic state and are still evolving is indicated by almost everything we observe.

Thus, Earth is evolving as demonstrated by the formation of the rocks, the huge varieties of biological species in various states of development, the drift of the continental plates, the formations of the ocean basins, and the existence of heavy radioactive elements of long half-lives. Earth's age (about 4.5 billion years) as determined from the half-life of uranium-238 agrees well with the age of the sun as determined from the theory of stellar evolution.

Outside the solar system we find abundant evidence that in the past the universe was younger and quite different from the way it is now. The existence of galaxies, globular clusters, open clusters, and clusters of galaxies is a very strong argument for the evolution of the universe from an original undifferentiated, very dense state to its present ordered state in which a remarkable hierarchy of different structures, from vast clusters of galaxies down to individual free atoms, exists.

Accepting this picture of the evolution of the universe, we can go back in time and trace events from the initial big bang some 15 billion years ago to the present low-temperature state. Since in its initial highly condensed, very hot state, shortly after the big bang, all the matter in the universe was in the form of a perfect gas, we may apply to it the gas laws and the laws of thermodynamics. From these we deduce that the initial temperature of the universe was extremely high because the absolute temperature of a sphere of gas varies inversely as the radius of the sphere and the initial radius of the universe was very small. We do not know how small because our understanding of the geometry of highly compressed matter depends on a proper synthesis of the general theory of relativity and the quantum theory. But nobody has yet discovered such a synthesis. If the quantum theory is not taken into account, the general theory of relativity by itself indicates that the initial radius should have been zero. Mathematicians refer to such a state as a singularity so that some cosmologists argue that the universe began as a singularity of infinitely high density, which is the ultimate state of a black hole. However, one cannot neglect quantum effects when particles of matter (protons) are squeezed very close together. The Heisenberg uncertainty principle tells us that if the volume to which a particle is confined is decreased, the motion (momentum) of the particle must increase, becoming ever more violent as the volume decreases. This violent quantum motion resists the confinement of the particle, and so it may well be that the initial radius of the universe could not have been zero but must have been some finite value owing to the quantum pressure.

In any case, the radius of the initial universe was so small that immediately following the big bang the temperature was many trillions of degrees. Under these conditions the state of the universe at this stage (the fire-ball stage as it is called) was dominated entirely by the intense radiation that was present because the density of the radiation far exceeded the density of the matter that was present. The reason for this is that the density of radiation varies directly as the fourth power of the absolute temperature, which, as we have noted, varies inversely as the radius. Hence, the density of the radiation in the universe in the past varied inversely as the fourth power of the radius of the universe. On the other hand, the density of matter varies inversely as the cube of the radius. Hence, when the universe was very small, radiation dominated and matter could not have existed in its present form.

If the universe was small enough shortly after the big bang, the temperature was so high that protons themselves were torn apart and only the constituent particles of protons (called quarks or unitons) existed. This state, however, could not have existed for more than a millionth of a second because the rapid expansion of the universe cooled the radiation drastically so that protons, neutrons, and other heavy particles were formed. Even then the density of the radiation was larger than the density of the matter by a factor of about 10 billion. After about 1 second the temperature dropped to a few billion degrees, which was cool enough for neutrons to decay into protons, electrons, and neutrinos. This state, consisting of a mixture of protons and electrons, continued for some thousands of years until the temperature had dropped to a few million degrees. The protons and electrons were still moving about too energetically to unite to form neutral hydrogen, but protons were coalescing (fusing) to form heavy hydrogen (deuterium) and helium. An analysis of the distribution of helium in the population-II stars today indicates that about 25% of the initial hydrogen was transformed into helium and a percent or so of hydrogen was transformed into deuterium. This fusion of hydrogen to form helium could not have lasted for more than about half an hour because the right temperature conditions for this process lasted only for that length of time.

With the end of the era of the formation of deuterium and helium, the radiation-dominated stage of the evolution of the universe
had not yet ended, and the matter-dominated stage had not yet
begun, because even though the radiation was no longer energetic
enough to make protons move fast enough to form helium, it was
still too energetic for the free electrons to be captured by the protons, and for the deuterons (the nuclei of heavy hydrogen), and the
alpha particles (helium nuclei) to form neutral hydrogen and helium.
But after some additional tens or hundreds of thousands of years,
the temperature had dropped below 10,000°K and neutral hydrogen
and helium were formed. Even so, the rapidly expanding gas was
still in too agitated a state to succumb to the force of gravity and
to begin to coalesce into separate self-gravitating clouds from which
the stars and galaxies were to arise. It is difficult to say how long
the universe expanded before this gravitational fragmentation occurred; estimates range from a million to a quarter of a billion years.
In any case the stage was set for gravitational fragmentation into
clouds and collapse of localized regions in these clouds into groups
of stars that we now observe as globular clusters and galaxies.

The precise way in which the rapidly expanding stream of primordial hydrogen and helium, some millions or hundreds of millions
of years after the big bang, was transformed into isolated concentrations of stars, gas, and dust is not known, but we know from
hydrodynamical theory that a liquid or a gas cannot continue to flow
in straight lines over a very long path if the speed of the gas exceeds
a certain value. The law that governs this was discovered by Os-
borne Reynolds many years ago; he showed that for a given density
of a fluid, and a given path length, the motion of the fluid changes
suddenly from uniform motion in a straight line to violently turbulent
motion as soon as the speed of the fluid becomes larger than a critical
value. This applies to the rapidly expanding gas in the early stages
of our universe; instead of just expanding radially, the gas broke
up into many large and small eddies that continued to take part in
the overall expansion, but with each eddy keeping its vortex structure and beginning to contract gravitationally. Moreover, the eddies
began to separate from each other.

Not all the eddies continued to contract uniformly because
many of them were too large. James Jeans showed, early in this
century, that if a large mass of gas is spread over too large a volume
of space, it becomes gravitationally unstable and breaks up into

smaller units; his figures, applied to the gas in our expanding universe, indicate that galaxies are just about the right size for stability. If an eddy of gas was originally much larger than a galaxy, it broke up into smaller eddies, thus forming a cluster of galaxies. This hypothesis, then, can account for the individual galaxies and the clusters of galaxies.

We can extend these ideas to account for the formation of the stars within the individual galaxies and also for the origin of planetary systems such as our solar system. In the early stages of galactic formation, the gaseous material composing the galactic cloud was about 99% hydrogen and helium. As contraction went on, new and smaller eddies were set up near the center of the cloud where the density was highest. These eddies finally contracted to form the population-II stars that are found in the center of galaxies and in the globular clusters. Ultimately, all the hydrogen in the center of the cloud was used up to form the nucleus (core) of stars that is observed at the center of each spiral nebula. The globular clusters were probably formed at about the same time as separate gravitational condensations in the outer regions of the contracting clouds. From this point on, the development proceeds in accordance with the picture we have already drawn of the evolution of population-II and population-I stars. The importance of the birth and death of population-II stars for our present state of the universe and, in particular, for the emergence of biological systems and intelligent life is now clear. Without such stars, in the deep interiors of which a few percent of the primordial hydrogen and helium was transformed into the heavy elements, second-generation stars (population-I stars) with their retinues of planets could not have originated.

In the description we have given above of the initial stages in the birth of the universe, we have used a time scale of about 15 billion years for the age of the universe (the time since the big bang); but this is an estimate, based on a value of 55 km/sec per megaparsec for the Hubble constant, a value that is uncertain and probably incorrect. Estimates of the Hubble constant have ranged from 30 to 75 km/sec per megaparsec, and, in accordance with such estimates, our estimates of the age of the universe have ranged from some 20 billion to about 10 billion years, with the smaller value of the Hubble constant assigned to the greater age of the universe.

The origin of our solar system, consisting of the sun and its remarkable retinue of planets, is also tied up with the evolution of the universe, for, as we have seen, population-I stars like our sun could not have been formed unless population-II stars preceded them, and these came on the scene only when conditions in the universe were appropriate. But once the scene was set for the formation of population-II stars, the emergence of population-I stars was inevitable, as was the formation of our solar system, which fits into the general picture quite nicely. After the early massive population-II stars exploded and ejected their metal-rich material into the galaxy to form vast interstellar clouds, new stars began to condense from these clouds. In the initial stages of this condensation, the entire cloud from which a star was to be formed extended over a very large volume and was rotating slowly. As the cloud contracted, the speed of rotation increased and the cloud began to flatten out in one plane. This process (bearing a superficial similarity to the original nebular hypothesis of Kant and Laplace) has been thoroughly studied by various astronomers and physicists, and they have shown that an equilibrium configuration existed during the last stages of this contraction in which the central nucleus (the sun) was surrounded by a thin rotating disk of gas and dust extending out to the extremities of our solar system.

Turbulent motions arose within this disk, which then broke up into eddies and vortices of various sizes that revolved in stable orbits around the central star at definite distances from the center (in agreement with Bode's law). As these eddies moved about, the friction between neighboring ones caused material to collect into lumps at these contact points, and these lumps then became the planets of our solar system. The stability, the masses, and the densities of the gravitationally coalescing turbulences that were to become planets were governed by the tidal action of the central star. Originally, there must have been much more material present in this cloud than there is now, most of it hydrogen. However, as the cloud warmed as the result of the contraction, the hydrogen atoms acquired speeds that enabled most of them to escape from the outer regions, so that we find very little hydrogen in our solar system outside the sun and the four massive outer planets. As time went on, most of the material left in the original clouds, after the escape of the hydrogen, was swept up by the sun and by the lumps of matter formed by the eddies. Here and there, however, smaller lumps were formed in the neighborhood of the larger condensations and we now see these as the satellites of the various planets.

But our story is by no means at an end even though our telling of it must stop at this point. Never before in the history of science has there been so much activity as there is at the present time. Never before has there been a promise of a richer harvest for the astronomer. As the flow of data from the great optical and radio telescopes and from space probes increases to a vast torrent, the astronomers can paint their portrait of the universe with bolder strokes. What now appears as a faint shape here, a poorly defined line there, a distortion of perspective elsewhere, will stand out sharply to complete the harmonious picture as we study, even more thoroughly, the model revealed to us by the stars.

At this point we complete our story with an account of some of the most recent cosmological observations and the theories that have been presented to explain them. The picture drawn above, based on the Einstein cosmological equations for the radius of the universe and its rate of change, is called the standard model. The manner in which cosmologists have handled the Einstein equations and applied them to our evolving universe leaves unanswered a number of perplexing questions and does not explain an array of puzzling observations. We present these cosmological puzzles in the order of their importance. Most critical is the so-called "initial singularity," which states that at the initial moment of the life of the universe (at its origin), the size of the universe was zero; it was a mathematical point in which all matter and energy in the universe were concentrated. Obviously, this is a physically inadmissible solution of Einstein's equations, telling us, as its does, that everything was infinite at the universe's initial moment. Such a situation is designated as a singularity in the mathematical equations that describe the dynamics of the universe. The question that this situation presents is whether the equations are at fault or the particular solutions of these equations that contemporary cosmologists use to describe the evolution of the universe are to blame.

The second problem stems from the observed large-scale structure of the universe. As we have already noted in our discussion of the distribution of clusters of galaxies, the universe seems to be essentially empty, with most of the visible matter (in the form of clusters of galaxies) spread thinly in a lacy pattern, over vast bubbles of space. Taking only this visible matter into account, we find that the visible density of the universe is smaller than the critical density (the density that just closes the universe gravitationally) by a factor of about 100.

The third question is presented by the nature of the hidden or dark matter that we know must be present in the universe, since, as we have already noted, the visible matter alone cannot produce the gravitational attraction (gravitational glue) that is required to keep galaxies, clusters of galaxies, and other large aggregates of matter from breaking up as a result of their large internal dynamics. The solution of the problem of the dark matter is important for two other cosmological reasons: does this matter generate the vast energetic explosive phenomena we see throughout the universe; and is the quantity of this hidden matter large enough to close the universe (i.e., does it increase the density of the universe above the critical density)?

Three other minor questions are presented by the observations that a correct model of the universe and of the basic particles (protons and electrons) that constitute it should answer. The first arises from the discrepancy between the observed number of neutrinos from the sun (solar neutrinos) and the calculated number. Why is the observed number only one-third of the calculated number? The second stems from the observation that the information brought to us by light from opposite directions in space shows the same physical conditions even though these two different regions are separated by such vast distances that they could never have communicated with each other. How did these two and, in fact, numerous such regions "know how" to evolve to the same state without the same set of initial conditions (without being in contact with each other at some time in the past)? The third question concerns the geometry of the universe, which, on the basis of the observed matter, has a density smaller than but very close to the critical density so that the geometry of the universe appears to be flat. How could this be so unless the initial geometry of the universe was set, with incredible precision, exceedingly close to flatness?

Cosmologists today, using the standard theory, have tried to answer these questions as though they were unrelated to each other. We believe that this is an unrealistic approach to these problems; they are all interrelated and, hence, soluble with a single assumption, as shown by one of the authors (L. Motz) in a series of scientific papers.

We consider first the initial singularity of the universe, which blocks any consideration of the universe at the moment of its birth. Since such a moment is unphysical in the sense that it is accompanied by infinities (e.g., infinite density, infinite temperature), we eliminate it by showing that such a state could never have existed at any time in the past if protons and neutrons now in the universe consist of triplets of particles (quarks or unitons) that are extremely tightly bound to each other by gravity. If we now imagine going back in time, with the universe growing ever smaller and the photons of the cosmic background radiation becoming bluer and more energetic, we reach a point in the past when protons and neutrons could not exist because the very energetic photons ripped them apart. In ripping these nucleons apart, the photons themselves disappeared and were replaced by free unitons. With the disappearance of the photons, the temperature of the universe dropped to zero. This may then be pictured as the initial state of the universe even though its radius was not zero. In other words, according to this picture, the universe, at some time in the past, was a cold, small sphere of free units with no radiation. But this state existed only for a fraction of a second owing to the very large mass (10^{-5} g) of each uniton; the mutual gravitational attraction of these free unitons caused them to recombine in triplets to form protons and neutrons with most of the mass of the unitons released in a vast explosion. This event was the "big bang" and the beginning of the expansion of the universe. According to this picture, the initial state of the universe was not a point of infinite temperature and density but a small, cold compact sphere of massive unitons dominated entirely by gravity.

We can now account for the dark matter in the universe today by postulating it to be the free unitons that, owing to the large separations from each other, had no chance to combine to form gravitationally bound triplets—protons and neutrons. One uniton per hundred thousand trillion (10^{17}) protons supplies enough mass to give the universe a density greater than the critical density—10^{-29} g/cm^3—and thus to close the universe and make it a pulsating rather than an ever expanding one.

The free unitons also account for the large-scale structure of the universe. The observed bubbles, with clusters of galaxies distributed over their surfaces, are not empty but gravitational bubbles in which unitons are held together by their mutual gravitational interactions. A typical bubble, with a radius of about 150 million light-years, contains about 1.4×10^{55} unitons or about one uniton per trillion trillion (10^{24}) cubic centimeters of space. They are thus about 1000 km apart. Space is uniform as far as the distribution of matter goes; it only appears lumpy if we look at the visible matter, which really accounts for very little as far as the dynamics are concerned.

Although we can offer a model of the evolution of the universe from a certain point in time, we cannot say where these unitons came from. But we are as ignorant of that as everyone else; it is one mystery that we cannot solve within the framework of scientific inquiry at the present time, and so we leave it to all who have followed us this far to speculate about the origin of the primordial matter.

Reference Notes

CHAPTER 1

1. Stephen F. Mason, *A History of the Sciences*. New York: Macmillan, 1962, p. 53.
2. Bertrand Russell, *A History of Western Philosophy*. New York: Simon & Schuster, 1972, p. 146.
3. Mason, *op. cit.*, p. 53.
4. *Ibid.*
5. *Ibid.*, p. 52.
6. Arthur Berry, *A Short History of Astronomy*. New York: Dover, 1961, pp. 40–41.
7. Morris Kline, *Mathematics and the Physical World*. New York: Dover, 1959, p. 92.
8. W. W. Rouse Ball, *A Short Account of the History of Mathematics*. New York: Dover, 1960, pp. 87–88.
9. Morris Kline, *Mathematics in Western Culture*. London: Oxford University Press, 1953, pp. 72–73.
10. Berry, *op. cit.*, p. 61.

CHAPTER 3

1. J. L. E. Dreyer, *A History of Astronomy from Thales to Kepler*. New York: Dover, 1953, p. 308.
2. J. D. Bernal, *Science in History*. Cambridge: MIT Press, 1985, pp. 407–408.
3. Quoted in James R. Newman, *The World of Mathematics*. New York: Simon & Schuster, 1956, p. 221.
4. Dreyer, *op. cit.*, p. 373.
5. *Ibid.*, p. 375.
6. Newman, *op. cit.*, p. 225.
7. *Ibid.*, pp. 229–230.
8. *Ibid.*, p. 231.

9. *Ibid.*, p. 232.
10. *Ibid.*, p. 233.

CHAPTER 10

1. Many of the details in this biography are drawn from: James R. Newman, "James Clerk Maxwell," *Scientific American* (June 1955). See also L. Campbell and W. Garnett, *The Life of James Clerk Maxwell.* London, 1882, 2nd ed.; J. J. Thomson, *James Clerk Maxwell 1831–1931.* Cambridge, 1931; J. G. Crowther, *British Scientists of the Nineteenth Century,* London, 1932; R. V. Jones, "James Clerk Maxwell at Aberdeen 1856–1860," in *Notes and Records. Royal Society of London,* 28 (1973), 57–81.
2. Max Planck, *Scientific Autobiography.* New York: Philosophical Library, p. 14.
3. Planck, *op. cit.*, p. 19.
4. See Max Born, *Obituary Notices of Fellows of the Royal Society,* 6, 1948, pp. 161–180.
5. Planck, *op. cit.*, p. 45.

CHAPTER 11

1. This biography is drawn from several sources including: K. A. Strand, "Ejnar Hertzsprung," *Dictionary of Scientific Biography.* New York: Scribner's, Vol. 6, 1972, pp. 350–353; Axel V. Nielsen, "Ejnar Hertzsprung—Measurer of Stars," *Sky and Telescope,* 35 (January 1968), 4–6; A. J. Wesselink, "Ejnar Hertzsprung," *Quarterly Journal of the Royal Astronomical Society,* 9 (1968), 337–341.
2. Much of this biographical material is drawn from: Bruce C. Cogan, "Henry Norris Russell," *Dictionary of Scientific Biography.* New York: Scribner's, Vol. 12, 1975, pp. 17–23; F. M. J. Stratton, *Biographical Memoirs of the Royal Society,* 3 (1957), 173–191; Harlow Shapley, *Biographical Memoirs. National Academy of Sciences,* 32 (1958), 354–378.

CHAPTER 14

1. A. Vibert Douglas, "Arthur Stanley Eddington," *Dictionary of Scientific Biography.* New York: Scribner's, Vol. 4, 1971, p. 278.

CHAPTER 17

1. Several books provided useful information for this biography including: Angus Armitage, *William Herschel.* London, 1962; J. B. Sidgwick, *William Herschel.* London, 1953; M. A. Hoskin, *William Herschel, Pioneer of Sidereal Astronomy.* London, 1953.

2. Isaac Asimov, *Asimov's Biographical Dictionary of Science and Technology*. New York: Doubleday, 2nd ed., 1982, p. 808.

3. *Ibid.*

4. Some valuable materials about Harlow Shapley's life include: Bart J. Bok, "Harlow Shapley, Cosmographer," *American Scholar*, 40 (1971), 470–474; Hudson Hoagland, "Harlow Shapley—Some Recollections," *Publications of the Astronomical Society of the Pacific*, 77 (1965), 422–430; Kirtley Mather, "Harlow Shapley, Man of the World," *American Scholar*, 40 (1971), 475–481.

CHAPTER 18

1. This biography is based on: G. J. Whitrow, "Edwin Powell Hubble," *Dictionary of Scientific Biography*. New York: Scribner's, Vol. 6, 1972, pp. 528–533; N. U. Mayall, "Edwin Hubble—Observational Cosmologist," *Sky and Telescope*, 13 (1954), 78–85; H. P. Robertson, "Edwin Powell Hubble: 1889–1953," *Publications of the Astronomical Society of the Pacific*, 66 (1954), 120–125.

CHAPTER 19

1. This biography is based on A. Deprit's "Monsignor Georges Lemaître," which appeared in A. Berger's *The Big Bang and Georges Lemaître*. Dordrecht, Holland: Reidel, 1984, pp. 363–392.

Index